International Conference on Big Data Analytics, Data Mining and Computational Intelligence (BigDaCI2024), Connected Smart Cities (CSC2024), and e-Health (EH2024)

Held at the Multi-Conference on Computer Science and Information Systems (MCCSIS 2024)

Budapest, Hungary
13-15 July 2024

ISBN: 979-8-3313-0302-0

Printed from e-media with permission by:

Curran Associates, Inc.
57 Morehouse Lane
Red Hook, NY 12571

International Association for Development of the Information Society (IADIS)
Rua Sao Sebastiao Da Pedreira
No. 100, 30, 1050-209
Lisbon, Portugal

Phone: 351 21 3151373
Fax: 351 21 3151244

secretariat@iadis.org

Additional copies of this publication are available from:

Curran Associates, Inc.
57 Morehouse Lane
Red Hook, NY 12571 USA
Phone: 845-758-0400
Fax: 845-758-2633
Email: curran@proceedings.com
Web: www.proceedings.com

PROCEEDINGS OF THE INTERNATIONAL CONFERENCES ON

BIG DATA ANALYTICS, DATA MINING AND COMPUTATIONAL INTELLIGENCE 2024

CONNECTED SMART CITIES 2024

and

E-HEALTH 2024

Budapest, Hungary
JULY 13 - 15, 2024

Organised by

iadis

international association for development of the information society

As a member of Crossref (a non-profit membership organization for scholarly publishing working with the purpose to make content easy to find, link, cite and assess) each published paper in this book of proceedings will be allocated a DOI (Digital Object Identifier) number for its fast and easy citation and indexation.

Volume Editors:
Ajith Abraham, Guo Chao Peng and Pedro Isaías

Computer Science and Information Systems Series Editors:
Piet Kommers and Pedro Isaías

Associate Editor: Luís Rodrigues

TABLE OF CONTENTS

FOREWORD ix

PROGRAM COMMITTEE xiii

KEYNOTE LECTURE xvii

FULL PAPERS

BIG DATA ANALYTICS, DATA MINING AND COMPUTATIONAL INTELLIGENCE

STRUCTURAL DAMAGE DETECTION BASED ON HYBRID ALGORITHM WITH SPARSE OBJECTIVE FUNCTION 3
Guangcai Zhang, Wenlong Zhao, Chunfeng Wan, Xiaonan Zhang, Xiujia Liu and Youliang Ding

GRAPH NEURAL NETWORKS FOR 3D OBJECT RECOGNITION AND RECONSTRUCTION: A COMPARATIVE STUDY 11
Rajaa Miftah and Mostafa Hanoune

FEATURE SELECTIONS AND EXPLORATORY ANALYSIS FOR ML STOCK PREDICTIONS USING SET50 19
Gridaphat Sriharee

RESEARCH ON KNOWLEDGE EXTRACTION OF CHINESE SHIP TECHNICAL SPECIFICATION 27
Kexin Cheng and Zuhua Jiang

FACE RECOGNITION ATTENDANCE SYSTEM 35
Zirije Hasani, Jakup Fondaj, Rrezart Kallaba, Xhafer Muja and Endrit Jelliqi

UNVEILING THE VALUE OF USER REVIEWS ON STEAM: A PREDICTIVE MODELING OF USER ENGAGEMENT APPROACH USING MACHINE LEARNING 43
María Olmedilla, Leonardo Espinosa-Leal, José Carlos Romero-Moreno and Zhen Li

CONNECTED SMART CITIES

REAL-TIME TRAFFIC MONITORING SYSTEM BASED ON COMPUTER VISION AND DEEP-LEARNING 50
Jaime Villa, Arturo de la Escalera and José Mª Armingol

TRANSFORMING ROME: STRATEGIC INITIATIVES FOR DEVELOPING A SMART CITY 57
Vincenzo Gattulli and Spartaco Paris

IS TOYOTA'S WOVEN CITY A SMART CITY? 65
Hiroshi Ito and Kazuhiko Kato

A REVIEW ON DATA PROTECTION AND REGULATION IN SMART CITY CONCEPT 73
Andrej Ignjatić

INVESTIGATING THE IMPACT OF TREES AS NOISE BARRIER ON URBAN NOISE POLLUTION THROUGH 3D SCENE MODELLING 81
Seanglidet Yean, Zheng Ping Heng, Guan Wei Leow, Bu Sung Lee and Peter Edwards

TOURISM SUPPORT SYSTEM TO UTILIZE VIRTUAL REALITY SPACE AND DYNAMIC INFORMATION IN REAL TIME 91
Yuro Koga and Kayoko Yamamoto

BEHAVIORAL BASED DECISION-MAKING SCHEME FOR COOPERATIVE CONNECTED AUTOMATED MOBILITY 99
David Yagüe-Cuevas, Pablo Marín-Plaza, María Paz-Sesmero and Araceli Sanchis

E-HEALTH

A DIGITAL HEALTH PLATFORM FOR SHARED EHR SERVICES: REAL WORLD CASE STUDIES 108
Ricardo Gonçalves, André Toffanello and Ricardo Puttini

TOWARDS PRECISE DETECTION OF PERSONAL INFORMATION LEAKS IN MOBILE HEALTH APPS 118
Alireza Ardalani, Joseph Antonucci and Iulian Neamtiu

INTEROPERABILITY PLATFORMS: A COMPONENT BASED APPROACH TO DERIVE FUNCTIONAL CLASSIFICATION 126
Martin Staemmler

VERIFICATION OF BODY SURFACE AREA-CALCULATING ANDROID APPS 133
Ian Almutawa and Iulian Neamtiu

TOPOLOGICAL DATA ANALYSIS FOR SELECTION OF MACHINE LEARNING MODELS IN CEREBRAL STROKE DETECTION 141
Aleksandra Vatian, Natalia Gusarova, Ivan Tomilov, Vladimir Shilonosov, Julia Valitova, Vladimir Sashin, Maria Bobrova, Stanislav Mashechkov, Tatiana Polevaya and Alexey Zubanenko

AI-BASED APPROACH FOR EARLY DIAGNOSIS SUPPORT IN HEMORRHAGIC STROKE 150
Athanase Sawadogo, Lydie Simone Tapsoba, Rodrique Kafando, Abdoul Kader Kabore, Aminata Sabane and Tegawende François d'Assise Bisyande

TEST AMPLIFICATION FOR MEDICAL APPS IMPLEMENTING LINEARLY-APPROXIMATABLE FUNCTIONS 157
Ian Almutawa, Muyeed Ahmed, Iulian Neamtiu and Peter Flis

MEAL-TAKING ACTIVITY MONITORING IN THE ELDERLY BASED ON SENSOR DATA: COMPARISON OF UNSUPERVISED CLASSIFICATION METHODS 165
Abderrahim Derouiche, Damien Brulin, Eric Campo and Antoine Piau

INFORMATION SYSTEM FOR CALCULATING THE HIGHEST ALLOWED WHOLESALE PRICE OF MEDICINAL PRODUCT IN CROATIA 173
Maja Fatiga and Sanja Grčić Plečko

ENABLERS OF STAKEHOLDER INVOLVEMENT IN THE DIGITALISATION OF THE UK HEALTHCARE SECTOR: AN ACTIVITY THEORY PERSPECTIVE 181
Boroto Hwabamungu and Paul Shepherd

USING DESIGN METHODOLOGIES TO EXPLORE A PRE-SURGICAL SYSTEM 190
Marko Cigljarev, Yefri Ventura, Christopher Pandolfi, Xavier Massé, Ana Rita Morais and Angela Jerath

SHORT PAPERS

BIG DATA ANALYTICS, DATA MINING AND COMPUTATIONAL INTELLIGENCE

IMPLEMENTING BUSINESS INTELLIGENCE IN HIGHER EDUCATION: A ROADMAP FOR DASHBOARD DEVELOPMENT 201
Romeu Sequeira, Paulo Alves, Arsénio Reis and Frederico Branco

LSTM NETWORK MAPPING MODEL FOR TEMPERATURE INDUCED-CRACK WIDTH VARIATION OF CONCRETE CABLE-STAYED BRIDGE 206
Xujia Liu, Youliang Ding, Chunfeng Wan and Xiaonan Zhang

DIGITAL MAPPING OF VEHICLE LOADS TO STRUCTURAL STRAINS BASED ON M-H SAMPLING AND BILSTM NETWORK 210
Xiaonan Zhang, Youliang Ding and Chunfeng Wan

CYBER-PHYSICAL CLOUD TESTBED ON RASPBERRY PI MICROCOMPUTERS 215
Adolf Kamuzora, Zakarya Farou, Tamás Jursonovics, Peter Pogrzeba and Imre Lendák

COMPARATIVE ANALYSIS OF CHANGE DETECTION METHODS IN CYBER-PHYSICAL SYSTEMS 219
Máté Hekfusz, Vrushali Mahajan, Adolf Kamuzora and Imre Lendák

CONNECTED SMART CITIES

LEVERAGING AI FOR ENHANCING DISASTER RESILIENCE IN SMART COASTAL CITIES: STRATEGIES FOR TOMORROW'S URBAN CHALLENGES 224
Rouzbeh Nazari, Mujungu Lawrence Museru and Maryam Karimi

E-HEALTH

CONDITIONAL DIFFUSION MODEL FOR GENERATING BIOLOGIC DATA 229
Denis Sidorenko and Anatoly Shalyto

MANAGEMENT BY DIGITAL RESPONSIBILITY IN HEALTH ORGANIZATIONS 233
Jutta Juliane Meier, Christian Weigand, Jochen Bauer and Björn Eskofier

AI-ENHANCED ANALYSIS OF BRAIN MORPHOMETRY LINKED TO FACTORS INFLUENCING MENTAL AND NEUROLOGICAL HEALTH OUTCOMES 239
Paula Arandia, Christian Núñez, Christian Stephan-Otto, Raul Benitez and Christian Mata

KNOWLEDGE GRAPH CONSTRUCTION VIA PRETRAINED END-TO-END TRANSFORMER MODEL 244
Denis Sidorenko and Anatoly Shalyto

REFLECTION PAPER

E-HEALTH

ADVANCING USER-CENTRIC DESIGN AND TECHNOLOGY ADOPTION FOR AGING POPULATIONS: A MULTIFACETED APPROACH 251
Andreea Stamate, Magda Velciu, Cosmina Paul, Luiza Spiru and Mircea Dan Marzan

POSTERS

E-HEALTH

USING EYE-GAZE TO EVALUATE E-HEALTH DECISION SUPPORT SYSTEMS 259
Elizabeth A. Krupinski

ENHANCING HEALTH CARE THROUGH REMOTE SERVICES AND INTERACTIVE TECHNOLOGIES 262
Ssu-Hsuan Lu

DOCTORAL CONSORTIUM

BIG DATA ANALYTICS, DATA MINING AND COMPUTATIONAL INTELLIGENCE

IMAGE SEGMENTATION APPROACHES: A COMPARATIVE STUDY 267
Rajaa Miftah, Mostafa Hanoune and Mohssine Bentaib

AUTHOR INDEX

FOREWORD

These proceedings contain the papers and posters of the 9th International Conference on Big Data Analytics, Data Mining and Computational Intelligence (BigDaCI 2024), the 10th International Conference on Connected Smart Cities (CSC 2024) and of the 16th International Conference on e-Health (EH 2024), which were organised by the International Association for Development of the Information Society, from 13 - 15 July, 2024. These conferences are part of the Multi Conference on Computer Science and Information Systems 2024, 13-15 July, which had a total of 603 submissions.

The aim of the Big Data Analytics, Data Mining and Computational Intelligence (BigDaCI'24) Conference is to serve as a forum to present current and future work as well as to exchange research ideas in this field. BigDaCI'24 invites authors to submit their original and unpublished work that demonstrate current research using big data analytics, computational intelligence and other intelligent computing techniques and their applications in science, technology, business and commerce. Submissions were accepted under the following areas and topics:
- Big Data Algorithms and Architectures
- Computational Intelligent Frameworks for Big Data Processing
- Data Mining Topics and Applications
- Big Data Applications
- Multi-Agent Systems: Models, Architectures and Applications

The Connected Smart Cities (CSC 2024) conference aims to provide a premier and exciting forum for the exchange of scientific ideas, research findings, and industrial practices among academics and practitioners, who are from different disciplines (e.g. information systems, computer science, telecommunication engineering, urban studies, and business and economics) but have common interests in smart cities. It welcomes all forms of relevant research, including conceptual, theoretical, quantitative, qualitative, and mixed-method studies. Papers with an interdisciplinary nature are particularly encouraging. In order to offer wide-ranging discussion opportunities, the conference seeks the most recent and relevant research on the following main topics:
- Wireless Connectivity in Smart Cities;
- New Business Concerns in Smart Cities;
- Data and Software Applications in Smart Cities;
- User-Centric Smart Cities;
- Security, Privacy and Identity Issues in Smart Cities;
- Greenness and Sustainability of Smart Cities;
- Smart City as an Integrated Ecosystem.

The use of ICTs (Information and Communication Technologies) in Healthcare Services is the main mechanism to improve efficiency and effectiveness. Nowadays ICTs are being developed to achieve the following objectives:
– To integrate heterogeneous systems;
– To develop frameworks to make all data meaningful, accessible and available everywhere and permanently;
– To develop AIDC (Automatic Identification and Data Collectors) systems;
– To develop intelligent systems to support clinical and management decisions;

The use of these technologies also improves the quality of patient care and reduces clinical risk. At the same time, the patient will be part of the healthcare process, having more information about diseases and access to his/her electronic health record.

The e-Health (EH) 2024 conference aims to draw together information systems, practitioners and management experts from all quadrants involved in developing computer technology to improve healthcare quality.

Submissions were accepted under the following 3 main areas in the field of e-Health within specific topics:

-Research Issues
-Management Issues
-Applications

These conferences received 152 submissions from more than 23 countries. Each submission has been anonymously reviewed by an average of four independent reviewers, to ensure that accepted submissions were of a high standard. Consequently, only 24 full papers were approved which means an acceptance rate of 16%. A few more papers were accepted as short papers and reflection paper, posters and doctoral. An extended version of the best papers may be published in the IADIS International Journal on Computer Science and Information Systems (ISSN: 1646-3692), IADIS International Journal on WWW/Internet (ISSN: 1645-7641), and also in other selected journals.

Besides the presentation of full, short and reflection papers, posters and a doctoral paper, these conferences also included one keynote presentation from an internationally distinguished researcher. We would therefore like to express our gratitude to Prof. Dr. habil. Bernd Blobel, Medical Faculty, University of Regensburg, Germany, for being our keynote speaker.

This volume has taken shape as a result of the contributions from a number of individuals. We are grateful to all authors who have submitted their papers to enrich the conference proceedings. We wish to thank all members of the organizing committee, delegates, invitees and guests whose contribution and involvement are crucial for the success of the conference.

Last but not least, we hope that everybody enjoyed the presentations, and we invite all participants for next year's edition of these conferences.

Ajith Abraham, Bennett University, India
BigDaCI 2024 Program Chair

Guo Chao (Alex) Peng, Sun Yat-Sen University, China
CSC 2024 Program Chair

Pedro Isaias, Universidade Aberta (Portuguese Open University), Portugal and
The University of New South Wales (UNSW – Sydney), Australia
EH 2024 Program Chair

Piet Kommers, University of Twente, The Netherlands
Pedro Isaias, Universidade Aberta (Portuguese Open University), Portugal and
The University of New South Wales (UNSW – Sydney), Australia
MCCSIS 2024 General Conference Co-Chairs

July 2024

PROGRAM COMMITTEE

BIG DATA ANALYTICS, DATA MINING AND COMPUTATIONAL INTELLIGENCE 2024
PROGRAM CHAIR

Ajith Abraham, Bennett University, India

CONNECTED SMART CITIES 2024
PROGRAM CHAIR

Guo Chao (Alex) Peng, Sun Yat-Sen University, Guangzhou, China

E-HEALTH 2024
PROGRAM CHAIR

Pedro Isaias, Universidade Aberta (Portuguese Open University), Portugal and
The University of New South Wales (UNSW – Sydney), Australia

MCCSIS 2024 GENERAL CONFERENCE CO-CHAIRS

Piet Kommers, University of Twente, The Netherlands
Pedro Isaias, Universidade Aberta (Portuguese Open University), Portugal and
The University of New South Wales (UNSW – Sydney), Australia

BIG DATA ANALYTICS, DATA MINING AND COMPUTATIONAL INTELLIGENCE 2024
COMMITTEE MEMBERS

Adina Florea, University Politehnica of Bucharest, Romania
Alessandro Giuliani, University of Cagliari, Italy
Álvaro Garcií-Barragán, Universidad Politécnica de Madrid, Spain
Anand Nayyar, School of Computer Science, Duy Tan University, Viet Nam
Angel Moreno, University of Alcala de Henares, Spain
Anirban Kundu, West Bengal University of Technology, India
Archil Maysuradze, Lomonosov Moscow State University, Russia
Arnulfo Alanis, Instituto Tecnológico de Tijuana, Mexico
Carmen Bustillo, Universidad Autónoma Metropolitana, Mexico
Carson K. Leung, University of Manitoba, Canada
Christos Kalloniatis, University of the Aegean, Greece
Christos Makris, University of Patras, Greece
Dan Wu, University of Windsor, Canada
Dariusz Krol, Wrocław University of Science and Technology, Poland
Egons Lavendelis, Riga Technical University, Latvia
Eva Sciacca, University of Turin, Italy

Felipe Ortega, University Rey Juan Carlos, Spain
Filippo Sciarrone, Universitas Mercatorum, Italy
Gerard Murray, Seventeen - 29, Australia
Gianni Costa, ICAR-CNR, Italy
Horia-Nicolai Teodorescu, Technical University of Iasi, Romania
I-Hsien Ting, Kaohsiung University, Taiwan
Jayesh Soni, Florida International University, USA
John Easton, University of Birmingham, United Kingdom
Jorge Adolfo Ramirez Uresti, Tecnologico de Monterrey, Mexico
Jose Molina Lopez, Universidad Carlos III de Madrid, Spain
Jose Alfredo Ferreira Costa, Federal University, UFRN, Brasil
Jose Antonio Iglesias Martinez, Carlos III University of Madrid, Spain
Kauser Ahmed, VIT University, India
Ki Chan, Hong Kong University of Science and Technology, Hong Kong
Marco Villani, University of Modena and Reggio Emilia, Italy
Maria de los Angeles Constantino-Gonzalez, Tecnologico de Monterrey, Mexico
Maria Dolores Rodriguez-moreno, University of Alcala, Spain
Maricela Quintana, Tecnológico de Monterrey, Mexico
Marius Silaghi, Florida Institute of Technology, USA
Michael Vassilakopoulos, University of Thessaly, Greece
Mohamed Ben Halima , REGIM-Lab: REsearch Groups in Intelligent Machines, Tunisia
Nicholas Beliz Osorio, Universidad Tecnológica de Panamá, Panamá
Panayiotis Bozanis, University of Thessaly, Greece
Pavel Osipov, Riga Technical University, Latvia
Rafael Leon, Universidad Politécnica de Madrid, Spain
Riccardo Ortale, ICAR-CNR, Italy
Salvatore Rinzivillo, KDDLab, ISTI, CNR, Italy
Silvia Parusheva, University of Economics - Varna, Bulgaria
Sviatoslav Braynov, University of Illinois, USA
Tibebe Beshah, Addis Ababa University, Ethiopia
Tomonobu Ozaki, Nihon University, Japan
Victor Lopez, Universidad Tecnologica de Panama, Panama
Victor Manuel Landassuri-Moreno, Universidad Autónoma del Estado de México, Mexico
Vincent Hilaire, Université de Technologie de Belfort-montbeliard, France
Wiem Chebil, Monastir University, Tunisia
Zahid Akhtar, State University of New York Polytechnic Institute, USA

CONNECTED SMART CITIES 2024

COMMITTEE MEMBERS

Alessandro Bocci, University of Pisa, Italy
Andy Hudson-smith, University College London, United Kingdom
Antonio Jara, Libelium, Spain
Assim Sagahyroon, American University of Sharjah, United Arab Emirates
Aurora Gonzalez, University of Murcia, Spain
Carlos Cruz, Universidad de Alcalá, Spain
Chafik Mahaya, University of Biskra, Algeria
Chiara Pugliese, Unviersity of Pisa, Italy
Felipe Araya, Universidad Tecnica Federico Santa Maria, Chile
Giandomenico Spezzano, ICAR-CNR, Italy
Giovanni Merlino, University of Messina, Italy

E-HEALTH 2024

COMMITTEE MEMBERS

KEYNOTE LECTURE

MANAGING HEALTHCARE TRANSFORMATION TOWARDS INTELLIGENT, ETHICAL AND SUSTAINABLE 5P MEDICINE ECOSYSTEMS

Prof. Dr. habil. Bernd Blobel,
FACMI, FACHI, FHL7, FEFMI, FIAHSI
Medical Faculty, University of Regensburg, Germany

ABSTRACT

Healthcare systems around the world are undergoing an organizational, methodological and technological transformation towards personalized, preventive, predictive, participative precision (5P) medicine ecosystems. These ecosystems consider individual health status, conditions, genetic and genomic dispositions in personal social, occupational, environmental, and behavioral contexts. For designing and managing the resulting highly interdisciplinary, complex, distributed and dynamic ecosystem, we must formally and consistently represent the system and its components at necessary granularity levels from the perspective of all actors including the subject of care. As those actors from different domains have different education, skills, and experiences, using different methodologies, languages and terminologies, communication and cooperation, i.e. interoperability, must advance from the data level (data sharing) to the knowledge level (knowledge sharing). To understand the business system, it must formally represent each considered use case structurally and functionally. Therefore, the design, implementation and management of intelligent and ethical transformed ecosystems must be realized, using a system-theoretical, architecture-centered, ontology-based and policy-driven approach, developed by the author over the last 30 years. The related model and framework has been meanwhile standardized as the ISO 23903 Interoperability and Integration Reference Architecture, defined as mandatory for any specification or project at ISO, CEN, IEEE, OMG, etc., addressing more than one domain. Thereby, it manages also security, privacy and trust in detail. The Keynote introduces necessary standards and methodologies for designing and managing 5P medicine ecosystems as well as practical examples.

Full Papers

STRUCTURAL DAMAGE DECTION BASED ON HYBRID ALGORITHM WITH SPARSE OBJECTIVE FUNCTION

Guangcai Zhang, Wenlong Zhao, Chunfeng Wan, Xiaonan Zhang, Xiujia Liu and Youliang Ding
Key Laboratory of concrete and prestressed concrete structure of Ministry of Education, Southeast University
No. 2, Southeast University Road, Jiangning District, Nanjing, China

ABSTRACT

Correlation function of acceleration response faces the difficulty in identifying local or minor structural damages, and correlation function of strain responses usually acquire unsatisfactory performance due to the insufficient global damage information. To deal with this issue, a novel objective function is proposed based on auto/cross-correlation function among acceleration and strain data with the purpose of simultaneously capturing the global and local damage information. $L_{0.5}$ sparse regularization is added into objective function to refine the ill-condition of the inverse problem and improve the robustness to noise. A new hybrid algorithm, integrating local search operator of Broyden-Fletcher-Goldfarb-Shanno (BFGS) into global search optimizer of Jaya algorithm, is developed to implement optimization. During the optimization process, the most promising domain of unknown parameters is initially explored with Jaya algorithm, and then the identified best-so-far solution is fine-tuned by BFGS algorithm. Numerical studies on a simply-supported beam structure are conducted to validate the applicability and effectiveness of the proposed method. Two different objective functions and four optimization algorithms including sequential quadratic programming, genetic algorithm, Jaya, Jaya-BFGS are compared. Identification results demonstrate that the proposed output-only method with Jaya-BFGS algorithm and objective function based on correlation function among different types of measurements and sparse regularization can accurately and efficiently detect, locate, quantify damages with limited sensors and 20% noise polluted measurements.

KEYWORDS

Output-Only Method, Jaya Algorithm, BFGS, Hybrid Algorithm, Sparse Regularization

1. INTRODUCTION

During the long-term service life, continuous health monitoring and damage assessment on the existing civil engineering structures are remarkably important to evaluate health status and predict the failure of the structures (Fan et al., 2011). In the past few decades, substantial attention has been attracted in the field of structural damage identification, especially for the vibration-based damage identification approaches (Avci et al., 2021). Some classical time domain methods have been proposed. These methods have solid background of mathematical theories, while a good initial guess of the unknown parameters and proper function gradient are required. Furthermore, inherent point-to-point search mechanism makes these methods easier to fall into the local optimum.

By contrast, computational intelligence approaches. e.g., machine learning methods, heuristic algorithms, can not only avoid the abovementioned limitations, but also nicely solve complicated optimization problems with multi-modality, nonlinearity, discontinuity. In recent years, various non-classical methods have been proposed and adopted due to their advantages of simplicity, flexibility, and robustness. It is found that computational intelligence approaches show effective performance in structural damage identification, while they still face some challenges, such as

(a) For heuristic algorithms, trial-and-error procedures are generally implemented to tune proper algorithm-specific parameters before addressing diverse optimization problems, which would inevitably waste demanding computational resources. In addition, considerable evaluations of objective function are still required after the neighborhood of the best solution is approached.

(b) In previous studies, adverse situations, such as high level of measurements noise, small or minor level of local damages, and limited sensors are rarely considered but they are quite practical in the real engineering applications, which needs to be further investigated.

To address the first challenge, some efforts have been made. A novel population-based stochastic optimization algorithm was proposed by Rao to solve complex optimization problems, named Jaya algorithm (Rao, 2016). However, as an emerging swarm intelligence algorithm, Jaya algorithm may exhibit an unfavorable convergence rate owing to heavy reliance on local search and fall into local optima because of excessive neglect of global search. In this study, a reasonable gradient-based quasi-Newton method, Broyden-Fletcher-Goldfarb-Shanno (BFGS) algorithm (Wu et al., 2020), is introduced to guide the best-so-far solution to the promising regions of the search space. In the proposed Jaya-BFGS method, the balance between exploration (global search capability) and exploitation (local search capability) is well realized. High convergence rate and good quality solution could be achieved with the aid of BFGS by making the Jaya algorithm escape from local optima, which is the first contribution of this article.

For addressing the second challenge, different types of sensors, e.g., accelerometers, strain gauges, are simultaneously employed considering their complementary characteristics. Accelerometers and strain gauges reflect damage information at global and local level, respectively (Sun et al., 2024). In this light, it is reasonable to combine local and global measurements to enhance the performance of single sensors. Considering the merits of correlation function-based damage identification method and data fusion, in this study, the correlation function among hybrid acceleration and strain is proposed and utilized to identify damages subjected to ambient excitation (Zhang et al., 2022a). Furthermore, in consideration of the sparsity of damage locations, $L_{0.5}$ sparse regularization technique (Ding et al., 2020) is introduced into objective function. It is found that the usage of the sparse regularization can partially mitigate the ill-condition and increase the identification accuracy (Zhang et al., 2023). Finally, a new objective function based on auto/cross-correlation function among hybrid measurements and sparse regularization is developed, which is the second contribution of this paper.

In this paper, two different objective functions, i.e., f_{Obj1} based on correlation function among acceleration and strain data, f_{Obj2} based on correlation function among hybrid data and sparse regularization, and four optimization algorithm namely sequential quadratic programming (SQP), GA, Jaya, Jaya-BFGS are compared. Numerical studies demonstrate the superiority of the proposed Jaya-BFGS algorithm and correlation function among hybrid measurements and sparse regularization based objective function.

2. RESEARCH METHOD

In this section, a hybrid scheme Jaya-BFGS is proposed by integrating local search operator of BFGS into global search optimizer of Jaya algorithm to improve the convergence speed and identification accuracy.

2.1 Jaya Algorithm

A novel swarm intelligence algorithm was proposed in 2016, named Jaya algorithm, to solve the constrained and unconstrained optimization problems (Rao, 2016). The behind concept of Jaya algorithm is that the offspring would try to move towards success (approaching the best solution) and avoid failure (moving away from the worst solution). There are no algorithm-specific parameters for Jaya algorithm, which significantly improves its applicability, efficiency, robustness. There are four main steps of Jaya algorithm, namely, initialization, individual updating, greedy selection, result output. The detailed introduction about Jaya algorithm can be found in (Zhang et al., 2022b).

2.2 Jaya-BFGS

The Broyden-Fletcher-Goldfarb-Shanno (BFGS) algorithm (Zhang et al., 2010) is a popular gradient-based quasi-Newton method, widely used in solving unconstrained local optimization problems, and its iteration equation is expressed as

$$\theta_{k+1} = \theta_k + \Delta\theta_k = \theta_k + \beta_k d_k \tag{1}$$

$$d_k = -G_k \nabla v(\theta_k) \tag{2}$$

where β_k means the step length; d_k represents the search direction; G_k stands for the Hessian inverse approximation; $\nabla v(\theta_k)$ is the gradient of objective function.

The gradient change Δg_k is calculated by

$$\Delta g_k = \nabla v(\theta_{k+1}) - \nabla v(\theta_k) \tag{3}$$

The Hessian approximation in BFGS is derived by

$$G_{k+1} = \left(I_n - \frac{\Delta\theta_k \Delta g_k^T}{\Delta\theta_k^T \Delta g_k} \right) G_k \left(I_n - \frac{\Delta\theta_k \Delta g_k^T}{\Delta\theta_k^T \Delta g_k} \right)^T + \frac{\Delta\theta_k \Delta\theta_k^T}{\Delta\theta_k^T \Delta g_k} \tag{4}$$

Swarm intelligence algorithms have the merits of efficient derivation-free mechanism, loose initial conditions, ease of implementation, while they may still face the problems of slow convergence speed or premature convergence. To improve the performance of Jaya algorithm, in this study, a new algorithm Jaya-BFGS is proposed by integrating local search operator of BFGS into global search optimizer of Jaya algorithm to accelerate the convergence rate to the optimal solution. The proposed Jaya-BFGS scheme, initially exploring the most promising search space of unknown parameters with Jaya algorithm, and then providing a good initial point to be quickly refined by BFGS algorithm, is shown in Figure 1. After implementing every *Nw* iterations for Jaya algorithm, the identified current best solution from Jaya is further refined as initial value with BFGS method. When the maximum iterations *Mw* of BFGS are satisfied, the updated solution is compared with its initial value, and the better one will be transferred back to the Jaya algorithm as the best-so-far solution. As presented in Figure 1, the flowchart of the proposed Jaya-BFGS algorithm has clear structure and simple operation.

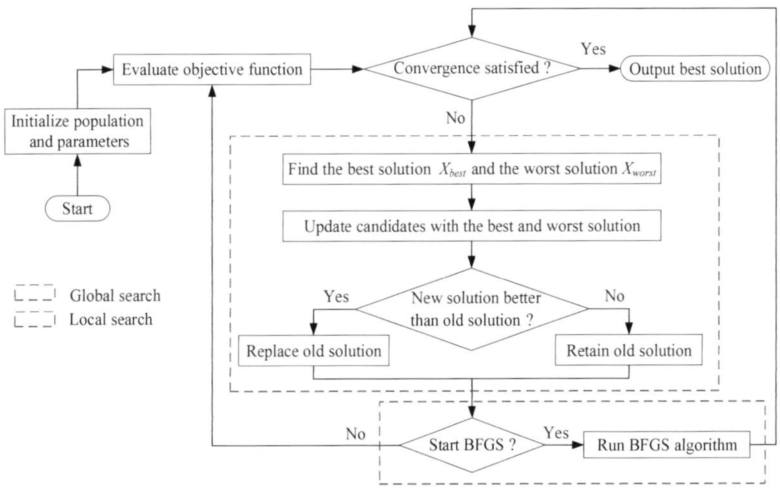

Figure 1. The flowchart of the proposed Jaya-BFGS algorithm

2.3 Sparse Objective Function

The dynamic equation of motion for a multiple degrees of freedom structural system can be expressed as

$$M\ddot{u}(t) + C\dot{u}(t) + Ku(t) = Bf(t) \tag{5}$$

where $\ddot{u}(t)$, $\dot{u}(t)$ and $u(t)$ represent the structural responses of displacement, velocity and acceleration vectors, respectively; *M*, *C*, *K* mean the mass, damping and stiffness matrices; *B* denotes the mapping vector with the value of 1 corresponding to the force location; $f(t)$ is the external ambient excitation.

It is known that accelerometers can effectively monitor the structural health state at a global level, while the nature of damage is typically a local phenomenon. In other words, the small or minor damages may not be successfully identified using acceleration responses only. In contrast, strain responses are more sensitive to small damages and they can capture local information if the damage lies in the vicinity of strain gauges. However, considerable strain gauges need to be installed and maintained to obtain the point-to-point local damage information. As a matter of fact, different types of measurements are available for the health monitoring system installed on the structure with multiple sensors. Therefore, in consideration of the complementary characteristic between acceleration and strain responses, herein, the cross-correlation function between acceleration and strain is derived.

The acceleration response at the μ-th DOF $\ddot{u}_\mu(t)$ under single excitation $f(\tau)$ is written as (Li et al., 2014)

$$\ddot{u}_\mu(t) = \int_{-\infty}^{+\infty} \ddot{h}_\mu(t-\tau) f(\tau) d\tau \tag{6}$$

where $\ddot{h}_\mu(t)$ represents the unit impulse response function, and it can be calculated by numerical integration methods, for example, Newmark-β method, assuming the structural system is initially in static equilibrium

$$\begin{cases} M\ddot{h}(t) + C\dot{h}(t) + Kh(t) = 0 \\ h(0) = 0, \dot{h}(0) = M^{-1}B \end{cases} \tag{7}$$

where $h(t)$, $\dot{h}(t)$, $\ddot{h}(t)$ are the unit impulse displacement, velocity, acceleration vectors, respectively.

The cross-correlation function between acceleration response $\ddot{u}_\mu(t)$ and strain response $\varepsilon_p(t)$ under ambient excitation $f(\tau)$ can be expressed as

$$\begin{aligned} R_{\ddot{u}_\mu \varepsilon_p}(\tau) &= E\left[\ddot{u}_\mu(t)\varepsilon_p(t+\tau)\right] = E\left\{\int_{-\infty}^{t} \ddot{h}_\mu(t-\sigma_1) f(\sigma_1) d\sigma_1 \int_{-\infty}^{t+\tau} h_p^\varepsilon(t+\tau-\sigma_2) f(\sigma_2) d\sigma_2\right\} \\ &= \int_{-\infty}^{t}\int_{-\infty}^{t+\tau} \ddot{h}_\mu(t-\sigma_1) h_p^\varepsilon(t+\tau-\sigma_2) \times E\left(f(\sigma_1) f(\sigma_2)\right) d\sigma_1 d\sigma_2 = S\int_0^{+\infty} \ddot{h}_\mu(t) h_p^\varepsilon(t+\tau) dt \end{aligned} \tag{8}$$

where S is a constant related to the applied ambient excitation.

If taking the beam-like structure as example, the convolution of the unit impulse response functions of acceleration and strain $H_{\ddot{u}_\mu \varepsilon_p}$ for candidate solution θ can be obtained by

$$H_{\ddot{u}_\mu \varepsilon_p}(\theta) = \int_0^{+\infty} \ddot{h}_\mu(t) h_p^\varepsilon(t+\tau) dt = \int_0^{+\infty} \ddot{h}_\mu(t) \begin{pmatrix} \dfrac{h_{e4}(t) - h_{e1}(t)}{l} + \left(\dfrac{6y}{l^2} - \dfrac{12xy}{l^3}\right) h_{e2}(t) + \left(\dfrac{4y}{l} - \dfrac{6xy}{l^2}\right) h_{e3}(t) \\ + \left(-\dfrac{6y}{l^2} + \dfrac{12xy}{l^3}\right) h_{e5}(t) + \left(\dfrac{2y}{l} - \dfrac{6xy}{l^2}\right) h_{e6}(t) \end{pmatrix} dt \tag{9}$$

where $\ddot{h}_\mu$ and $h_{e1}(t), h_{e2}(t), h_{e3}(t), h_{e4}(t), h_{e5}(t), h_{e6}(t)$ can be calculated with the Newmark method.

Thus, the acceleration-strain cross-correlation function $R_{\ddot{u}_\mu \varepsilon_p}$ is expressed as

$$R_{\ddot{u}_\mu \varepsilon_p}(\tau) = H_{\ddot{u}_\mu \varepsilon_p}(\theta) S \tag{10}$$

The auto/cross-correlation function among hybrid acceleration and strain data can be obtained as $\bar{R} = \left[R_{\ddot{u}_y, \ddot{u}_1}, R_{\ddot{u}_y, \ddot{u}_2}, \cdots, R_{\ddot{u}_y, \ddot{u}_{na}}, R_{\ddot{u}_y, \varepsilon_1}, R_{\ddot{u}_y, \varepsilon_2}, \cdots, R_{\ddot{u}_y, \varepsilon_{ns}}\right]^T$ if the γ-th acceleration measurement is selected as reference point.

A new objective function, denoted as f_{Obj1}, is proposed based on the auto/cross-correlation function among acceleration and strain measurements

$$f_{Obj1} = \sum_{i=1}^{nm}\sum_{j=1}^{np} \frac{\left|\bar{R}_{mea}(i,j) - \bar{R}_{est}(i,j)\right|^2}{E\left(\bar{R}_{mea}^{\ 2}(i)\right)} \tag{11}$$

where $\bar{R}_{mea}$ and $\bar{R}_{est}$ are the measured and estimated correlation function among acceleration and strain; *nm* and *np* mean the total number of sensors and data points.

In fact, damages usually occur at a few local positions in real engineering structures, which is sparse in physical space. $L_{0.5}$ regularization technique is utilized by adding a sparse penalty item, which makes the most of the damage indexes equal to zero or at least close to zero. As a result, objective function based on auto/cross-correlation function among acceleration and strain data and sparse regularization is presented as

$$f_{Obj2} = \sum_{i=1}^{ns}\sum_{j=1}^{np} \frac{\left|\bar{R}_{mea}(i,j) - \bar{R}_{est}(i,j)\right|^2}{E\left(\bar{R}_{mea}^{\ 2}(i)\right)} + \lambda\|\alpha\|_{0.5} \tag{12}$$

where λ denotes the regularization parameter; $\|\alpha\|_{0.5}$ represents the $L_{0.5}$ norm of the solution.

Compared with the L_1 and L_2 regularization, the $L_{0.5}$ regularization is more suitable to stimulate the sparsity of structural damages. Introducing the sparse penalty item can refine the ill-condition of the inverse problem and improve robustness to noise. The effectiveness of the proposed sparse objective functions will be tested and compared in numerical and experimental studies.

2.4 Procedures of Damage Identification

The proposed output-only damage identification method is carried out by the procedure as follows

Step 1: measure the structural acceleration and strain responses under unknown white noise excitation with preinstalled *na* accelerometers and *ns* strain gauges. Then, define the γ-th acceleration as reference point, and compute the measured correlation function among acceleration and strain.

Step 2: randomly initialize the structural parameters in the predefined upper and lower search space limits.

Step 3: for each candidate solution in the population, calculate the convolution of unit impulse response function and the constant vector. Compute the estimated correlation function among acceleration and strain.

Step 4: evaluate the hybrid objective function using Eq. (11) or Eq. (12) for each candidate solution.

Step 5: update new solutions with the proposed Jaya-BFGS algorithm.

Step 6: Step 3 to Step 5 is performed until the maximum iterations or convergence criterion is satisfied.

3. NUMERICAL STUDIES

To validate the effectiveness of the proposed identification approach, a series of numerical studies on a simply-supported beam structure are conducted. As presented in Figure 2, the length, width, section height of simply supported beam are 1600 mm, 50 mm, 3 mm, respectively. There are 17 nodes and 16 identical elements in total, so the length of each element is 100 mm. Intermediate nodes, i.e., from node 2 to node 16, have two degrees of freedom, namely vertical translation and rotation. The boundary connection is modelled as a pin support at node 1 and a roller support at node 17. Euler-Bernoulli beam theory is employed due to its large ratio of length to height. Young's modulus and mass density of steel material are $E = 2.1\times10^{11}$ N/m^2 and $\rho = 7850$ kg/m^3, respectively. An unknown ambient excitation with zero mean and unit standard deviation is vertically applied at nodes 12.

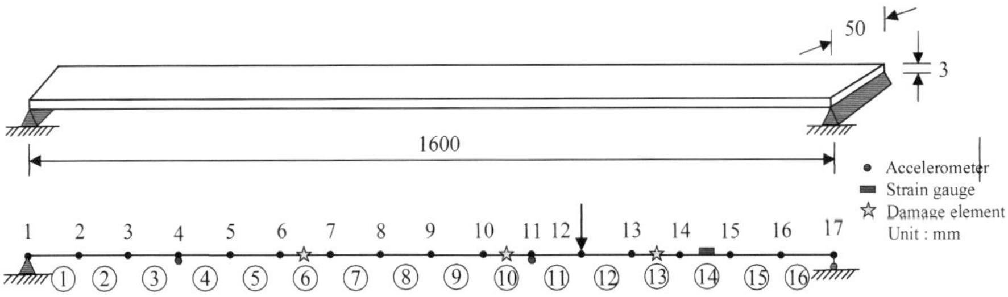

Figure 2. Numerical model of a simply-supported beam structure

Sensors are installed on the structure to record the vertical acceleration or strain responses for duration of 1800 s with sampling rate of 400 samples/s. To investigate capacity of proposed approach to identify different damage levels, large damage and small damage are both considered. It is just assumed as large damage if the damage extent is greater than 10%, and small damage if the damage extent is not more than 5%. As for the parameters of Jaya-BFGS, population size and maximum iterations are set as 100 and 200, respectively, $Nw = 20$, $Mw = 20$. In the light of stochastic nature of the optimization process, the average value based on 20 independent runs is used to provide more reliable identification results. To simulate the measured signal with noise, 0%, 10% or 20% white Gaussian noise is added into clean acceleration measurements.

3.1 Comparison on Different Objective Functions

Both of large damage and small damage are considered, $\alpha_6 = 0.2$, $\alpha_{10} = 0.1$, $\alpha_{13} = 0.05$, to study the performance of the proposed objective function f_{obj1} and f_{obj2}. Three sensors, including two accelerometers and one strain gauges, are employed, thus there are two acceleration-acceleration auto/cross-correlation functions $R_{\ddot{u}_{11},\ddot{u}_4}$, $R_{\ddot{u}_{11},\ddot{u}_{11}}$ and one acceleration-strain cross-correlation function $R_{\ddot{u}_{11},\varepsilon_{14}}$ involved in the objective function evaluation. The regularization parameter λ for the f_{obj2} is set as 1×10^{-4}. Figure 3 presents the identified damage results using f_{obj1} and f_{obj2}. It can be found that favorable results are achieved using f_{obj1} with maximum error of identified parameters less than 3% when measurements are contaminated with 20% noise, while some minor false identifications of intact elements still can be observed at elements 4, 5, 9, 11 and 15. By contrast, with the aid of sparse regularization term in f_{obj2}, the stability and accuracy of the identification results are significantly improved. A more accurate results are obtained with average error of 0.32% and maximum error of 1.52% even under 20% noise. The identified damage extents agree well with the true values and almost no false identification is observed, which demonstrates the superiority of the proposed f_{obj2} based on the correlation function among hybrid measurement data and sparse regularization.

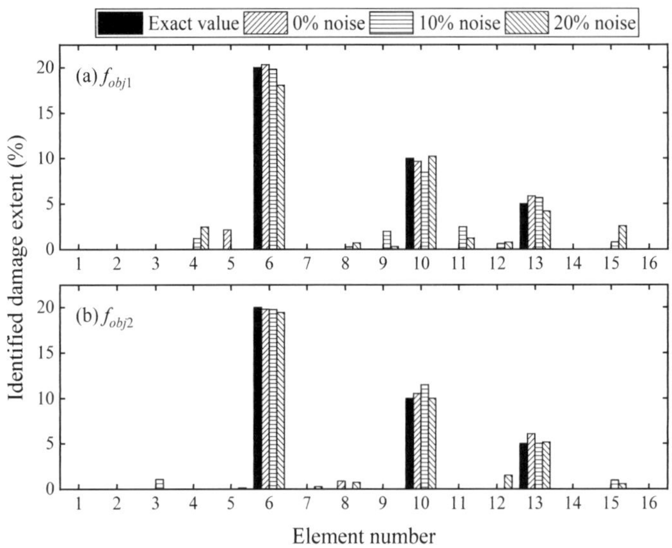

Figure 3. Identified damage results with objective function: (a) f_{obj1}; (b) f_{obj2}

3.2 Comparison with Other Optimization Algorithms

In this section, the superiority of the proposed Jaya-BFGS algorithms would be verified by comparing with other three optimization approaches, including sequential quadratic programming (SQP) (Mazzotti et al., 2019), GA, and Jaya. SQP, GA, Jaya are the typical representatives of traditional gradient-based optimization algorithm, classical heuristic algorithm, and modern swarm intelligence algorithm, respectively. Population size and convergence criterion are 100 and 10^{-3}. Maximum iterations for GA, Jaya, Jaya-BFGS are 400, 200, 100. It is assumed that there are 5% and 15% stiffness reduction in elements 6 and 13, $\alpha_6 = 0.05$, $\alpha_{13} = 0.15$. Two accelerometers are placed at the nodes 4 and 11, and one strain gauge is installed at middle position of element 14. The acceleration response at node 11 is set as reference point, thus there are two acceleration-acceleration correlation functions $R_{\ddot{u}_{11},\ddot{u}_4}$, $R_{\ddot{u}_{11},\ddot{u}_{11}}$ and one acceleration-strain cross-correlation function $R_{\ddot{u}_{11},\varepsilon_{14}}$ involved in the objective function evaluation. The objective function f_{obj2} is utilized to identify large and small damages with the regularization parameter set as $\lambda = 1\times10^{-4}$.

Figure 4 presents the final damage identification results with SQP, GA, Jaya, Jaya-BFGS under 0%, 10% and 20% noise. It can be observed that SQP and GA fail to detect the relatively small damage at element 6 when contaminated with 10% or 20% noise. Jaya algorithm is able to successfully identify the damage locations but suffers from the difficulty in quantifying the severities of stiffness reduction. In contrast, the proposed Jaya-BFGS algorithm provides the most satisfactory results with the mean errors of 0.23%, 0.58%, 0.60% and maximum errors of 1.09%, 1.73%, 1.98% with 0%, 10%, 20% noise contaminated responses.

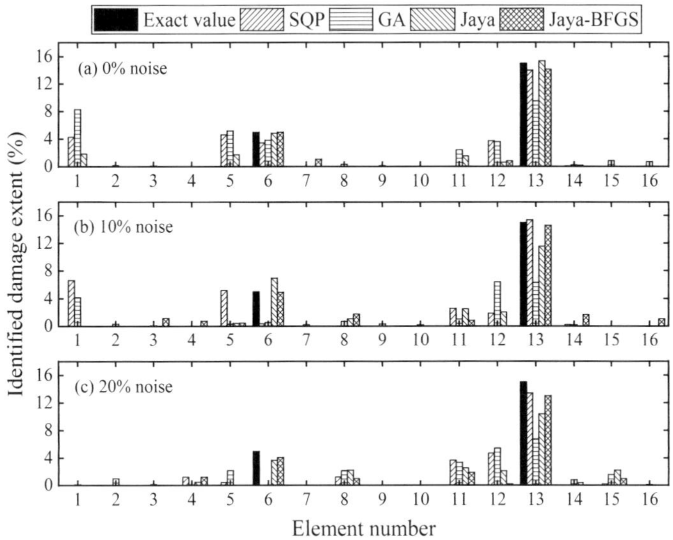

Figure 4. Identification results for SQP, GA, Jaya, Jaya-BFGS: (a) 0% noise; (b) 10% noise; (c) 20% noise

4. CONCLUSION

This paper proposes an output-only structural damage identification approach using Jaya-BFGS algorithm and sparse objective function. The numerical results on a simply-supported beam structure demonstrate that the proposed output-only method based on Jaya-BFGS and auto/cross-correlation function among hybrid measurement data and sparse regularization can simultaneously identify the location and severity of large and small damages with limited sensors and serious noise contaminated measurements. The proposed Jaya-BFGS algorithm can effectively combine the merits of Jaya and BFGS. Taking the identification accuracy and computational efficiency into consideration, the proposed Jaya-BFGS algorithm presents better performance than SQP, GA, Jaya, with the mean errors of 0.23%, 0.58%, 0.60% and maximum errors of 1.09%, 1.73%, 1.98% with 0%, 10%, 20% noise contaminated responses. Some aspects are not considered, modelling error, boundary stiffness alternation, optimal sensor placement, etc. The proposed method would be extended to other structure types, such as plates and trusses in the future study.

ACKNOWLEDGEMENT

This research has been supported by the National Natural Science Foundation of China (52378288), National Key R&D Program of China (2021YFE0112200), the Japan Society for Promotion of Science (Kakenhi No. 18K04438), the Tohoku Institute of Technology research Grant and the Postgraduate Research&Practice Innovation Program of Jiangsu Province (KYCX23_0273). The first author acknowledges China Scholarship Council (CSC 202306090271) to support visiting study in the School of Civil and Mechanical Engineering at Curtin University.

REFERENCES

Avci O., et al, (2021). A review of vibration-based damage detection in civil structures: From traditional methods to Machine Learning and Deep Learning applications. *Mech Syst Signal Process*, 147: 107077.

Ding Z. H., et al, (2020). Non-probabilistic method to consider uncertainties in structural damage identification based on Hybrid Jaya and Tree Seeds Algorithm. *Engineering Structures*, 220: 110925.

Fan W., et al, (2011). Vibration-based damage identification methods: a review and comparative study. *Structural health monitoring*, 10(1): 83-111.

Li J., et al, (2014). Substructure damage identification based on wavelet-domain response reconstruction. *Structural Health Monitoring*, 13(4): 389-405.

Mazzotti M., et al, (2019). A multiplicative regularized Gauss-Newton method with trust region Sequential Quadratic Programming for structural model updating. *Mechanical Systems and Signal Processing*, 131: 417-433.

Rao R. (2016). Jaya: A simple and new optimization algorithm for solving constrained and unconstrained optimization problems. *International Journal of Industrial Engineering Computations*, 7(1): 19-34.

Sun L. M., et al, (2024). Hybrid monitoring methodology: A model-data integrated digital twin framework for structural health monitoring and full-field virtual sensing. *Advanced Engineering Informatics*, 60: 102386.

Wu J. Y., et al, (2020). On the BFGS monolithic algorithm for the unified phase field damage theory. *Computer Methods in Applied Mechanics and Engineering*, 360: 112704.

Zhang F. L., et al, (2023). Structural damage detection based on fundamental Bayesian two-stage model considering the modal parameters uncertainty. *Structural Health Monitoring*, 22(4): 2305-2324.

Zhang G. C., et al, (2022a). Output-only structural damage identification using hybrid Jaya and differential evolution algorithm with reference-free correlation functions. *Measurement*, 199: 111591.

Zhang G. C., et al, (2022b). Structural system identification and damage detection using adaptive hybrid Jaya and differential evolution algorithm with mutation pool strategy. *Structures*, 46: 1313-1326.

Zhang Z., et al, (2010). Uniformly sampled genetic algorithm with gradient search for structural identification–Part I: Global search. *Computers & structures*, 88(15-16): 949-962.

GRAPH NEURAL NETWORKS FOR 3D OBJECT RECOGNITION AND RECONSTRUCTION: A COMPARATIVE STUDY

Rajaa Miftah and Mostafa Hanoune
LTIM laboratory. Faculty of Science Ben M'sik
Hassan II University of Casablanca
Casablanca, Morocco

ABSTRACT

This paper presents a comprehensive analysis and comparative study of existing approaches utilizing Graph Neural Networks (GNNs) for 3D object recognition and reconstruction, alongside other methods existing in the literature. Various architectures, and training strategies proposed in the literature are systematically analyzed, highlighting their strengths, weaknesses, and performance metrics. Through this comparative study, we aim to provide insights into the current state-of-the-art techniques and identify potential avenues for future research in this rapidly evolving field. By delving into the intricacies of GNN applications for 3D object recognition and reconstruction, we aim to contribute to the advancement of intelligent systems capable of navigating the complexities of 3D data. Our findings not only enrich understanding but also serve as a catalyst for innovation, driving forward the frontiers of GNN technology in computer vision and beyond.

KEYWORDS

Graph Neural Networks, 3D Object Recognition, 3D Object Reconstruction, Object Detection, Deep Learning, Computer Vision

1. INTRODUCTION

The field of artificial intelligence has witnessed remarkable progress in recent years, with Graph Neural Networks (GNNs) emerging as a transformative paradigm for processing complex data structures. Within the domain of computer vision, the challenges of three-dimensional (3D) object recognition and reconstruction have long been prominent. Traditional approaches often struggle to effectively capture the intricate spatial relationships and structural complexities inherent in real-world environments, especially when dealing with unstructured and irregular 3D data such as point clouds.

Graph Neural Networks offer a promising avenue for addressing these challenges. By directly operating on the relational information encoded in graph representations, GNNs excel at modeling complex dependencies and interactions within data, making them particularly suitable for tasks involving spatial data like 3D object recognition and reconstruction.

In this paper, we embark on a comprehensive exploration of the application of Graph Neural Networks to the tasks of 3D object recognition and reconstruction. Our primary objective is to survey the existing literature, compare various methodologies, architectures, and training strategies, and provide insights into the current state-of-the-art in this rapidly evolving field. By doing so, we aim to not only contribute to the body of knowledge in this area but also to identify potential directions for future research.

The remainder of the paper is organized as follows: Section II provides background information and reviews related work on GNNs and their applications in 3D computer vision. In Section III, we present a comparative analysis of GNN architectures for 3D object recognition and reconstruction. Subsequent sections delve into leveraging GNNs for 3D object reconstruction, shedding light on the performance metrics and evaluation, followed by future direction, and conclusions.

Through this endeavor, we aim to shed light on the potential of Graph Neural Networks in advancing the field of 3D object recognition and reconstruction and pave the way for future advancements in this exciting area of research.

2. BACKGROUND AND RELATED WORK

First, In this section, we provide essential background information on Graph Neural Networks (GNNs) and their applications in computer vision, particularly focusing on 3D object recognition and reconstruction. We begin by elucidating the fundamental concepts of GNNs, highlighting their unique ability to process graph-structured data and model complex relational information. Key architectures and methodologies have been proposed in the literature, ranging from graph convolutional networks to message passing networks, showcasing their suitability for tasks involving spatial data.

Furthermore, we review related work in the field of 3D object recognition and reconstruction, emphasizing prior studies that have leveraged GNNs or similar graph-based approaches. For example, Shi et Rajkumar in Shi and Rajkumar (2020) introduced a novel GNN architecture for 3D object recognition from point clouds, achieving state-of-the-art performance on benchmark datasets. Additionally, recent work by Wang et al. Wang et al. (2021) explored the use of graph-based representations for 3D object reconstruction from multi-view images, demonstrating promising results in capturing fine-grained object details. Moreover, Park et al. Park et al. (2019) proposed a graph-based method for 3D object detection and tracking in point clouds, achieving competitive performance compared to traditional approaches.

By surveying the terrain of existing research, we aim to establish a comprehensive understanding of the state-of-the-art in GNN-based approaches to 3D object recognition and reconstruction. This background knowledge serves as a foundation for our subsequent methodology and comparative analysis, enabling a contextualized interpretation of our findings and insights.

3. COMPARATIVE ANALYSIS OF GNN ARCHITECTURES

In this section, we delve into the comparative analysis of Graph Neural Network (GNN) architectures "Figure 1" for 3D object recognition and reconstruction. Through a systematic review and synthesis of existing literature, we compare and contrast various approaches based on their methodologies, architectures, performance outcomes, and key findings.

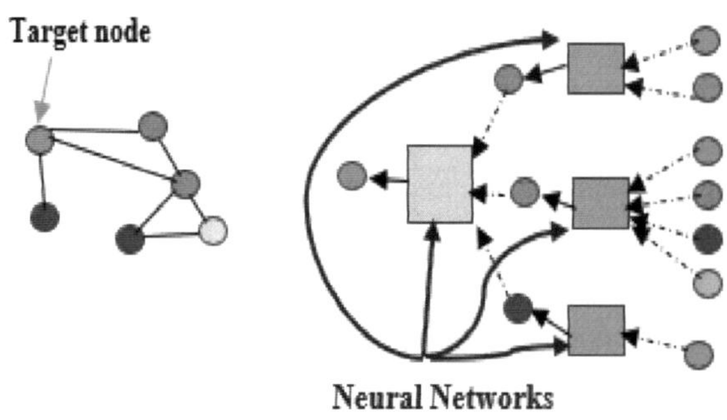

Figure 1. Illustration of GNNs architecture

3.1 Graph Convolutional Networks

Graph Convolutional Networks (GCNs) are a type of neural network that operates directly on graphs, enabling the model to learn from the graph structure itself. They generalize convolutional operations from regular grids to graphs, allowing for the incorporation of node connectivity into the learning process. This capability makes them particularly effective for processing data that is naturally represented as graphs, such as social networks, molecular structures, and, relevantly, 3D point clouds. The landmark study by Kipf and Welling in 2017 Kipf and Welling (2017) introduced semi-supervised classification using GCNs, setting a foundational framework for their application in 3D object recognition. Their method, focusing on the aggregation of neighboring information, proves highly effective for analyzing the irregular and complex nature of point clouds.

3.2 Graph Attention Networks

Graph Attention Networks (GATs) Innovating within the graph neural network domain, GATs incorporate attention mechanisms to prioritize information in the network, enhancing the model's focus on relevant parts of the graph for predictive accuracy. This is particularly valuable in environments where connections carry varying degrees of significance. Introduced by Velickovic et al. in 2018 Velickovic et al. (2018), GATs have elevated the capabilities of GCNs by dynamically weighting the significance of each node's contributions, a feature that has shown to be advantageous in dynamic environments like 3D object detection, enabling nuanced feature learning across complex point clouds.

3.3 Dynamic Graph CNNs

Dynamic Graph CNNs (DGCNNs) mark a significant advancement in graph-based learning by introducing a dynamic graph architecture, which adapts during the learning process to better capture complex spatial relationships. This innovation allows for a more accurate representation of 3D structures, as seen in the work of Wang et al. in 2019 Wang et al. (2019). Their approach, through the EdgeConv operation, dynamically updates graph connections to reflect local geometric structures more accurately, proving invaluable in tasks like 3D object reconstruction where detail preservation is key.

3.4 Hierarchical Graph Networks

This approach leverages multi-resolution data processing to create a scalable and efficient means of handling extensive datasets, as demonstrated by Guibas et al. in 2017 Guibas et al. (2017). By constructing hierarchical representations, these networks can process information at various granularities, allowing for a comprehensive understanding of complex 3D environments. This method has proven particularly effective in enhancing the efficiency and accuracy of 3D object reconstruction tasks. Hierarchical graph networks facilitate the modeling of objects at different levels of detail, enabling the network to capture both global shapes and fine-grained features. This dual focus is crucial for tasks that require a detailed understanding of 3D objects, from broad structural outlines to specific surface characteristics.

3.5 Performance and Comparison

Comparing these architectures reveals distinct advantages depending on the application scenario. GCNs and GATs offer foundational frameworks with broad applicability but might lack the specificity required for complex 3D geometries. DGCNNs and hierarchical methods excel in capturing local to global structures, proving invaluable for detailed reconstruction tasks.

GCNs and GATs generally provide a robust starting point for feature extraction in 3D objects, suitable for a wide range of applications but may fall short in capturing fine-grained details. DGCNNs, with their dynamic edge updates, are particularly adept at adapting to the intricate variability in point cloud data, making them superior for tasks requiring high fidelity in local geometric structures. Hierarchical approaches offer a scalable solution to processing large 3D datasets, enabling effective data management and processing efficiency Chen et al. (2019).

The choice of GNN architecture in 3D object recognition and reconstruction tasks is influenced by the specific requirements of the task at hand, including the need for fine-grained detail, scalability, and adaptability to complex structures. As the field progresses, hybrid and enhanced versions of these architectures could emerge, further pushing the boundaries of what is achievable in 3D computer vision tasks.

4. LEVERAGING GNNS FOR 3D OBJECT RECONSTRUCTION

In this part, we explore the burgeoning field of 3D object reconstruction leveraging Graph Neural Networks (GNNs). Recent advancements in GNNs have opened up exciting new possibilities for reconstructing complex 3D structures from sparse or unstructured data, revolutionizing traditional approaches to object reconstruction.

4.1 GNN-Based Approaches to Object Reconstruction and Graph Representation of 3D Scenes

Graph Neural Networks (GNNs) offer a novel paradigm for object reconstruction by directly operating on graph representations of data. For instance, recent works like the study by Wang et al. Wang et al. (2019) have demonstrated the effectiveness of GNNs in reconstructing 3D shapes from point cloud data. By encoding spatial relationships between points as edges in a graph, GNNs can reason about the underlying structure of the scene and reconstruct objects with remarkable accuracy.

Central to the success of GNN-based object reconstruction is the representation of 3D scenes as graphs. Pioneering works, such as that of Simonovsky and Komodakis Simonovsky and Komodakis (2017), have championed the use of graph-based representations for 3D shape reconstruction. These approaches highlight the advantages of graph convolutions in capturing both local and global features of 3D shapes. By leveraging graph representations, GNNs can effectively model the complex spatial relationships and dependencies within the data, enabling more accurate and detailed reconstructions.

Together, these advancements underscore the transformative potential of GNNs in the field of object reconstruction. By operating directly on graph representations of 3D scenes, GNNs offer a powerful framework for understanding and reconstructing complex spatial structures with unprecedented accuracy and fidelity.

4.2 Applications and Challenges

GNN-based approaches to object reconstruction have found applications across various domains. For example, in cultural heritage preservation, works like that of San et al. San et al. (2018) have utilized GNNs to reconstruct damaged artifacts and historical sites from sparse archaeological data. Similarly, in robotics, studies such as the work by Guibas et al. Guibas et al. (2017) have leveraged GNNs to reconstruct 3D maps of indoor environments, enabling robots to navigate and interact with their surroundings autonomously.

However, challenges persist in the realm of GNN-based object reconstruction. Works such as the study by Li et al. Li et al. (2018) highlight issues such as scalability with large-scale scenes, handling occlusions and clutter, and ensuring geometric accuracy in the reconstructed models. Addressing these challenges requires further research and innovation in GNN architectures, training methodologies, and data representation techniques.

5. LEVERAGING GNNS FOR 3D OBJECT RECONSTRUCTION

In this section, we assess the performance of Point-GNN, a type of Graph Neural Network (GNN) specifically designed for processing point cloud data in 3D object detection tasks. Point-GNN operates on graph representations of point clouds, leveraging the inherent spatial relationships and structural dependencies encoded within the data.

To provide a comprehensive assessment, we compare this GNN type to other state-of-the-art methods for 3D object detection. We briefly introduce the competing methods, including AVOD-FPN, F-PointNet, VoxelNet, PointRCNN, UberATG-ContFuse, UberATG-MMF, SECOND, and PointPillars, followed by an overview of the comparative results obtained on the KITTI Geiger et al. (2012) test dataset.

5.1 Competing Methods

- AVOD-FPN: Anchor-based Voting 3D Fully Convolutional Network with Feature Pyramid Networks (AVOD-FPN) utilizes anchor-based detection combined with feature pyramid networks for 3D object detection, typically using LiDAR dataKu et al. (2018).
- *F-PointNet*: An extension of PointNet, introducing improvements in feature learning and contextual modeling, typically operating on LiDAR data.
- UberATG-ContFuse: a fusion-based method developed by Uber Advanced Technologies GroupXu et al. (2018), combining data from multiple sensors, such as LiDAR and cameras, for comprehensive scene understanding.
- UberATG-MMF: Another method developed by Uber ATG, UberATG-MMF (Multi-Modal Fusion) integrates information from LiDAR and images using multimodal fusion techniques for improved object detection and localization Liang et al. (2019).
- *PointPillars*: is a state-of-the-art LiDAR-based object detection method that utilizes a pillar-based encoding scheme for point clouds, followed by a 2D convolutional network for feature extraction and object detection.
- VoxelNet: Operates on voxelized representations of point clouds, leveraging volumetric convolutions for feature extraction and object detection, typically using LiDAR data Zhou and Tuzel (2018).
- SECOND: Sparsely Embedded Convolutional Detection (SECOND) is a LiDAR-based 3D object detection framework that operates directly on raw point clouds, utilizing sparsely embedded convolutional layers for efficient feature extraction Yan et al. (2018).
- *PointRCNN*: A region-based 3D object detection framework that incorporates both 2D and 3D information for accurate localization and classification of objects in point clouds, typically using LiDAR data.

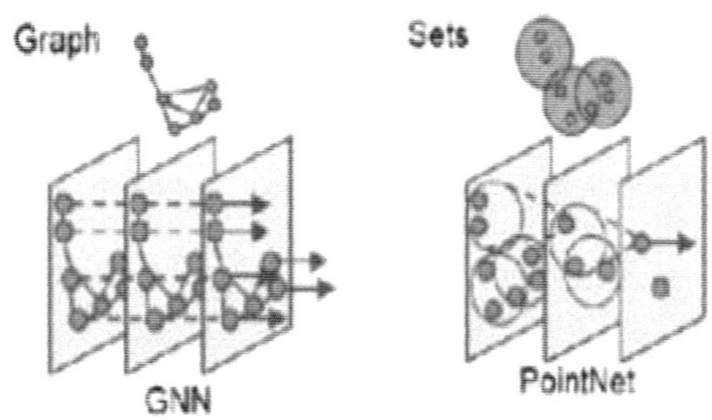

Figure 2. Example of Point Cloud representations in GNN and PointNet

5.2 Comparative Analysis

To assess the performance of Point-GNN in relation to these competing methods, we present a comparative analysis based on evaluations conducted on the KITTI test dataset. The modalities used by these methods primarily consist of LiDAR or a combination of LiDAR and image data, denoted as (LiDAR or LiDAR+Image). This indicates the reliance on point cloud data captured by LiDAR sensors, with some methods augmenting this information with additional imagery for enhanced object detection capabilities. The comparative table below summarizes the results in terms of Average Precision (AP) obtained for different categories of objects (e.g., cars, cyclists) across varying difficulty levels (e.g., easy, moderate, hard).

Table 1. Average Precision (AP) comparison of 3D object detection on the KITI test dataset

Method	Modality	Car			Cyclist		
		Easy	*Moderate*	*Hard*	*Easy*	*Moderate*	*Hard*
AVOD-FPNKu et al. (2018)	LiDAR + Image	81.94	71.88	66.38	64.00	52.18	46.61
F-PointNetQi et al. (2018)	LiDAR + Image	81.20	70.39	62.19	71.96	56.77	50.39
UberATG-ContFuseXu et al. (2018)	LiDAR + Image	82.54	66.22	64.04	N/A	N/A	N/A
UberATG-MMFLiang et al. (2019)	LiDAR + Image	86.81	76.75	68.41	N/A	N/A	N/A
PointPillarsLang et al. (2019)	LiDAR	79.05	74.99	68.30	75.78	59.07	52.92
VoxelNetZhou and Tuzel (2018)	LiDAR	81.97	65.46	62.85	67.17	47.65	45.11
SECONDYan et al. (2018)	LiDAR	83.13	73.66	66.20	70.51	53.85	53.85
PointRCNNShi et al. (2019)	LiDAR	85.94	75.76	68.32	73.93	59.60	53.59
Point-GNNShi and Rajkumar (2020)	LiDAR	**88.33**	**79.47**	**76.06**	**78.60**	**63.48**	**57.08**

5.3 Results

This comparative analysis offers valuable insights into the efficacy of Point-GNN and its competitors across different object categories and difficulty levels. By thoroughly examining their performance metrics, it aims to provide a clear understanding of the strengths and limitations of each method in real-world 3D object detection scenarios. Notably, the presented table showcases the effectiveness of Graph Neural Networks, particularly highlighting the remarkable performance of Point-GNN. This underscores the growing prominence and applicability of GNNs in the field of 3D object detection, emphasizing their potential to significantly advance the capabilities of modern detection systems.

The choice of architecture depends on the specific requirements of the application and the nature of the input data. For scenarios with limited sensor modalities or computational resources, simpler architectures like PointPillars may be preferable. However, for tasks demanding high precision and robustness, more complex architectures such as Point-GNN offer superior performance.

While our comparative analysis focused on performance metrics, it's essential to acknowledge the impact of data quality on model performance. Future research should explore how different models handle noise, occlusions, and other challenges present in real-world datasets. Strategies for data preprocessing, augmentation, and robust feature extraction will be critical in improving the resilience of object recognition systems to environmental variations.

6. FUTURE DIRECTIONS

The future of Graph Neural Networks (GNNs) in the domains of 3D object recognition and reconstruction holds immense potential for innovation and advancement. One promising avenue for future exploration involves the development of hybrid architectures that seamlessly integrate GNNs with other deep learning frameworks. By combining the strengths of GNNs in capturing spatial dependencies with the robust feature learning capabilities of convolutional neural networks (CNNs), as demonstrated by Wang et al. Wang et al. (2019), researchers can potentially achieve more comprehensive and accurate representations of 3D objects.

Furthermore, the application of attention mechanisms within GNN architectures represents an exciting direction for enhancing object recognition and reconstruction performance. Recent studies, such as the work by Velickovic et al. Velickovic et al. (2018), have highlighted the effectiveness of attention mechanisms in GNNs for capturing salient features and refining predictions. Incorporating attention mechanisms into

GNN-based models could enable more focused and context-aware processing of 3D scene data, leading to improved object recognition accuracy and reconstruction quality.

Another area ripe for exploration is the integration of GNNs with multimodal sensor data for enhanced perception in complex environments. By leveraging data from multiple sensors, including LiDAR, cameras, and inertial measurement units (IMUs), researchers can create more robust and versatile object recognition and reconstruction systems. Works such as that by Uber ATG Chai et al. (2019) have demonstrated the benefits of multimodal fusion techniques in improving object detection performance, and extending these approaches to incorporate GNN-based processing could further enhance system capabilities.

Moreover, the development of GNN-based frameworks for real-time object recognition and reconstruction represents a crucial research direction for practical deployment in autonomous systems and robotics. By optimizing model architectures and leveraging efficient inference techniques, researchers can enable GNN-based systems to operate in real-world scenarios with low latency and high throughput. Studies such as that by Qi et al. Guibas et al. (2017), which propose lightweight GNN architectures for point cloud processing, provide valuable insights into the feasibility of real-time GNN applications.

In summary, the future directions for GNNs in 3D object recognition and reconstruction encompass a wide range of research avenues, including hybrid architectures, attention mechanisms, multimodal fusion, and real-time deployment. By exploring these directions, researchers can advance the state-of-the-art in intelligent perception systems and pave the way for transformative applications in various domains.

7. CONCLUSION

In conclusion, this paper has provided a comprehensive overview of Graph Neural Networks (GNNs) and their applications in 3D object recognition and reconstruction. Through a systematic review of existing methodologies and approaches, we have highlighted the transformative potential of GNNs in capturing spatial relationships and dependencies within 3D data, leading to accurate and detailed reconstructions.

From the foundational principles of GNNs to the latest advancements in architecture design and methodology, we have surveyed the landscape of GNN-based object recognition and reconstruction. Additionally, through comparative analyses and discussions, we have examined the performance of Point-GNN, a specific type of GNN architecture, alongside other state-of-the-art methods on the KITTI test dataset. This analysis provides valuable insights into the effectiveness of GNNs, particularly Point-GNN, in practical applications of 3D object recognition and reconstruction.

As we look ahead, the performance on the KITTI test dataset offers valuable benchmarks for assessing the efficacy of GNNs in real-world scenarios. Continued research and innovation in this domain hold great promise for revolutionizing object reconstruction across various domains, including robotics, autonomous systems, and medical imaging. By embracing interdisciplinary collaboration and pushing the boundaries of innovation, we can unlock new frontiers in understanding and reconstructing complex 3D environments with unprecedented accuracy and fidelity.

This paper serves as a foundational resource for researchers and practitioners interested in leveraging GNNs, especially Point-GNN, for object reconstruction, inspiring further exploration and discovery in this exciting and rapidly evolving field.

REFERENCES

Shi and Rajkumar (2020) W. Shi and R. Rajkumar, "Point-GNN: Graph neural network for 3D object detection in a point cloud," in Proceedings of the IEEE Computer Society Conference on Computer Vision and Pattern Recognition, 2020, pp. 1708–1716, doi: 10.1109/CVPR42600.2020.00178.

Wang et al. (2021). Y. Wang, W. Sun, J. (Judy) Jin, Z. (James) Kong, and X. Yue, "MVGCN: Multi-View Graph Convolutional Neural Network for Surface Defect Identification Using Three-Dimensional Point Cloud," J. Manuf. Sci. Eng., vol. 145, no. 3, pp. 1–16, 2023, doi: 10.1115/1.4056005.

Park et al. (2019). M. Park, H. Seong, W. Jang, and E. Kim, "Graph-Based Point Tracker for 3D Object Tracking in Point Clouds," in Proceedings of the 36th AAAI Conference on Artificial Intelligence, AAAI 2022, 2022, vol. 36, no. 1, pp. 2053–2061, doi: 10.1609/aaai.v36i2.20101.

Kipf and Welling (2017). T. N. Kipf and M. Welling, "Semi-supervised classification with graph convolutional networks," in 5th International Conference on Learning Representations, ICLR 2017 - Conference Track Proceedings, 2017, pp. 1–14.

Velickovic et al. (2018) P. Veličković, A. Casanova, P. Liò, G. Cucurull, A. Romero, and Y. Bengio, "Graph attention networks," in 6th International Conference on Learning Representations, ICLR 2018 - Conference Track Proceedings, 2018, pp. 1–12, doi: 10.1007/978-3-031-01587-8_7.

Wang et al. (2019). Y. Wang, Y. Sun, Z. Liu, S. E. Sarma, M. M. Bronstein, and J. M. Solomon, "Dynamic graph Cnn for learning on point clouds," ACM Trans. Graph., vol. 38, no. 5, 2019, doi: 10.1145/3326362.

Guibas et al. (2017). C. R. Qi, H. Su, K. Mo, and L. J. Guibas, "PointNet: Deep learning on point sets for 3D classification and segmentation," in Proceedings - 30th IEEE Conference on Computer Vision and Pattern Recognition, CVPR 2017, 2017, vol. 2017-Janua, pp. 77–85, doi: 10.1109/CVPR.2017.16.

Chen et al. (2019). J. Chen, B. Lei, Q. Song, H. Ying, D. Z. Chen, and J. Wu, "A Hierarchical Graph Network for 3D Object Detection on Point Clouds," 2020.

Simonovsky and Komodakis (2017) M. Simonovsky and N. Komodakis, "Dynamic edge-conditioned filters in convolutional neural networks on graphs," in Proceedings - 30th IEEE Conference on Computer Vision and Pattern Recognition, CVPR 2017, 2017, vol. 2017-Janua, pp. 29–38, doi: 10.1109/CVPR.2017.11.

San et al. (2018) W. Liu, J. Sun, W. Li, T. Hu, and P. Wang, "Deep learning on point clouds and its application: A survey," Sensors (Switzerland), vol. 19, no. 19. pp. 1–22, 2019, doi: 10.3390/s19194188.

Li et al. (2018). Y. Li, R. Bu, and X. Di, "PointCNN : Convolution On X -Transformed Points," 2017.

Geiger et al. (2012). A. Geiger, P. Lenz, C. Stiller, and R. Urtasun, "Vision meets robotics: The KITTI dataset," Int. J. Rob. Res., vol. 32, no. 11, pp. 1231–1237, 2013, doi: 10.1177/0278364913491297.

Ku et al. (2018). J. Ku, M. Mozifian, J. Lee, A. Harakeh, and S. L. Waslander, "Joint 3D Proposal Generation and Object Detection from View Aggregation," in IEEE International Conference on Intelligent Robots and Systems, 2018, pp. 5750–5757, doi: 10.1109/IROS.2018.8594049.

Xu et al. (2018). M. Liang, B. Yang, S. Wang, and R. Urtasun, "Deep Continuous Fusion for Multi-sensor 3D Object Detection," in Lecture Notes in Computer Science (including subseries Lecture Notes in Artificial Intelligence and Lecture Notes in Bioinformatics), 2018, vol. 11220 LNCS, pp. 663–678, doi: 10.1007/978-3-030-01270-0_39.

Liang et al. (2019) M. Liang, B. Yang, Y. Chen, R. Hu, and R. Urtasun, "Multi-task multi-sensor fusion for 3D object detection," in Proceedings of the IEEE Computer Society Conference on Computer Vision and Pattern Recognition, 2019, vol. 2019-June, pp. 7337–7345, doi: 10.1109/CVPR.2019.00752.

Zhou and Tuzel (2018) Y. Zhou and O. Tuzel, "VoxelNet: End-to-End Learning for Point Cloud Based 3D Object Detection," in Proceedings of the IEEE Computer Society Conference on Computer Vision and Pattern Recognition, 2018, pp. 4490–4499, doi: 10.1109/CVPR.2018.00472.

Yan et al. (2018). Y. Yan, Y. Mao, and B. Li, "Second: Sparsely embedded convolutional detection," Sensors (Switzerland), vol. 18, no. 10, pp. 1–17, 2018, doi: 10.3390/s18103337.

Qi et al. (2018). C. R. Qi, W. Liu, C. Wu, H. Su, and L. J. Guibas, "Frustum PointNets for 3D Object Detection from RGB-D Data," in Proceedings of the IEEE Computer Society Conference on Computer Vision and Pattern Recognition, 2018, pp. 918–927, doi: 10.1109/CVPR.2018.00102.

Lang et al. (2019) A. H. Lang, S. Vora, H. Caesar, L. Zhou, J. Yang, and O. Beijbom, "Pointpillars: Fast encoders for object detection from point clouds," in Proceedings of the IEEE Computer Society Conference on Computer Vision and Pattern Recognition, 2019, vol. 2019-June, pp. 12689–12697, doi: 10.1109/CVPR.2019.01298.

Shi et al. (2019) S. Shi, X. Wang, and H. Li, "PointRCNN: 3D object proposal generation and detection from point cloud," in Proceedings of the IEEE Computer Society Conference on Computer Vision and Pattern Recognition, 2019, vol. 2019-June, pp. 770–779, doi: 10.1109/CVPR.2019.00086.

Chai et al. (2019) Uber Advanced Technologies Group, "No Title," Research Publications [White paper].

FEATURE SELECTIONS AND EXPLORATORY ANALYSIS FOR ML STOCK PREDICTIONS USING SET50

Gridaphat Sriharee
Department of Computer and Information Science, Faculty of Applied Science,
King Mongkut's University of Technology North Bangkok
1518 Pracharat 1, Wonsawang, Bangsur, Bangkok, 10800, Thailand

ABSTRACT

Stock prediction using machine learning is an interesting topic for investors. However, the performance of the prediction relies on different techniques and the train data set. Feature selection is an important factor in creating the model. In this paper, feature selection based on correlation was proposed, and exploratory analysis of SET50 from Thailand Stock market was experimented. The fundamental data – OHLC, percent change, and three technical indicators – SMA, RSI, and EMA were applied in the experimentation. The fundamental ML models: linear regression, logistic regression, ANN, decision tree, random forest, and XGBoost were conducted to evaluate the model's performance based on the proposed features. The result of the experiments showed that multiple-parameter model performed better than single-parameter model. RSI was the most powerful indicator, and different periods can be a critical factor for different ML models.

KEYWORDS

Feature Selection, Stock Analysis, Technical Indicator, Machine Learning

1. INTRODUCTION

Stock analysis can be classified broadly into two types: fundamental analysis and technical analysis. However, other techniques are addressed widely, for example, algorithm trading and quantitative analysis. Quantitative analysis is one of the techniques for stock analysis that applies mathematical and statistical principles together. It also includes machine learning or artificial intelligence as the technique for stock prediction. Therefore, these become the challenging topics among different perspectives of the investors. Algorithmic trading addresses trading using an autonomous agent – the agent that performs bidding and offering into the trading platform by running the script automatically when conditions are satisfied. Algorithmic trading can combine machine learning techniques and the strategy for buy/sell signals using technical indicators.

The machine learning model relies on feature selections. The model requires training data on which features can predict the target, and the features should be independent of each other, but they should define the target at the same time. However, stock data relates to each other as a basis, for example, open, high, low, and close prices are highly relevant. Reduction of irrelevant or redundant features results in better prediction accuracy and decreases the computational complexity. Feature selection requires OHLC information, technical indicators (e.g., RSI, Moving average), and fundamental indicators (e.g., interest rates, consumer price index) (Htun et al., 2023). Stock trading has both internal factors (e.g., business, factsheet, business performance) and external factors (e.g., news, other stock markets, events) that cannot be controlled, and these affect the stock price. Thus, analysis for risk management was chosen for the investors.

Technical indicators are used commonly for buy/sell signals, monitoring stock trends, and movement analysis. Investors can use the indicators to support his/her decision in taking action. The technical indicators can be used as features in the machine learning model. However, the performance and accuracy can be varied according to different techniques. Stock prediction using machine learning can enhance algorithmic trading where bids/offers are executed automatically.

This paper proposes exploratory analysis for feature selection to investigate which technical indicators are powerful for which ML models. Some contributions are as follows.

(1) Exploratory analysis was explored to study the existence of stock and to investigate fundamental data of SET and SET50 correlations in the period (2022-2023).
(2) To find better features using correlation analysis. Each stock was explored to examine its technical indicators values and evaluated on ML models for stock predictions.
(3) To evaluate the ML model that used the proposed features for stock predictions. Algorithms: linear regression, logistic regression, decision tree, random forest, XGBoost, and artificial neural network (ANN) were focused.

Section 2 is the background of this paper. Section 3 is related work. Section 4 is an exploratory analysis of SET and SET50 index. Correlation analysis of feature selection is explained in this section. Section 5 is the experimentation using the machine learning models and various technical indicators, and Section 6 is the conclusion and future work.

2. BACKGROUND

2.1 Stock Analysis

Stock analysis can be classified into two types: fundamental analysis, and technical analysis. However, there are many other analyses, for example, quantitative analysis, and algorithmic analysis which can be a part of quantitative analysis. Fundamental analysis focuses on fundamental business information, for example, balance sheets, reports, and business revenue. This technique relies on public information however, this information can be disclosed from the facts because of business reasons.

Technical analysis relies on technical indicators, for example, the Simple Moving Average (SMA), Relative Strength Index (RSI), and Exponential Moving Average (EMA). These indicators present the movement of stock price over a period of time (see formula 1-3). SMA is the average of price in the n period. RSI is calculated from gain/loss (RS). EMA is the average of price which has a weight given by a constant value.

$$SMA_n = \sum_{n=1}^{n-1} \frac{P_t + \cdots + P_n}{n} \tag{1}$$

$$\mathrm{RSI} = 100 - \frac{100}{1+\mathrm{RS}} \tag{2}$$

$$EMA_t = (P_t \times \frac{Constant}{1+n}) + (EMA_{t-1} \times \left(1 - \frac{Constant}{1+n}\right)) \tag{3}$$

Fundamental data for analysis can be return, log return, and percent change. The technical indicators revealed the value according to periods (short and long periods) and the average within those periods. The indicators of a long-term period usually give significant signals such as buy/sell signals and reversion of the stock. The investor and the analyst hence use technical indications wildly in stock prediction. Quantitative analysis applies mathematical and statistical techniques for modeling, analyzing, and predicting stock prices. Statistical values such as mean, standard deviation, correlation, and probability are factors for exploring the data. Because stock price may be affected by internal factors (management, strength of the business), and external factors which cannot be controlled. Therefore, there are no analysis techniques that can deal with the uncertainty. However, analysis is still necessary for risk management to avoid loss.

2.2 Machine Learning and Stock Prediction

The technical analysis applies many indicators that support buy/sell signal and prediction. Fundamental indicators that are used for movement analysis are SMA, RSI, ROC, MACD, and EMA. These can be used with different periods, for example, SMA 14 days (Figure 1) is the average price 14 days before a day. The longer period may give a significant signal over a shorter period for example, when SMA 30 days crosses over SMA 14 days it may mean the stock is in the down trend.

Figure 1. SET prices period 2022-01-01 – 2023-12-31

Machine learning requires features and targets. The model can be used for different strategies, for example, buy signal only, sell signal only, and buy and sell signal. ML classification model supports buy/sell signal, and this can be analyzed by decision tree, random forest, and XGBoost. For stock prediction, the percentage change can be the target and the features of the model can be varied. Linear regression logistic regression, and ANN can be the prediction model.

3. RELATED WORK

Some research works experimented with SET50. Chaigusin, Chirathamjaree and Clayden (2008) proposed the use of neural networks in the prediction of the stock exchange of Thailand (SET). They experimented the neural network model construction which were three to five-layered neural networks. The better model for each model was reported and that was evaluated by the Mean Absolute Percentage Error (MAPE). Their experimentation used both internal factors (e.g., SET index) and external factors (e.g., the gold prices, the minimum load rate (MLR), and other stock indexes (Dow Jones, Nikkei, Hang Seng)) as the features in the constructed model. Sanboon, Keatruangkamala and Jaiyen (2019) experimented with a deep learning model for predicting Buy and Sell Recommendations in SET using Long-Short-Term Memory. They reported that the LSTM can achieve the highest accuracy of all SET50 stocks. The performance of the proposed model is evaluated and compared to SVM, Logistic Regression, Random Forest, Decision tree, KNN, and MLP. Rungruang et al. (2019) proposed prediction accuracy of the direction of the SET50 index in the Stock Exchange of Thailand movement by using the Support Vector Machines (SVM) model. Inthachot, Boonjing and Intakosum (2016) presented a technique that was a combination of ANN and GA models for predicting SET50 stock index movement. Their experiment showed that the proposed techniques achieve a better prediction accuracy than their previous work that implemented only ANN. Sopipan, Kanjanavajee and Sattayatham (2012) predicted SET50 Index by using a multiple regression based on PCA. Their model included some stock markets index, gold, and currency market. Correlation matrix of SET50 index and explanatory variables and the experiments of the models using different PCAs were reported.

Regarding feature detection, some related works are as follows. Peng et al. (2021) investigated a set of 124 technical analysis indicators from seven stock markets. They proposed feature selection using LASSO, TS, and SFFS techniques. They focused on neural network with different setting of hidden layers and dropout rate. Kumari, Patnaik and Swarnkar (2023) mentioned that a suitable model may require input variables from four to ten. They suggested that the features can be classified into fundamental, technical features and macroeconomic features. The technical features were suitable for analysis on a particular stock.

Prediction techniques may vary in accordance with the feature set. The most widely used methods for feature selection are PCA, GA and decision trees. Another approach for feature selection is ensemble-based feature selection. Aloraini (2015) proposed the combination of stock features and gold features for stock prediction. Tumay, Aydin and Kozat (2024) presented a hierarchical stacking involves training an initial machine learning model using a subset of the features, and then updating the model's output using another machine learning algorithm that takes the remaining features or a subset of them to adjust the first layer predictions while minimizing a custom loss. Gunduz, Çataltepe and Yaslan (2017) used different types of feature selection and classification methods together. Logistic regression and gradient boosting machine were conducted.

4. EXPLORATORY ANALYSIS AND FEATURE SELECTION

4.1 SET and SET50 Index

Correlation analysis, and data distribution can be the techniques for investigating the relations between the SET50 index and the SET market. The SET market in the period of conducting research shows downtrend (Figure 1). SET announced the SET50 index in 2 periods (end of June, and end of December). Most of the stocks in SET50 for two periods were the same and only a few securities were removed in a period and may come back in another period subsequently. The SET50 index available in January 2024 was used in the proposed experiment. Figure 2 depicts the return of the SET50 index using the retrieve period 2022-01-01 to 2023-12-31. There were some stocks which had better and poor performance in this period. There were 38% of SET50 stocks which have positive (log) returns. The high returns in top-5 ranks were DELTA (79.6895%), TRUE (59.06363%), BH (48.78397%), WHA (47.45137%), and SCB (34.96678). Table 1 depicts the correlation scores of SET and SET50 stocks. Correlation scores are ranked from highest to lowest and 25 stocks are given here. For example, SET close price and SET50 close price, SET percent change and SET50 percent changes, and SET volume and SET50 volume. The volume and volatility of SET and SET50 has a very low correlation score in each stock.

Table 1. Correlations of SET and SET50 index

Stocks	Close	Stock	Change	Stock	Volume	Stock	Volatility
RATCH	0.93802	GULF	0.61281	BBL	0.49435	CPF	0.480143
EA	0.91301	EA	0.59177	TRUE	0.48930	PTT	0.456681
GPSC	0.89368	KTC	0.58778	GULF	0.46083	BEM	0.392176
KTC	0.88344	SAWAD	0.58677	SCC	0.42970	OR	0.372512
SCGP	0.86771	GPSC	0.57184	KBANK	0.39216	TLI	0.322013
HMPRO	0.85355	CPALL	0.55251	PTTEP	0.39198	SCB	0.315708
EGCO	0.83975	MTC	0.54729	TISCO	0.36783	IVL	0.303838
IVL	0.83371	PTTGC	0.54286	HMPRO	0.36505	TOP	0.303676
PTTGC	0.82753	CPN	0.54279	CPN	0.36159	TISCO	0.284421

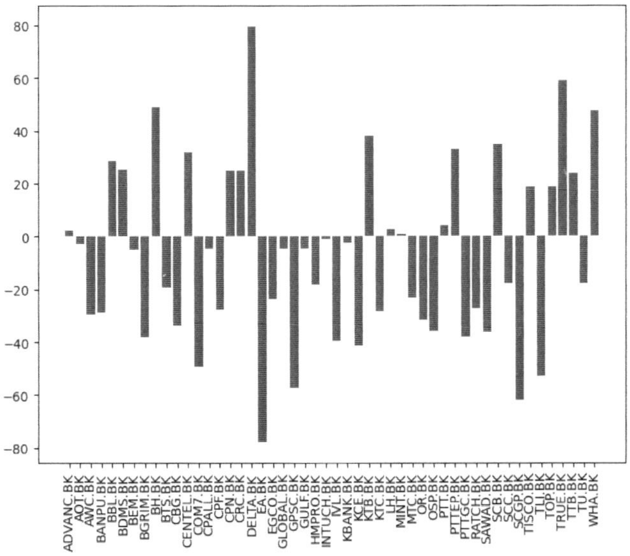

Figure 2. Log return of SET50 period 2022-01-01 – 2023-12-31

4.2 Feature Selection

The experiment to explore each stock in SET50 was conducted. Technical indicators – SMA, EMA, RSI, and percent change for the periods 14, 30, and 60 days were used for correlation analysis. The technical indicator with a similar period has higher correlations, for example, percent change 14 days has a high correlation with SMA-14, RSI-14, and EMA-14 (see Figure 3 x-label 6 and 7). Between indicators, SMA-14 has a high correlation with EMA-14 (see Figure 3 x-label 10). Volume has a negative correlation with percent change one day (see Figure 3 x-label 11). An example of the correlation scores of each stock and its movements regarding technical indicators is depicted in Table 2. RSI has highest correlation score to percent change.

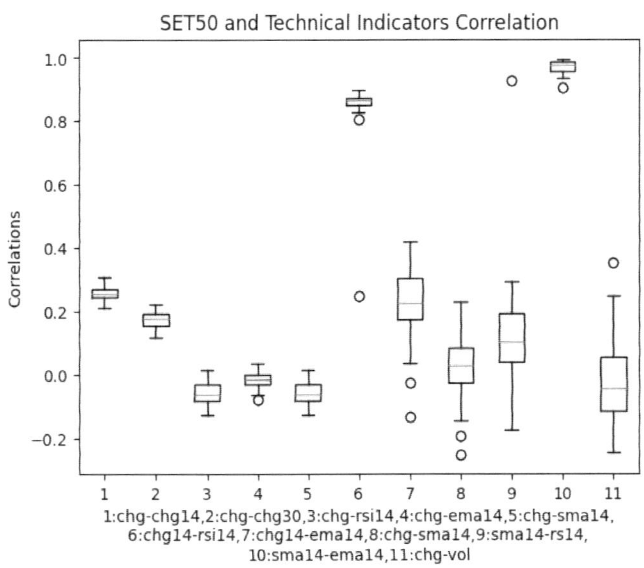

Figure 3. Correlations of technical indicators

Table 2. Correlations of percent change 14 day and others Technical indicators (14 days) of stocks

Stocks	CHG-SMA	CHG-RSI	CHG-EMA
ADVANC	0.01425	0.84387	0.18337
AOT	0.18878	0.84710	0.42070
AWC	0.23116	0.86810	0.33206
BANPU	0.22363	0.89918	0.35972
BBL	-0.05734	0.86641	0.06736

5. STOCK PREDICTION

5.1 Predicting Percent Change

The experimentation was designed for stock prediction with two purposes: to find buy signals, and to predict percent change. For the former, the target can be 1 (buy) or 0 (do nothing), and for the latter, the target is a continuous value that represents a percent change of movement. The machine model was designed into single-parameter model (only one feature) and multiple-parameter model. In this experiment, 14 days and 30 days were used. The single-parameter model was experimented with RSI-14, SMA-14, and EMA-14. The multiple-parameter model was designed into two scenarios: a group of RSI, SMA, and EMA for 14 days and another for 30 days.

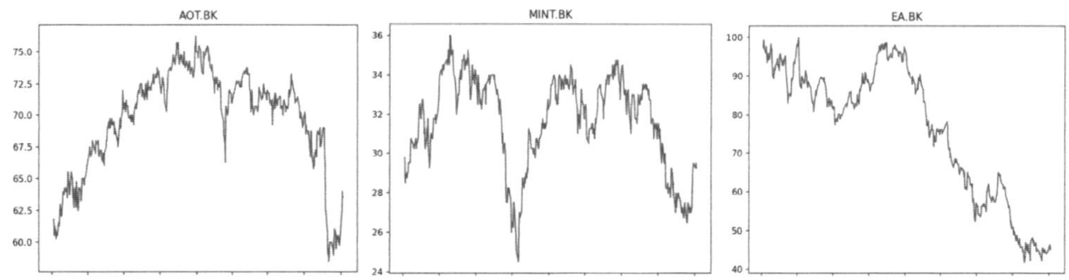

Figure 4. AOT, MINT, EA trend

SET50 stocks are ranked from high to low correlations. RSI correlation score has a high score, therefore the stocks in higher rank were chosen for the experiment. In stock selection, the stock can be seen as a stock with high, neutral, and low correlation scores, one from each group is chosen for the experimentation. They were AOT, MINT, and EA, respectively. Figure 4 depicts the trend of each stock. Table 3 depicts the R-squared value of linear regression with single parameter, and with multiple parameters. The RSI 14 days has better performance than other indicators in a single-parameter model. There were 470 and 453 observation values (data from 2022-01-01 – 2023-12-31) for 14 days and 30 days period. The R-squared values are very low, which means that the independent parameter may not be able to explain the dependent parameter.

Logistic regression (Table 3) was experimented for stock prediction using the maximum depth as 30 and a minimum sample is three. MINT and EA have similar results in which the period 30 days produced better performance than other indicators, but EMA-14 produced better performance on AOT. For multiple-parameter model, the model performed well with the 30 days period. The experiment using ANN was executed in 1000 epochs using 4 dense layers. The ReLU was used for activation. The period of 30 days produced a better performance for AOT and EA. Using single parameter, ANN produced poor results.

Table 3. R-squared of logistic regression

Models	RSI30	RSI14	EMA-14	SMA14	Multiple 30	Multiple 14
Linear regression						
AOT	0.001	0.005	0.000	0.000	0.147	0.190
MINT	0.002	0.005	0.000	0.000	0.229	0.244
EA	0.001	0.001	0.004	0.004	0.041	0.097
Logistic regression						
AOT	0.43063	0.37978	0.47935	0.37949	0.63109	0.57445
MINT	0.48032	0.43847	0.40616	0.45016	0.64832	0.64816
EA	0.47842	0.45866	0.45191	0.36394	0.60882	0.58383
Neural Network						
AOT	0.26379	0.20738	0.08915	0.09014	0.69221	0.65010
MINT	0.12065	0.11058	0.09430	0.09763	0.30619	0.92346
EA	0.09621	0.05507	0.07869	0.04347	0.54837	0.52474

5.2 Predicting Buy/Sell Signal

In this experiment, the buy signal strategy was conducted. The target was determined by the buy signal. For example, if the percent change of the next day is higher than 0 the signal is *buy* (1) otherwise, *do nothing* (0). Decision tree, random forest, and XGBoost model were experimented. All models were specified 5 maximum depths. Table 4 depicts the results of the experiment. EMA produced better performance in the random forest and XGBoost model but poor performance in decision trees. Using multiple parameters produces better accuracy than using single parameter. The train data has a better performance than the test data in all models. Ranking technical indicator according to ML model is in Table 5. All indicators were judged from all stocks at least 2 of 3. Some ranks cannot be judged (indicated by -).

Table 4. Accuracy of decision tree, random forest, and XGBoost

Models	RSI30	RSI14	SMA14	EMA14	Multiple 30	Multiple 14
Decision Tree						
AOT	0.65745	0.63466	0.63031	0.61968	0.69889	0.65333
MINT	0.64088	0.62133	0.63297	0.66755	0.68508	0.66933
EA	0.65745	0.66133	0.62765	0.65957	0.66574	0.65066
Random forest						
AOT	0.64364	0.69333	0.71542	0.72340	0.79281	0.77066
MINT	0.70441	0.70400	0.71542	0.71809	0.80939	0.82666
EA	0.69613	0.71466	0.69946	0.73670	0.79005	0.79733
XGBoost						
AOT	0.85359	0.85333	0.81383	0.87500	0.95027	0.98133
MINT	0.83425	0.83466	0.82978	0.85638	0.95027	0.97600
EA	0.85635	0.84800	0.82180	0.86170	0.94475	0.95466

Table 5. Rank of indicators of AOT, MINT, and EA

ML models	Single-parameter model				Multiple-parameter model	
	(1)	(2)	(3)	(4)	(1)	(2)
Linear regression	RSI14	RSI30	-	-	14 days	30 days
Logistic regression	RSI30	-	-	SMA14	30 days	14 days
ANN	RSI30	RSI14	SMA14	EMA14	30 days	14 days
Decision tree	-	-	-	-	30 days	14 days
Random forest	EMA14	SMA14	RSI14	RSI30	14 days	30 days
XGBoost	EMA14	RSI30	RSI14	SMA14	14 days	30 days

Because the conducted experiment revealed different ranks with different stocks, therefore each stock was explored further. The result of prediction using SET50 stocks with logistic regression and random forest is in Table 6. Each stock in SET50 performed different with different indicators. The table shows the number of stocks with the indicators in the first rank. For example, RSI 30 days period in single-parameter model is the most powerful indicators in 16 stocks with logistic regression, and EMA 14 days period performed better in 17 stocks with random forest. Using multiple-parameter model with 30 days period performed well in most stocks.

Table 6. Rank of indicators of SET50 stocks

ML Model	Single-parameter model				Multiple-parameter model	
	RSI30	RSI14	SMA14	EMA14	30 days	14 days
Logistic regression	16	12	11	11	31	19
Random forest	13	8	12	17	36	14

6. CONCLUSION

Feature selection was experimented using correlation analysis. Technical indicator RSI is the most significant indicator for stock prediction. XGBoost is the most efficient model for buy/sell signal prediction. Linear regression, logistic regression, and artificial neural networks performed well in multiple-parameter model. A period is a critical factor for feature selection. Using different periods, the model produced different outcomes.

In this paper, the correlation between SET50 and the SET market was reported to present the existence of SET50 stock in the stock market. The proposed feature selection method is a single method which means only technical indicators were focused. The ensemble feature selection method can achieve better single-feature selection. Thus, other methods will be explored in future work as well as the multiple layers for stock prediction using a combination of machine learning models will be considered in future work.

REFERENCES

Aloraini, A. (2015). Penalized ensemble feature selection methods for hidden associations in time series environments case study: equities companies in saudi stock exchange market. *Evolving Systems*, *6*(2), 93-100.

Chaigusin, S., Chirathamjaree, C. and Clayden, J., (2008. December). The use of neural networks in the prediction of the stock exchange of Thailand (SET) Index. *Proceedings of 2008 International Conference on Computational Intelligence for Modelling Control & Automation*, pp. 670-673.

Gunduz, H., Çataltepe, Z., & Yaslan, Y., (2017). Stock daily return prediction using expanded features and feature selection. *Turkish Journal of Electrical Engineering and Computer Sciences*, Vol. 25: No. 6, Article 32.

Htun, H. H., Biehl, M., & Petkov, N., (2023). Survey of feature selection and extraction techniques for stock market prediction. *Financial Innovation*. Springer;Southwestern University of Finance and Economics, vol. 9(1),pp. 1-25.

Inthachot, M., Boonjing, V., & Intakosum, S., (2016). Artificial neural network and genetic algorithm hybrid intelligence for predicting Thai stock price index trend. *Computational intelligence and neuroscience*, Vol. 2016, pp. 1-8.

Kumari, B., Patnaik, S., & Swarnkar, T., (2023). Feature selection for stock price prediction: a critical review. *International Journal of Intelligent Enterprise*, *10*(1), pp. 48-72.

Peng, Y., Albuquerque, P. H. M., Kimura, H., & Saavedra, C. A. P. B., (2021). Feature selection and deep neural networks for stock price direction forecasting using technical analysis indicators. *Machine Learning with Applications*, *5*, 100060.

Rungruang, C., Srichaikul, W., Chanaim, S., & Sriboonchitta, S., (2019). Prediction the direction of SET50 index using support vector machines. *Thai Journal of Mathematics*, pp. 153-165.

Sanboon, T., Keatruangkamala, K., & Jaiyen, S., (2019. February). A deep learning model for predicting buy and sell recommendations in stock exchange of thailand using long short-term memory. *IEEE 4th International Conference on Computer and Communication Systems (ICCCS)*, pp. 757-760.

Sopipan, N., Kanjanavajee, W., & Sattayatham, P., (2012). Forecasting SET50 index with multiple regression based on principal component analysis. *Journal of Applied Finance and Banking*, SCIENPRESS Ltd, vol. 2(3), pp. 1-10.

Tumay, A., Aydin, M.E. and Kozat, S.S., (2024). Hierarchical Ensemble-Based Feature Selection for Time Series Forecasting. *arXiv preprint arXiv:2310.17544*.

RESEARCH ON KNOWLEDGE EXTRACTION OF CHINESE SHIP TECHNICAL SPECIFICATION

Kexin Cheng and Zuhua Jiang
School of Mechanical Engineering,
Shanghai Jiao Tong University
Shanghai, China

ABSTRACT

Extracting multiple kinds of knowledge from large ship technical specification texts plays an important role in improving the efficiency of designers in acquiring and applying knowledge. In view of the huge number of design specification texts and the need for deep information extraction, a knowledge extraction method for Chinese ship design specification based on coding rules and named entity recognition is proposed. First of all, morpheme analysis and grammar analysis are conducted on ship design specifications, which can automatically identify the knowledge type of sentences according to the type, quantity and grammar relationship of morphemes in each sentence. Combined with named entity recognition algorithm, corresponding contents are extracted according to different knowledge type extraction rules to fill in knowledge attributes, so as to complete the knowledge extraction of the design specification. The effectiveness of the proposed method is verified by experiments in the collected corpus of ship design specifications.

KEYWORDS

Deep Learning, Machine Learning, Neural Networks, Knowledge Extraction

1. INTRODUCTION

As the rules and regulations to be followed in the complex knowledge-intensive process of ship design, the design specifications contain a large number of different types of knowledge. Multiple kinds of knowledge extracted from ship technical specifications can reasonably expand the ship design knowledge base, and make it meet the knowledge requirements of multi-disciplinary and multi-sectoral collaborative design in the design and manufacturing process of large ships. In addition, they can also help automatically assist the generation of new design schemes, or help review whether the new design schemes meet the corresponding design specifications, providing great help to design experts, so as to ensure the efficient implementation of ship collaborative design and improve the design efficiency and quality of high-tech ships.

In the past 50 years, the complexity of various technical specifications has continued to rise significantly. The number of existing ship design specifications and the complexity of natural language expression have led to the error prone process of manual extraction, the complexity and inefficiency of the correction process. Numerous professional words and formulas contained in texts and the complex and inconsistent sentence structure bring difficulties to further acquiring and mining knowledge. It also has a negative impact on the subsequent computer automatic review work, which cannot play the due value of relevant design knowledge. Therefore, it is extremely urgent to construct a computer-aided automatic knowledge extraction method for design specification text.

2. RELATED WORK

In terms of knowledge extraction, experts and scholars initially adopted the methods based on expert system. The scholar Lee J.M(2010) used the model of expert system to simplify each specification into a production rule form (IF-THEN structure) and built the specification into a knowledge base; Lee H.(2016) transformed the natural language specification provisions into a computer-executable rule set file KBimCode through the management system KBimLogic based on logical rule mechanism . Guo (2016) established the coding rules of security specifications and BIM model design components, and corresponded the natural language specifications and design components one by one according to the coding . However, these methods are costly and not universally applicable in other fields. In addition, compared with English, the grammatical structure and expression of Chinese natural language are also very different. In terms of knowledge extraction of Chinese natural language specifications, the method based on expert system modeling requires domain experts to spend a lot of time on ontology reasoning or modeling, and it is easy to make mistakes.

Later, methods based on natural language processing (NLP) gradually became popular. Yu (2022) combines natural language processing (NLP) and context-free syntax (CFG) to propose a novel framework for general rule interpretation that can parse regulatory text like domain-specific languages. Based on term matching and semantic similarity analysis, Guo (2021) proposes a semantic method to map rule terms to keywords (concepts or attributes) in BIM data, which is used to extract rule terms and logical relationships between these terms from text specification documents. Compared to general non-technical texts such as news articles and general websites, domain-specific design specification texts are better suited to automatic NLP and can better explain and reduce ambiguity in automatic processing.

NLP knowledge extraction splits into rule-based, requiring manual rule refinement for improved text processing accuracy but high human input, and machine learning-based, offering full automation but with lower text processing performance in accuracy and recall.

For the knowledge extraction of ship design specifications studied in this paper, the number of texts is huge, the manual intervention should be as low as possible, and complex sentences need to be analyzed, which has high requirements for text processing performance; At the same time, it is necessary to identify the meaning of the text in a specific field, and carry out a deeper level of information extraction to facilitate the interpretation and understanding of engineers. It is necessary to extract not only terms or concepts, but also other information elements (such as restrictions) to extract all information expressed in sentences/requirements.

3. KNOWLEDGE EXTRACTION METHOD OF CHINESE SHIP DESIGN SPECIFICATION

3.1 Knowledge Extraction Strategy and Process of Chinese Ship Design Specification

The complexity of ship design specifications has surpassed manual extraction methods, highlighting the need for automated solutions. This study aims to improve the efficiency of the ship design process by building an innovative knowledge extraction method. Our approach combines morphemic and syntactic analysis with named entity recognition to systematically analyse and classify multifaceted knowledge in Chinese ship design texts. Its main goal is to automatically identify knowledge types and extract relevant attributes to build a robust knowledge base. The approach aims to support design experts by facilitating the generation and review of design solutions, ultimately improving the quality and efficiency of collaborative ship design work.

This paper combines a coding rule-based approach with a machine learning-based approach. Firstly, this paper analyses the lexemes and syntax of the cruise design specifications with the help of natural language processing methods. Then, the knowledge type of the article is automatically determined based on the type, number and syntactic relationship of morphemes in each standard. Combined with the named entity recognition algorithm, the corresponding content is extracted according to different knowledge type extraction rules and knowledge attributes are filled in. The knowledge extraction of design specification text is finally completed, which facilitates the research and production in the field of ship design.

Firstly, the text is preprocessed to make it more consistent with the training of the model, including cutting the long normative text containing multiple chapters, and each sentence will be extracted into one knowledge; Secondly, the sentences are classified. According to the corresponding rules, they are judged as rule, formula, taboo or small domain technical specification. The text sentences of each category will be extracted into corresponding knowledge forms. After that, the named entity recognition is carried out for each standard sentence, and the entities are filled into the object attributes. On this basis, the rest contents are extracted and filled into the corresponding knowledge attributes to realize the standardized representation of each knowledge.

3.2 Construction of Design Specification Data Set

So far, there are few researches on knowledge extraction of design specification text. In addition, there is a lack of typical data sets for experimental testing. Therefore, this paper downloaded some specification files related to ship design from the official website of China Classification Society, converted PDF into text format through optical character recognition (OCR) technology, and used it as the source of original text data for knowledge extraction. Data sets are self-built and annotated. Specifically, this paper collected eight ship design codes, including "Cruise Ship Code", "Domestic Marine Ship Construction Code 2018 (online edition)", "Ship Safety Management System Certification Code", "Steel Marine ship Classification Code", "Green Ecological Ship Code", "Smart Ship Code", "Cruise Cabin Noise Prediction Guide", "Hull Structure Fatigue Strength Guide 2021". The specific steps of data set self-construction are as follows.

Data preprocessing. The data preprocessing stage is mainly to clean and standardize the data in order to make the data more consistent with the training of the model and get better results. We have processed the ship specification text as follows: ①Remove the blank space and miscellaneous information caused by the OCR recognition technology problem; ②Remove the content in the form of tables and pictures. This paper does not consider table extraction and picture extraction; ③Replace some colloquial expressions with normative expressions, for example, "最小...为 (the minimum of... is)" to " ...不小于(... no less than)", " 不超过 (no more than)"→" 不大于 (no more than)".

Text segmentation. After data cleaning, the text should be segmented when text classification. ".", "!" and "?" punctuation marks are used to break sentences.

Title storage. Since the title number (for example, 2.3.2.1) will be automatically ignored after the specification file PDF converted into text, this work can only be done manually. The first level title (e.g. Chapter 2 Hull Structure), second level title (2.2 General Provisions), and third level title (e.g. 2.1.1 General Requirements) in the specification text file are stored in a specific table for the convenience of later filling in the "applicable scenario" attribute of the knowledge.

Establishment of entity classes. According to the requirement of knowledge extraction, self-annotated entities can be classified into five categories, including "concept term", "action term", "state term", "parameter term", and "parameter comparison", which are used for filling in the subsequent various knowledge attributes, as shown in Table 1.

Annotating system. The BIO mode is used for annotating all entities. "B" represents the starting position of the entity; while "I" and "O" represents other parts of the entity and the non-entity part, respectively. The decoder in the system could align the BIO sequence identified by the model with the original text, and take out the subsequence pairs corresponding to the BI annotations. The set of all subsequence pairs is the named entity set, which can be input into the deep neural network structure for training.

Table 1. Entitles of Design Specification

Entity category	Entity example	Entity category symbol
Name term	开孔角隅 (opening corner), 船体构件 (hull member), 开孔角隅 (opening corner), 船体构件 (hull member), 强力甲板 (strong deck)	NAME
Action term	细化分析 (detailed analysis), 屈服强度校核(yield strength check),	ACT
State term	动态平衡 (dynamic balance), 应力集中(stress concentration)	STATE
Parameter term	分布系数 (distribution coefficient), 校核总纵 强度 (checked total longitudinal strength), 静水切力 (hydrostatic shear force)	PARA

3.3 Text Classification

According to the actual project investigation and analysis, the design specification text is generally divided into four types: formula type, rule type, taboo type, and small domain technical standard type. They will be extracted into corresponding formula knowledge, rule knowledge, taboo knowledge, and standard knowledge. The formula text contains the formula used in the corresponding design stage (mostly algebraic formula); The rule text is in the "condition-conclusion" style, which describes the specific conclusion formed under specific conditions; The taboo text will restrict some subjects to make them conform to the standard; The small domain technical standard text is an empirical description of the terms to be followed in the corresponding design stage. The judgment basis of each type of text is shown in Table 2.

3.4 Named Entity Recognition

For sentences in the formula category, there is a "formula name" attribute in the corresponding knowledge representation, and for sentences in the rule and small domain technical standard categories, there is a "key concept" attribute in the corresponding knowledge representation. Here, the entire sentence will be populated with descriptive objects or entities. Since many standard sentences generally have a professional noun consisting of adjectives and nouns as the object, direct participle extraction of nouns or verbs is not effective. For example, a strong deck edge plate is the entity of interest in this paper. However, direct participle extraction of nouns can only extract the deck and cannot accurately express the meaning of knowledge. There are two ways to solve this problem. One is to establish a professional cruise ship specification vocabulary, which is difficult to achieve at present; the other is to use certain algorithms and rules to identify entities, find out the key concepts of each sentence, and then fill in the corresponding knowledge attributes.

Table 2. Judgment Basis of Four Text Types

Text type	Example	Judgment basis
Formula	计算中垂波浪弯矩时，分布系数 m 按下式的 cf 值确定 (when calculating the vertical wave bending moment,the distribution coefficient m is determined by the cf value cf of the following formula):$c_f = 0.2\ v/\sqrt{L} + \left(a_d - a_w\right)/l * h_f$	1.A string containing at least 5 consecutive non-Chinese characters and non-punctuation marks. 2.Contain the "=" Symbol.
Rule	当开孔角隅为光滑自由边时，应选取 s-n 曲线中的 c 曲线。 (when the opening corner is a smooth free edge, select c curve in s-n curve.)	There is one specific morpheme: " 如果 (if)" or " 当... 时 (when) " or " 对于 (for) ".
Taboo	细化分析采用的网格尺寸应不大于 50x50 毫米。 (mesh size used for detailed analysis shall not be greater than 50×50mm.)	There is one specific morpheme: " 大于 (more than) " or " 小于 (less than) " or " 等于 (equal to)" or ">" or "<" or "≥" or "≤".
Small domain technical standard	有限元模型采用船体构件的建造尺寸 (finite element model adopts the construction dimensions of hull components.)	The remaining texts after three levels of classification belong to this category.

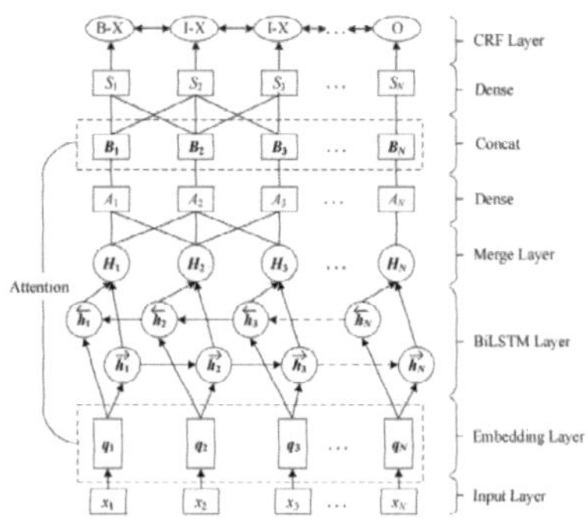

Figure 1. Architecture of QTCE-BiLSTM Attention CRF model

Aiming at the characteristics of large vocabulary of design specifications, high contextual relevance, incomplete corpus features and insufficient character information, this paper adopts the named entity design specification recognition model, named Quality Topic-Char Embeding-BiLSTM-Attention-CRF (Zhou &Jiang, 2022).

QTCE-BiLSTM-Attention-CRF model consists of input layer, embedding layer, BiLSTM layer, multi-layer fully connected layer and CRF layer. First, character vectors pre-trained by word2vec are combined with high-quality topic vectors of characters, which are used as character embedding vectors of the model. Then, they are input into BiLSTM to obtain long-distance dependent features of sentences. Next, the attention mechanism is used to implement a weighted redistribution strategy for the embedding layer, so that the influence of the character vector in the embedding layer is directly transmitted to the fully connected layer. Next, the fully-connected layer receives the output of the BiLSTM layer and the embedding layer via the attention mechanism. It performs multiple fully-connected operations to fully extract entity features. Finally, the label transfer function of CRF is used to decode the rational relationship among predicted labels and obtain the global optimal label sequence. The model architecture is shown in Figure 1,where $(x_1, x_2, .., x_n)$ is the input sequence;$(q_1, q_2, .., q_n)$ is the character embedding sequence; $(\overrightarrow{h_1}, \overrightarrow{h_2}, ..., \overrightarrow{h_N})$ is the pre-order output sequence of LSTM; $(\overleftarrow{h_1}, \overleftarrow{h_2}, ..., \overleftarrow{h_N})$ is the post-order output sequence of LSTM; $(H_1, H_2, ..., H_N)$ is the state sequence obtained by combining $(\overrightarrow{h_1}, \overrightarrow{h_2}, ..., \overrightarrow{h_N})$ and $(\overleftarrow{h_1}, \overleftarrow{h_2}, ..., \overleftarrow{h_N})$; A_i is the weight coefficient; $(B_1, B_2, ..., B_N)$ is the intermediate sequence; S_i is the CRF predicted score; $(B-X, I-X, ..., O)$ is the optimal predicted label sequence;"B" represents the starting position of the entity; "I" represents other parts of the entity; "O" represents the non-entity part; "X" represents the entity category symbol.

3.5 Knowledge Element Extraction and Knowledge Attribute Filling

For each sentence, a specific knowledge ID is generated for retrieval.

Formula knowledge:

Step 1: The "formula name" attribute is filled with the "concept term" identified by named entity recognition;

Step 2: Extract the first continuous non-Chinese and non-punctuation character string with "=" in each formula type sentence. The segmentation of the string is achieved by inserting special symbols (such as "*");

Step 3: From the formula expression on both sides of "=" in each formula sentence, the shorter string is extracted and filled into "output parameter", and the string on the other side is filled into "input expression";

· Step 4: The "formula meaning description" attribute is filled with the "action term" identified by named entity recognition;

· Step 5: The "Variable parameters" attribute is filled with the "parameter term" identified by named entity recognition.

Rule knowledge:

· Step 1: Extract the clause number (a.b.c.d) of each sentence, extract the title name numbered "a", "a.b" or "a.b.c" in the title table, and fill in the "applicable scenario";

· Step 2: Extract all the contents from the first word after the morpheme "if" or "when" or "for" in each rule-type sentence to the first punctuation mark, and fill in the "rule premise";

· Step 3: Extract all the content after the punctuation after the rule premise in each formula sentence and fill it into the "rule conclusion";

· Step 4: The "key concept" attribute is filled with "concept term" identified by named entity recognition.

Taboo knowledge:

· Step 1: The "parameter" attribute is filled with the "parameter term" identified by named entity recognition;

· Step 2: The "taboo content" attribute is filled with the "parameter comparison" identified by named entity recognition.

Normative knowledge:

· Step 1: Extract the clause number (a.b.c.d) of each sentence, extract the title name numbered "a", "a.b" or "a.b.c" in the title table, and fill in the "applicable scenario";

· Step 2: Extract all the contents of each normative sentence and fill them into the "content" attribute;

· Step 3: The "key concept" attribute is filled with "concept term" identified by named entity recognition.

3.6 Compliance Check and Adjustment

After the above extraction steps, the extraction results need to be checked for compliance and adjusted manually, mainly for the following problems:

Problems of rule knowledge extraction:

The interference of special symbols. For example, if the special symbol "①" appears in a clause, the sentence needs to be adjusted by deleting or modifying it in order to make the sentence sound.

Conditional sentences consist of more than one sentence. If there is more than one conditional message in a sentence, punctuation or conjunctions should be used to break it up and ensure that the condition is expressed smoothly.

Problems of formula knowledge extraction:

Some parameters and algebraic expressions may be confused. For example, the sound absorption coefficient is identified as "coefficient", and the parameter level "v = 3.5055" is also identified as a formula.

Uncontrollable problems, mainly the problem of letter recognition error or garbled code in PDF to WORD, such as part of letters garbled code, "m2" identified as unit "m/s", etc.

Problems of taboo knowledge extraction: As the knowledge template is more complete, the extracted elements are more distinctly characterized and have fewer errors.

Problem of standard knowledge extraction: There is no obvious error.

4. EXPERIMENT AND RESULT ANALYSIS

4.1 Text Classification Result

Table 3 shows the results of classifying the preprocessed text using the method described in section 3.3.

Table 3. Preprocessing Text Classification Results

File name	Document page	Number of knowledge	Formula	Rule	Taboo	Standard
Cruise ship specifications	68	576	20	114	79	363
Domestic seagoing ship construction specifications 2018 (online version)	636	10305	559	1779	1029	6821
Ship safety management system certification	43	520	0	111	0	409
Steel seagoing ship classification specifications	2672	21481	0	4241	1565	15675
Green ecological ship specifications	106	985	54	301	22	598
Smart ship specifications	98	817	0	200	1	616
Cruise ship cabin noise prediction guide	40	267	63	20	5	179
Hull structure fatigue strength guide 2021	152	691	5	99	18	569

4.2 Effect of Named Entity Recognition

As mentioned earlier, there is an obvious lack of literature on knowledge extraction from Chinese design documents, which prevents comparative analyses with existing methods. Therefore, the accuracy precision, recall recall and f-value, which are common to named entity models, are adopted as the evaluation metrics of entity recognition models to evaluate the recognition effect of five types of entities. The calculation process is as follows:

$$Precision = \frac{TP}{TP + FP} \quad (1)$$

$$Recall = \frac{TP}{TP + FN} \quad (2)$$

$$F - score = \frac{2 Precision \times Recall}{Precision + Recall} \tag{3}$$

where TP, FP and FN denote the number of true positive,the number of false positive and the number of false negative,respectively.

The recognition effect of QTCE-BilSTM-Attention-CRF model on different named entities is shown in Table 4.

Table 4. Recognition Effect of Different Entitles

Entity symbol	Precision	Recall	F-score
Name	88.56%	83.52%	85.97%
Act	88.83%	82.57%	85.59%
State	88.56%	82.45%	85.40%
Para	88.32%	82.68%	85.50%
Comp	88.58%	82.56%	85.46%

4.3 Knowledge Extraction Result

Examples of formula knowledge (Formula ID 35):

Formula name: 中垂弯矩分布系数 (vertical bending moment distribution coefficient).

Formula expression: $C_f = 0.2\,V/\sqrt{L} + (A_d - A_w)/L * h_f$.

Output parameter: C_f.

Input parameter: $0.2V/\sqrt{L} + (A_d - A_w)/L * h_f$.

Formula meaning description: 计算中垂波浪弯矩 (calculate vertical wave bending moment).

Variable parameters: V-最大服务航速 (maximum service speed); L-船长 (captain); A_d -干舷甲板投影面积 (projected area of freeboard deck); A_w -水线面面积 (waterline area); h_f -系数 (coefficient).

Examples of rule knowledge (Rule ID 47):

Applicable scenario: 船体构件尺寸(hull member size), 上层建筑及甲板室(superstructure and deckhouse).

Rule premise: 上层建筑、甲板室等长上层建筑的纵向结构上包含门、窗时(when the longitudinal structure of the superstructure (e.g. superstructure and deckhouse) includes window and door openings).

Rule conclusion: 应设置圆角或较厚嵌入板(fillet shall be set or thicker embedded plate shall be added).

Key concept: 上层建筑的纵向结构 (longitudinal structure of superstructure).

Examples of taboo knowledge (Taboo ID 294):

Parameter: 强力甲板边板在端部的宽度 (the width of the strong deck edge plate at the end).

Taboo content: 不小于船中部宽度的 65% (no less than 65% of the midship width).

Applicable objects: 船上剧院、中庭等处所的高应力区域 (high stress areas in theatres, atriums and other places on board).

Examples of standard knowledge (Standard ID 2433):

Applicable scenario: 整船直接计算 (direct calculation of the whole ship), 载荷分量及施加 (load components and application).

Content: 动压力通过波浪载荷直接计算得到，并以压力的形式作用在有限元模型上 (The dynamic pressure is directly calculated by wave load and acts on the finite element model in the form of pressure).

Key concept: 有限元模型 (finite element model).

It can be seen that the method proposed in this paper can effectively extract the four types of knowledge closely related to the design content in the unstructured design text, and can be able to completely and accurately extract the attributes corresponding to each type of knowledge, which is easy to store, retrieve and integrate in the knowledge base.

5. CONCLUSION

The design specification in the engineering field has a large number of articles and complex natural language expression, which makes the manual extraction process prone to errors, the correction process complex and inefficient, and has a negative impact on the subsequent computer automatic review work. The design of computer-aided automatic extraction process can greatly improve the accuracy and efficiency. On the basis of relevant research, this paper carries out morpheme analysis and grammar analysis of ship design specifications with the help of natural language processing methods, automatically identifies the knowledge type of articles according to the type, quantity and grammatical relationship of morphemes in each sentence. Combined with named entity recognition algorithm, corresponding contents are extracted according to different knowledge type extraction rules to fill in knowledge attributes. Finally, the knowledge extraction of the design specification text is completed. When the relevant code and data are used for experiments, the knowledge classification, feature extraction and entity extraction results of each sentence can be obtained by simply entering the PDF file containing the normative knowledge. The classification and extraction basis is only related to the text syntax, such as whether it contains formulas, whether it contains inequality relations, and does not involve the recognition of professional terms. Therefore, the code can be smoothly and widely transferred to the engineering fields of electronic chips, automobile manufacturing, and so on. It is a highly portable knowledge classification and extraction method. Combines coding rules with named entity recognition, this method reduces manual intervention, improves text processing performance, and extracts complete and accurate results.

ACKNOWLEDGEMENT

This research was supported by National Natural Science Foundation of China (No. 72271163).

REFERENCES

Guo, et al. (2016). *Bim and safety rules based automated identification of unsafe design factors in construction*. Procedia Engineering.

Guo, et al. (2021). *A Semantic Approach for Automated Rule Compliance Checking in Construction Industry*. IEEE ACCESS129648-129660.

Lee H. et al, (2016). *Translating building legislation into a computer-executable format for evaluating building permit requirements.* Automation in Construction, 71(pt.1), 49-61.

Lee, J. M. (2010). *Automated checking of building requirements on circulation over a range of design phases*. (Doctoral dissertation, Georgia Institute of Technology.;).

Yu C Z. et al, (2022). *Integrating NLP and context-free grammar for complex rule interpretation towards automated compliance checking*. Computers in Industry,142,103746,

Zhou & Jiang. (2022). *Named Entity Recognition of Design Specification Integrated with High-Quality Topic and Attention Mechanism,* Journal of Shanghai Jiaotong University (Science).

FACE RECOGNITION ATTENDANCE SYSTEM

Zirije Hasani[1], Jakup Fondaj[2], Rrezart Kallaba[1], Xhafer Muja[1] and Endrit Jelliqi[1]
[1]University "Ukshin Hoti" Prizren, Faculty of Computer Science,
Prizren, Kosovo
[2]Southeast European University, Faculty of Contemporary Science and Technology
Tetovo, North Macedonia

ABSTRACT

The paper explores the design, implementation, and evaluation of an innovative approach to streamline attendance tracking in various domains. Leveraging advancements in computer vision and facial recognition technology, the proposed system aims to automate the attendance process, eliminating the need for traditional methods such as manual sign-ins or card swipes. The system employs state-of-the-art deep learning algorithms to accurately detect and recognize facial features, ensuring reliable and secure identification of individuals. The paper discusses the system's architecture, data processing pipeline, and integration with existing infrastructure. Furthermore, the study evaluates the system's performance in real-world scenarios, considering factors such as accuracy, speed, and scalability. The results indicate the system's potential to revolutionize attendance management, offering a more efficient, accurate, and user-friendly alternative to conventional methods. The findings contribute to the ongoing discourse on the practical applications of facial recognition technology in enhancing organizational processes.

KEYWORDS

Face Recognition, Attendance Systems, Deep Learning Algorithms, Detect and Recognize Facial Features

1. INTRODUCTION

In the fast-paced and technologically evolving landscape of the 21st century, the demand for efficient and secure attendance management systems has become increasingly pronounced. Traditional methods, such as manual sign-ins and card-based systems, often prove to be time-consuming, susceptible to errors, and cannot adapt to dynamic environments. In response to these challenges, this paper introduces the "Face Recognition Attendance System," an innovative and automated solution harnessing the power of facial recognition technology.

Facial recognition, a subset of computer vision and artificial intelligence, has witnessed significant advancements in recent years, enabling accurate and rapid identification of individuals based on facial features. This paper delves into the design, development, and evaluation of a robust attendance management system that leverages state-of-the-art deep learning algorithms for facial recognition. By integrating this technology into the attendance tracking process, we aim to address the limitations of traditional methods and offer a more seamless and sophisticated alternative.

The proposed Face Recognition Attendance System seeks to streamline attendance management across various domains, including educational institutions, corporate settings, and public events. Through a comprehensive exploration of its architecture, functionality, and real-world performance, this paper aims to contribute to the growing body of knowledge surrounding the practical applications of facial recognition technology in organizational processes. As we embark on this exploration, the potential benefits, challenges, and ethical considerations associated with the implementation of such systems will be examined, paving the way for a more informed and nuanced discussion on the role of facial recognition in attendance tracking.

2. RELATED WORKS

Facial recognition technology has gained significant traction in recent years, offering a promising avenue for revolutionizing attendance management systems. This literature review explores the existing research on face recognition technology and its application in attendance tracking, shedding light on the advancements, challenges, and ethical considerations associated with implementing a Face Recognition Attendance System.

Several studies highlight the rapid advancements in facial recognition algorithms and deep learning techniques. Li et al. (2018) demonstrated the effectiveness of deep neural networks in achieving high accuracy rates in facial recognition, paving the way for robust attendance systems. Similarly, Zhang et al. (2020) explored the use of convolutional neural networks (CNNs) for feature extraction in facial recognition, contributing to improved identification performance.

Researchers have recognized the potential of integrating facial recognition into attendance management systems. Liang et al. (2019) proposed a hybrid attendance system combining facial recognition and radio-frequency identification (RFID) technology, enhancing accuracy and reliability. The study emphasized the seamless integration of facial recognition with existing infrastructure, addressing the practical challenges of deployment.

The evaluation of facial recognition systems involves various performance metrics. Zhang and Tan (2019) conducted a comprehensive review of evaluation metrics, emphasizing the importance of accuracy, speed, and robustness. Their work provides insights into the criteria for assessing the effectiveness of Face Recognition Attendance Systems in real-world scenarios.

Ethical concerns regarding facial recognition technology have been widely discussed in the literature. Smith et al. (2021) examined the ethical implications of using facial recognition in attendance tracking, emphasizing the importance of transparent policies, informed consent, and measures to address potential biases. The study contributes valuable insights into the ethical considerations associated with implementing such systems.

Ensuring the security and reliability of Face Recognition Attendance Systems is a crucial aspect of their deployment. Jiang et al. (2020) investigated security vulnerabilities and proposed encryption techniques to safeguard facial recognition data. The study underscores the need for robust security measures to protect sensitive information in attendance systems.

There has been great interest and development in the field of facial recognition and presence management systems. Some of the most important works related to the project are:

FaceNet's Florian et al. (2015) introductory paper presents a methodology for using a deep model to create a uniform representation of a person's face. Contributes to the challenges of face recognition, using a deep multidimensional algorithm to determine the distance between faces in space.

DeepFace Taigman et. Al. (2014) seeks to reduce the gap between the performance of face recognition algorithms and human capabilities. Uses a deeply targeted model to achieve a high level of performance in face verification.

A survey on face recognition techniques (Bhaskar et. al. (2018)) provides a broad overview of the techniques used in face recognition, including traditional and artificial intelligence-based ones. Discusses challenges and opportunities in the field, including the concerns of security and privacy.

The study (Nemade, Bhushankumar. (2015)) presents a concrete implementation of an attendance management system using face recognition. Discusses the advantages of using this technology compared to traditional attendance methods.

3. RESEARCH METHODOLOGY

This study aims to develop a facial recognition system, using computer vision technologies and artificial intelligence. The system is designed to automate and improve attendance registration processes in various environments, providing an accurate and efficient means of identifying individuals.

The objectives of this study are:

- Efficient attendance management: Replacing traditional registration methods with a modern and advanced system.
- Facial Recognition: Using facial recognition algorithms to identify and verify individuals.
- Real-time operation: Real-time facial recognition capability, providing instant information.

- User-friendly interface: Creating an intuitive and easy user interface for system management.

Technologies Used:

- Python: The main programming language for system development.
- OpenCV: Computer vision library used for face detection and analysis.
- Tkinter: GUI Toolkit for Creating Graphical User Interfaces.
- MySQL: Database management system for storing and organizing data.
- LBPH (Local Binary Pattern Histogram): Algorithm used for face recognition.

This system aims to improve and modernize the existing attendance registration processes, using facial recognition as an efficient and reliable tool.

3.1 Project Implementation

The project is implemented in Python and have used the Tkinter library (python, 2023). This library is used for GUI views of the project, as well as PIL, image (Pillow, 2023), and image TK. Where the latter are used to import images and visualize them in the project.

Figure 1. The main menu of the proposed face recognition system

Here in the project, you can see the GUI, where there are buttons and imported pictures where each button takes you to its page. The student, but also the button has its photo, making it possible to visualize them with the PIL Library of images (Pillow, 2023). When we open the program, this window appears (Figure 1) with the result (1530 by 790). It also continues by taking each button and placing it in the main label of the panel. In addition to the buttons that are all functional, we take for example clicks on the student page.

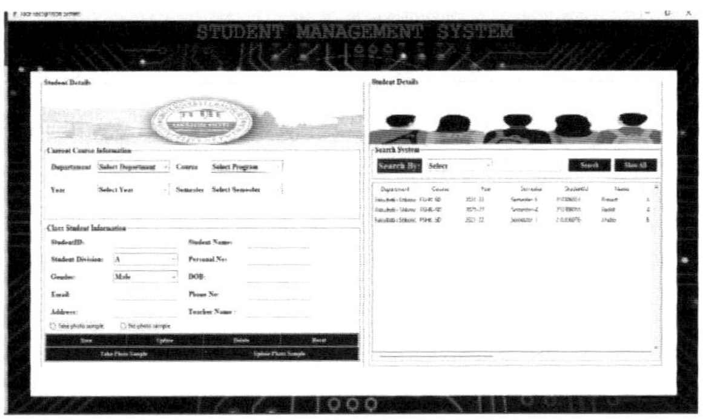

Figure 2. The proposed system for face recognition

This is the student's page (Figure 2) in this part of the GUI, a main FRAME is taken, where then two other frames are placed above the main frame. Frame for the left part and for the right part. The student section is also connected to a database where data is recorded for each student.

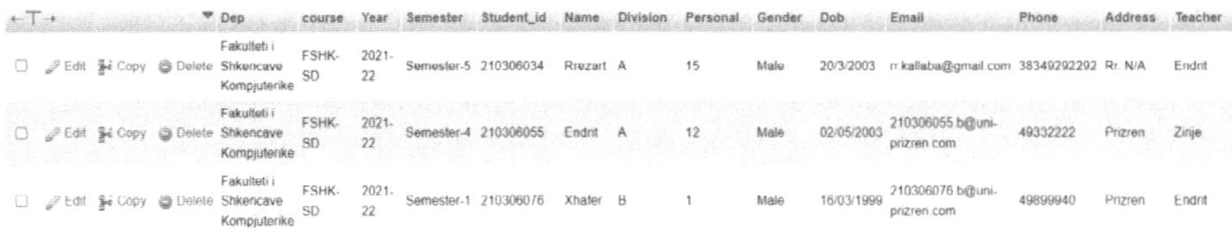

	Dep	course	Year	Semester	Student_id	Name	Division	Personal	Gender	Dob	Email	Phone	Address	Teacher
Edit Copy Delete	Fakulteti i Shkencave Kompjuterike	FSHK-SD	2021-22	Semester-5	210306034	Rrezart	A	15	Male	20/3/2003	rr.kallaba@gmail.com	38349292292	Rr. N/A	Endrit
Edit Copy Delete	Fakulteti i Shkencave Kompjuterike	FSHK-SD	2021-22	Semester-4	210306055	Endrit	A	12	Male	02/05/2003	210306055.b@uni-prizren.com	49332222	Prizren	Zinje
Edit Copy Delete	Fakulteti i Shkencave Kompjuterike	FSHK-SD	2021-22	Semester-1	210306076	Xhafer	B	1	Male	16/03/1999	210306076.b@uni-prizren.com	49899940	Prizren	Endrit

Figure 3. Database for proposed system

A student must be registered in MySQL- DBMS before we receive the model of his face. To create a dataset of photos, the user must take 500 photos to increase the accuracy in training the predictive model. In addition to these properties, we can register a new user where he is assigned his data, such as Name, ID, Faculty Program, and other data. It is also possible to pay off that student depending on the needs of the user. It is also worth noting that when a photo is taken of a person, we save the photos as (cv2.COLOR_BGR2GRAY) or in JPEG format to take up as little space as possible.

A photo is stored in a folder that takes up approximately 20kb, in total for 1500 photos it takes up approximately 40.5 MB, which means it takes up very little space and has little spatial complexity.

The code in the train, classifier method of the **Train class** aims to train a face recognition classifier through the LBPH (OpenCV, 2023) (Local Binary Pattern Histogram) algorithm in OpenCV.

Face_Recognition class is used for face recognition, this uses the OpenCV library (Path Sing, 2023) to perform face recognition based on a classifier. In addition, the class includes functionality to mark the presence of a person after facial recognition. Here are some important points about what this class contains:

- GUI Integration: The Face_Recognition class is integrated with Tkinter (Python, 2023) for the graphical user interface. The face recognition process is activated by clicking on the "Face Recognition" button.

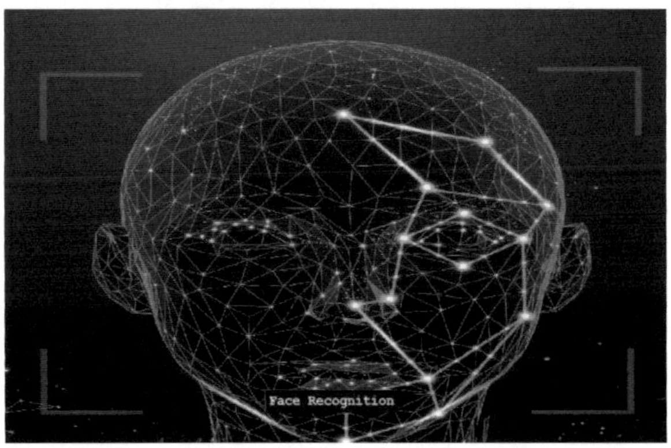

Figure 4. Face Recognition page

- Mark attendance: The mark_attendance method marks the attendance information in a CSV file, and verifies that the student details (ID, Personal Number, Name, Department) are not already present in the file before marking the information.

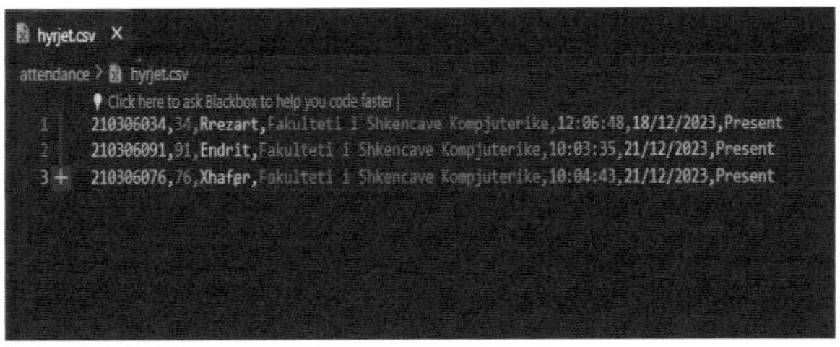

Figure 5. Marking of the detected user

As seen in the picture below, when a face is recognized with a confidence level of more than 80 %, the name appears in the terminal and shows the similarity:

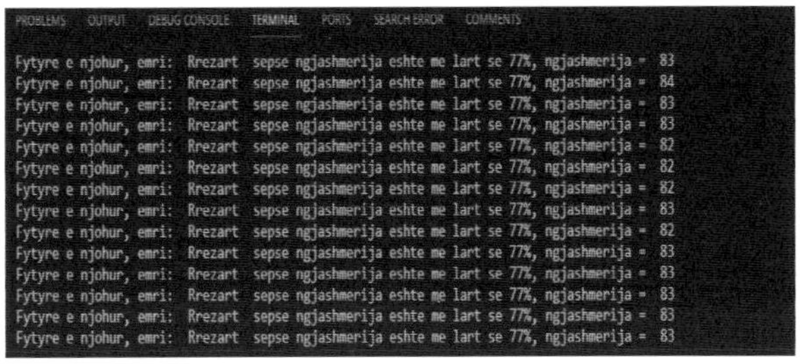

Figure 6. Recognition of the student if the similarity > 80%99

- Face Recognition Process: The face_recog method uses the Haar Cascade classifier (haarcascade_frontalface_default.xml) and the LBPH (OpenCV, 2023) (Local Binary Patterns Histograms) face recognizer (classifier.xml) to recognize faces in real-time using a camera (cv2.VideoCapture(0)).
- Drawing Boundaries: The draw_boundray functionality draws rectangles around detected faces and displays information like ID, Roll, Name and Department. If the confidence level is below 80%, it labels the face as "Unknown Face".

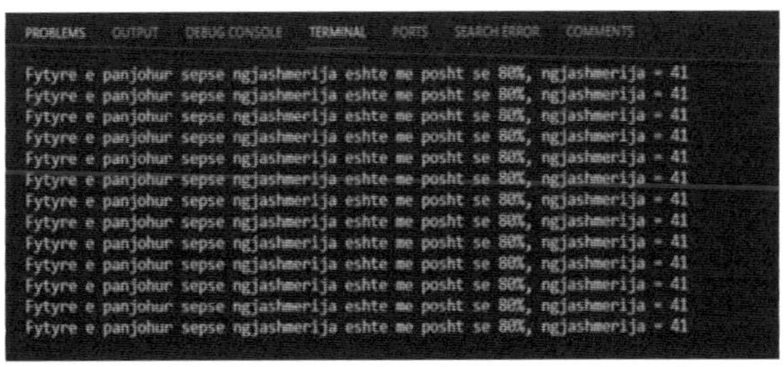

Figure 7. Not recognizing the student if the similarity < 80%

- Database Connection: The code connects to a MySQL database (MySQL, 2023) to retrieve student information about familiar faces.

4. FACE RECOGNITION BY USING THE LBPH (LOCAL BINARY PATTERN HISTOGRAM) ALGORITHM

LBPH (OpenCV, 2023) is a face recognition algorithm used to recognize a person's face. It is known for its performance and how it can recognize a person's face from both the front face and the side face.

12	15	18
5	8	3
8	1	2

Figure 8. LBPH algorithm

Let's select the pixel in the center here, value 8, and apply a condition. If the value is greater than or equal to 8, the result is '1' otherwise, if the value is less than eight, the result is zero. After applying the conditioner, the matrix will now look like this:

1	1	1
0	8	0
1	0	0

Figure 9. LBPH algorithm

Now we need to generate a binary value. Binary value = 11100010. The algorithm will start applying the condition from the element of the upper left corner going to the element 1 of the second row, as if it is making a circle as in the picture below:

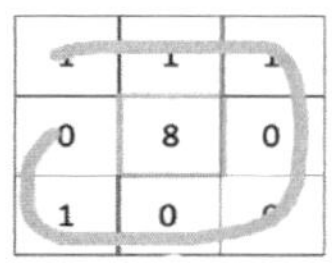

Figure 10. LBPH algorithm

After converting the Binary value to decimal value, we get **Decimal value = 226.** It shows that all these pixels around the center value are equal up to 226.

Let's consider this another image here. To better understand how the algorithm will recognize a person's face.

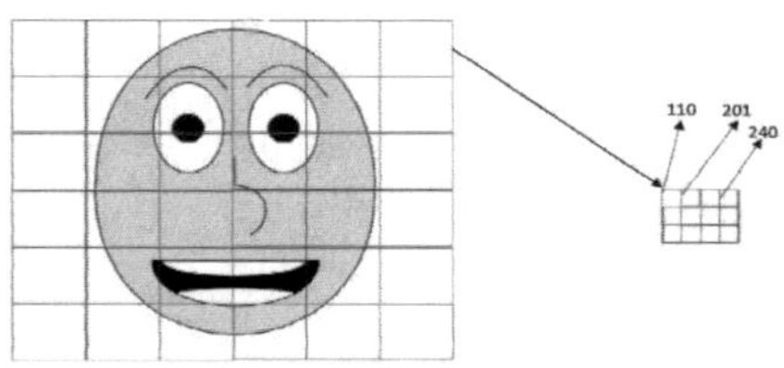

Figure 11. Operation of the LBPH algorithm on faces

Here we have the image of a face and what the algorithm will do is create some squares, as you can see here. Within each of these squares, we have the representation of the previous one is light. For example, this square here doesn't represent just one pixel, but is set with multiple pixels that are three rows and four columns. Three out of four equals twelve pixels total in these squares here in each of these squares which is twelve pixels. And then we apply that condition to each one. Considering the center pixel. This Attendance Details class code manages attendance in an organization or institution built with the help of the Tkinter Python library (Python, 2023).

5. RESULT AND DISCUSSION

Detecting people's faces can be used in various fields, such as law enforcement, justice systems, identity recovery, and biometrics. Facial recognition and identification technology integrates technological solutions to ensure student protection. For selected locations, such as university campuses, the main goal is to identify and record the presence of students through their faces, enabling a safe and reliable monitoring of the time spent by each student on campus.

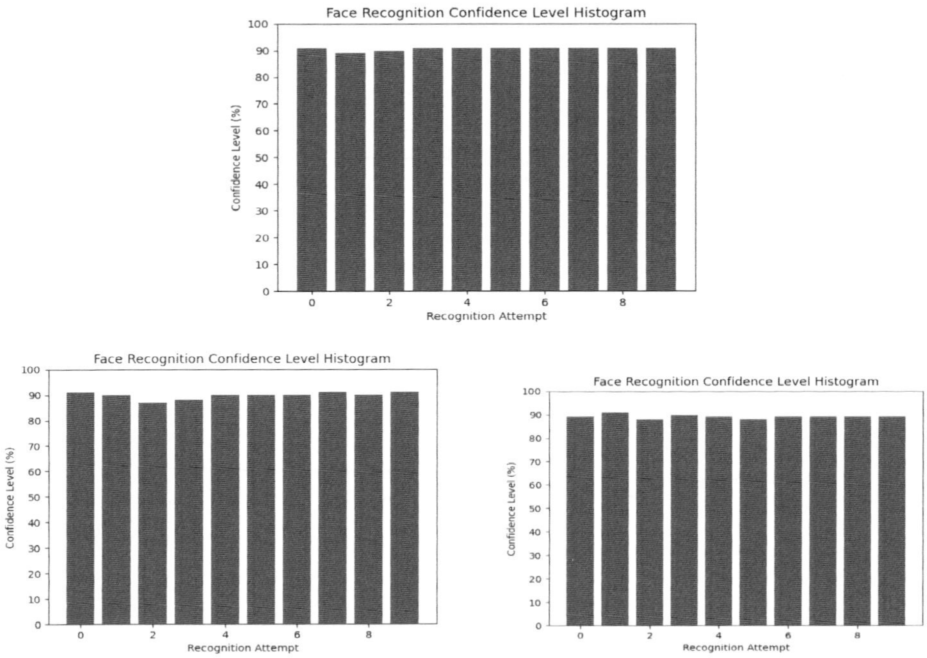

Figure 12. Acurency testing for three users

This project stands out for its high accuracy, which reaches an average of 91% for students. Modeling the dataset with 500 photos has resulted in a high efficiency, showing good data processing capabilities. It is estimated that the photos are stored in GRAYSCALE form, ensuring an efficient use of space with a low percentage of information retention. With an overall size of 13mb for 500 photos, this also improves data storage efficiency. Overall, the project is characterized by a successful effort in implementing and using models to handle the dataset efficiently and accurately.

6. CONCLUSION

In conclusion, the exploration of the Face Recognition Attendance System underscores the transformative potential of facial recognition technology in the realm of attendance management. The synthesis of advancements in deep learning algorithms, integration strategies, performance evaluation metrics, and ethical considerations paints a comprehensive picture of the current landscape and future prospects for this innovative system.

The research has highlighted the rapid strides in facial recognition technology, especially with the utilization of deep neural networks and convolutional neural networks, demonstrating its efficacy in achieving high accuracy rates. The amalgamation of facial recognition with existing attendance systems, as proposed by Liang et al. (2019), offers a promising solution to enhance reliability and streamline the attendance tracking process.

Performance evaluation metrics, as discussed by Zhang and Tan (2019), provide a critical framework for assessing the effectiveness of Face Recognition Attendance Systems. The emphasis on accuracy, speed, and robustness in real-world scenarios ensures a practical understanding of the system's capabilities and limitations, contributing to informed decision-making during deployment.

However, the implementation of facial recognition technology in attendance systems is not without its challenges, as illuminated by Smith et al. (2021). Ethical considerations, privacy concerns, and potential biases necessitate careful deliberation and the establishment of transparent policies to ensure responsible deployment.

The exploration of security measures by Jiang et al. (2020) reinforces the importance of safeguarding sensitive facial recognition data. Encryption techniques and robust security protocols play a pivotal role in addressing vulnerabilities and bolstering the overall reliability of Face Recognition Attendance Systems.

As we reflect on the findings of this research, it becomes evident that the Face Recognition Attendance System represents a paradigm shift in attendance management. The system's potential to offer a seamless, accurate, and secure alternative to traditional methods aligns with the evolving needs of diverse domains, including educational institutions, corporate settings, and public events.

In moving forward, it is imperative to acknowledge the continuous evolution of facial recognition technology and its integration into attendance systems. Future research should focus on addressing remaining challenges, refining ethical frameworks, and adapting the system to dynamic environments. As technology advances and societal expectations evolve, the Face Recognition Attendance System stands poised to play a pivotal role in shaping the future of attendance tracking.

REFERENCES

Bhaskar, Nuthanakanti & Das, Kapili & Ganashree, (2018). A SURVEY ON FACE RECOGNITION TECHNIQUES. 4. 2455-2585.

Florian Schroff, Dmitry Kalenichenko, James Philbin. FaceNet: A Unified Embedding for Face Recognition and Clustering. Proceedings of the IEEE Computer Society Conference on Computer Vision and Pattern Recognition 2015

Jiang, S., Zhang, Z., & Xie, X. (2020). A secure and efficient face recognition system for attendance management. Information Sciences, 537, 361-376.

Li, W., Li, X., & Lu, J. (2018). Deep learning for generic object detection: A survey. arXiv preprint arXiv:1809.02165

Liang, X., Xu, C., Shen, J., & Lin, L. (2019). Attend and recognize: A novel attention-based hybrid attendance system. IEEE Transactions on Image Processing, 28(2), 1001-1013.

MySQL. (2023, 12 28). MySQL. Retrieved from https://www.mysql.com/.

OpenCV. (2023, 12 28). Opencv-python 4.8.1.78. Retrieved from https://pypi.org/project/opencv-python/.

Parth Singh. (2023, 12 01). Understanding Face Recognition using LBPH algorithm. Retrieved from https://www.analyticsvidhya.com/blog/2021/07/understanding-face-recognition-using-lbph-algorithm

Pillow. (2023, 12 28). Pillow. Retrieved from https://pillow.readthedocs.io/en/stable/.

Python. (2023, 12 27). Python interface to Tcl/Tkl. Retrieved from https://docs.python.org/3/library/tkinter.html

Smith, A. N., Li, X., & Hengartner, U. (2021). Ethical issues in facial recognition for attendance tracking: A case study. Ethics and Information Technology, 23(3), 281-294.

Y. Taigman, M. Yang, M. Ranzato and L. Wolf, "DeepFace: Closing the Gap to Human-Level Performance in Face Verification," 2014 IEEE Conference on Computer Vision and Pattern Recognition, Columbus, OH, USA, 2014, pp. 1701-1708, doi: 10.1109/CVPR.2014.220.

Zhang, Y., Wang, S., & Phillips, P. (2020). A survey on deep learning in face recognition. Computer Vision and Image Understanding, 189, 102884.

Zhang, Z., & Tan, T. (2019). A review of face recognition evaluation. In 2019 8th International Conference on Multimedia Systems and Signal Processing (pp. 153-158). IEEE.

UNVEILING THE VALUE OF USER REVIEWS ON STEAM: A PREDICTIVE MODELLING OF USER ENGAGEMENT APPROACH USING MACHINE LEARNING

María Olmedilla[1], Leonardo Espinosa-Leal[2], José Carlos Romero-Moreno[3] and Zhen Li[2]
[1]*SKEMA Business School – Université Côte d'Azur, France*
[2]*Arcada University of Applied Sciences, Graduate Studies and Research, Finland*
[3]*Applied Computational Social Sciences-Institute, University of Paris-Dauphine-PSL, France*

ABSTRACT

In an era where user-generated content is both ubiquitous and influential, accurately evaluating videogame reviews' relevance becomes critical. The vast digital domain of videogames brims with user feedback, presenting the challenge of distinguishing genuinely helpful reviews. Our study, analyzing over a million videogame reviews from the Steam platform, employs cutting-edge machine learning techniques to ascertain review helpfulness. We applied both regression and binary classification models, revealing the latter's enhanced predictive prowess. Interestingly, our findings contradict the anticipated benefit of incorporating features from pre-trained NLP models into enhancing prediction accuracy. This paper not only highlights methods for assessing review helpfulness effectively but also promotes the application of computational techniques for the insightful analysis of user-generated content. Furthermore, it provides valuable perspectives on the elements influencing user engagement and the intrinsic value of feedback within the context of videogame consumption, marking a significant contribution to understanding digital user interaction dynamics.

KEYWORDS

User Reviews, Extreme Machine Learning, Steam, Binary Classification, NLP, Review Helpfulness Prediction

1. INTRODUCTION

Since Atari introduced the first home console version of Pong in 1977, the video game industry has undergone a remarkable evolution. This sector has experienced explosive growth, with annual global revenues now nearing $218 billion (Juniper Research, 2021). The recent global pandemic has further accelerated this growth, as consumers turned to video games for entertainment during lockdown periods. In this burgeoning online video game market, reviews have become a critical factor influencing consumer purchase decisions. Consequently, analyzing video game reviews is increasingly important for developers to understand user preferences and enhance game quality (Li et al., 2019).

This paper delves into the realm of online video game platforms, with a specific focus on the Steam platform—a leading digital distribution service. By examining nearly one million reviews of games on Steam, we aim to ascertain the feasibility of predicting a game's helpfulness score based on various features, including review text. Understanding the predictive power of these features could enable game developers to leverage review insights for making informed decisions, from shaping game development to optimizing marketing strategies.

Moreover, this research seeks to contribute to the burgeoning field of user-generated content analysis within the video game industry. By assessing the potential of predictive analytics in understanding review helpfulness, we aspire to provide developers with tools for harnessing online reviews in guiding game design, securing investor interest, targeting key markets, and refining marketing approaches.

The structure of this paper is as follows: Section 2presents the research background, offering an overview of the video game industry, the Steam platform, and existing literature on helpfulness ratings. Section 3 outlines the methodology employed in our analysis. Section 4 discusses the results obtained from our study. Finally, Section 5 concludes the paper, summarizing our findings and their implications for video game development and marketing.

2. RESEARCH BACKGROUND

2.1 Video Games and Steam Game Platform

The evolution of video games from simple arcade forms to complex, immersive experiences reflect significant technological and cultural shifts. With the advent of various gaming devices—ranging from dedicated consoles and personal computers to the ubiquitous presence of smartphones—access to video games has never been easier or more diverse. This accessibility is further enhanced by the digital distribution of games, which has become a dominant force in the industry over the past decade. Digital sales platforms have not only expanded the reach of video games but also significantly increased the industry's revenue potential. MarketLine (2015, 2021) notes that the shift toward digital formats aligns with broader retail trends favoring e-commerce, effectively reducing the market share of traditional physical outlets. In 2020, the revenue per user from digital games vastly outstripped that of physical copies, highlighting the economic advantages of digital distribution for the industry (Winter, 2021).

The digitization of video games has also facilitated innovations in game design and monetization strategies, allowing developers to experiment with new genres and gameplay mechanics. Digital platforms offer a more direct relationship between game creators and their audience, enabling rapid feedback and iterative development processes. This dynamic has, however, introduced the challenge of maintaining player engagement in an environment where new titles are constantly vying for attention.

Steam, developed by Valve Corporation, has been instrumental in shaping the digital distribution landscape. Launched in 2003 initially as a means to update games and combat piracy, Steam has grown into the premier platform for digital game sales, community engagement, and content creation among PC gamers. It serves as both a marketplace for purchasing games and a social hub where gamers can connect, share, and discuss gaming content. Steam's role extends beyond commerce and community; it provides valuable analytics on game performance, including reviews, which are crucial for developers and publishers in strategizing game development and marketing efforts (Lin et al., 2019).

2.2 Rating Helpfulness of Online Reviews

The digital era has seen an explosion in user-generated content, with online reviews becoming a critical source of information for consumers navigating digital marketplaces. In the context of video games, reviews on platforms like Steam play a significant role in influencing purchasing decisions. The ability to predict the helpfulness of these reviews can significantly enhance consumer experience by highlighting the most relevant and informative feedback.

Research in this domain has increasingly applied machine learning techniques to predict review helpfulness, underscoring the potential of computational models to extract insights from vast quantities of user-generated data. The authors in Olmedilla et al. (2022) provide a comprehensive summary of these efforts, indicating a growing academic interest in leveraging technology to enhance the utility of online reviews.

Despite advancements in helpfulness prediction, challenges remain, particularly in accurately assessing review quality based on intrinsic characteristics such as the reviewer's profile and the content of the review itself. Previous studies have attempted to correlate various features with perceived helpfulness, with mixed results (Wu, 2017). The complexity of this task is underscored by the diversity of online platforms and the varying criteria users may apply when rating a review's helpfulness (Krishnamoorthy, 2015).

In the specific arena of video game reviews, the Steam platform offers a rich dataset for analysis. Prior research has identified factors such as review length, the amount of time spent playing the game, and the game's genre as significant determinants of a review's helpfulness. However, there is a notable gap in understanding how the textual content of reviews—beyond simple keyword frequency—affects perceived helpfulness. This gap presents an opportunity to explore more sophisticated text analysis techniques, such as semantic analysis, to uncover deeper insights into what makes a review helpful to potential buyers (Eberhard et al., 2018; Kasper et al., 2019).

3. METHODOLOGY

We used a dataset openly available from Kaggle (Kamel, 2022). The original dimensions of the dataset are 1,124,903 rows and 10 columns. The dataset corresponds to reviews of 14,469 different games. The features of the dataset are *'game_id'*, *'review_id*, *'steamid'*, *'playtime_forever'*, *'playtime_at_review'*, *'pos_review'*, *'votes_relevant'*, *'review_text'*, *'weighted_vote_score'* and *'steam_purchase'*. We created three new features by binarizing both the *'pos_review'* and *'steam_purchase'* and by adding the length of the reviews. We estimated the character length of the reviews and filtered the ones longer than 512 characters, which is the typical limitation of most of NLP available models. The dataset after the removal contains only 898,326 rows. In Figure 1, a general description of the dataset is presented by plotting the frequency of the sentence length (top-left), the ordered number of reviews as a function of the Game ID for the top 100 games (top-right) and the frequency distribution of the helpfulness score (*weighted_vote_score*).

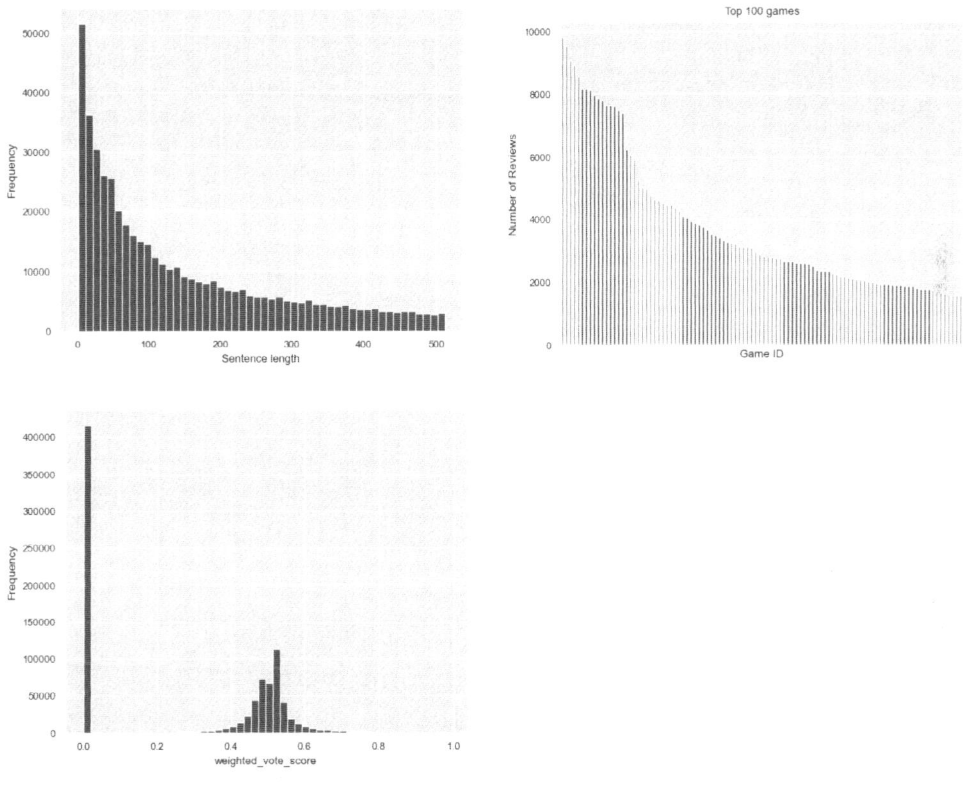

Figure 1. Characteristics of the original dataset used in this work. Top-left: Frequency of the sentence length. Top-right: Ordered number of reviews versus Game ID and Bottom-left: The frequency distribution of the helpfulness score (*weighted_vote_score*)

After inspecting the frequency distribution of the helpfulness score (*weighted_vote_score*), we noted almost 50% of the scores are equal to zero. We removed these rows in the dataset and the final dataset is of 482,460 rows. In Figure 2, the ordered number of reviews as a function of the Game ID for the top-100 games (top-left) shows a clear exponentially decaying behavior. The helpfulness score shows a clear Gaussian distribution centered around 0.5 (top-right). The frequency of the playtime at review (bottom-left) shows also a clear exponentially decaying behavior. The Pearson correlation matrix for the nine features used for training (bottom-right) shows that the following three features are the most important: votes_relevant (0.36), pos_review_bin (0.21) and length (0.13). The other six features have a correlation of less than 0.1: *review_id* (0.098), *game_id* (0.072), *playtime_forever* (0.05), *playtime_at_review_new* (0.047), *steam_purchase_bin* (0.046) and *steamid* (0.013).

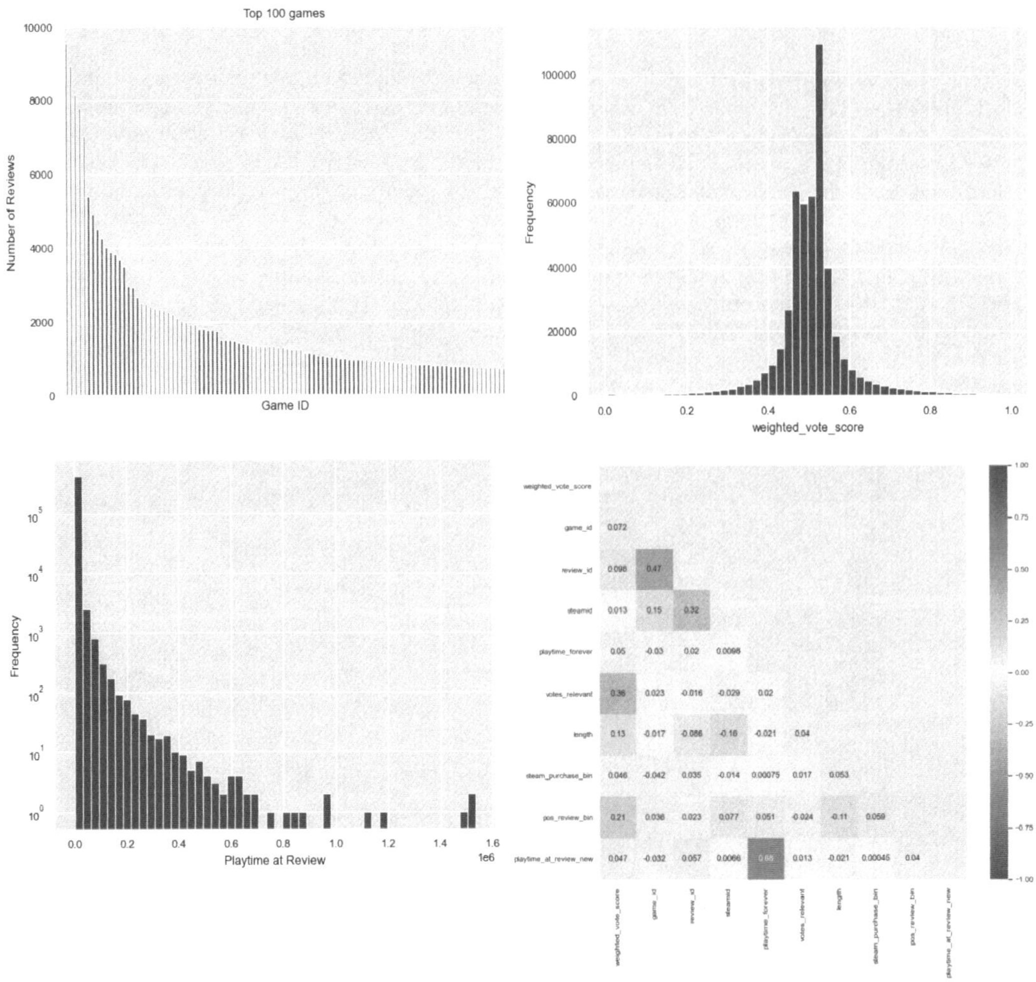

Figure 2. Characteristics of the dataset after filtering used in this work. Top-left: Ordered number of reviews versus Game ID. Top-right: The frequency distribution of the helpfulness score (*weighted_vote_score*). Bottom-left: Frequency distribution of the playtime at review and bottom-right: Pearson correlation matrix among the features of the dataset

4. RESULTS AND DISCUSSION

We have trained three different machine learning models to predict the helpfulness score of videogames online reviews, from two different approaches: (1) a regression problem and (2) a binary classification problem. In the regression problem, the models were Ridge, extreme learning machine (ELM) and extreme gradient boosting (XGBoost). For the calculations the well-known Python libraries: scikit-learn (Pedregosa et al., 2011) and scikit-ELM (Akusok et al., 2021). We trained all models using a randomized grid search of 50 steps for parameters optimization with a three-fold cross-validation and a 70/30 ratio for training and testing. For the Ridge and ELM, the features were normalized before training. The best results are consigned in Table 1.

Table 1. Summary of results for the three machine learning models for predicting the helpfulness score as a regression problem

Models	Optimized parameters	Best score (train/test) R^2
Ridge	alpha: 2020.2	0.22/0.21
ELM	alpha: 2.08e-05, density: 0.22, include_original_features: False, n_neurons: 3270, pairwise_metric: cityblock, ufunc: sigm	0.52/0.51
XGBoost	colsample_bytree: 0.51, learning_rate: 0.09, max_depth: 8, n_estimators: 861, scale_pos_weight: 3, subsample: 0.81	0.61/0.60

Table 2. Summary of results for the three machine learning models for predicting the helpfulness score as a binary classification problem

Models	Optimized parameters	Best score (train/test) Accuracy	F1-score / ROC-AUC
Logistic Regression	C: 98.99	0.79/0.78	0.77/0.71
ELM	alpha: 0.026, density: 0.41, include_original_features: True, n_neurons: 3417, pairwise_metric: cityblock, ufunc: relu	0.88 / 0.87	0.84 / 0.77
XGBoost	colsample_bytree: 0.76, learning_rate: 0.02, max_depth: 13, n_estimators: 990, scale_pos_weight: 2, subsample: 0.60	0.91/0.90	0.85 / 0.79

The regression results show that XGBoost presents the best score, then ELM and Ridge. However, the best ration performance and computational time is achieved by the ELM model. To model the helpfulness score as a binary classification model we changed the values in the range (0,0.5) as 0 and the values in the range [0.5,1.0] as 1. We trained three different models: Logistic regression, ELM and XGBoost with the same steps as the regression models. The obtained results for the best models are in Table 2. The obtained results for classification follow the same tendency as the regression problem. XGBoost is the best model, however in this case the ELM results have a similar performance.

NLP modelling

We included the features generated by the reviews from the text by using a state-of-the-art NLP model from the Huggingface library 0. We used the pre-trained DistilBERT model, that upon feeding with text, produces a vector of 768 of length. We added the dataset generated by this model to the original 9 features. The obtained results for regression and classification are consigned in Table 3 and Table 4.

Table 3. Summary of results for the ELM model for predicting the helpfulness score as a regression problem, including the NLP features

Models	Optimized parameters	Best score (train/test) R^2
ELM	alpha: 0.036, density: 0.65, include_original_features: True, n_neurons: 4025, pairwise_metric: cityblock, ufunc: sigm	0.51/0.50

Table 4. Summary of results for two machine learning models for predicting the helpfulness score as a binary classification problem, including the NLP features

Models	Optimized parameters	Best score (train/test) Accuracy	F1-score / ROC-AUC
Logistic Regression	C: 52.53	0.80/0.79	0.77/0.72
ELM	alpha: 0.00013, density: 0.135, include_original_features: True, n_neurons: 4040, pairwise_metric: cityblock, ufunc: lin	0.80 / 0.80	0.80 / 0.72

Our findings show that by adding the features from the NLP pre-trained model in the training step, there is not an apparent improvement in the scoring metrics for both regression and classification strategies, except in the case of the logistic regression model that shows a minor improvement.

5. CONCLUSIONS

In this work, we predicted the helpfulness score for a dataset of videogames from the Steam platform using two approaches: (1) as a regression problem and (2) as a binary classification problem. For each approach, we trained three different machine learning models on a dataset of around 500 rows. The best model in both cases is the XGBoost, and the next is ELM. Both linear models underperform in comparison when comparing the metrics in both train and test sets. Our results also show that using features from the text reviews does not improve the performance of the models when predicting the helpfulness score. Besides, these results show that the early prediction of review helpfulness can be useful in helping the users to make better purchase decisions and interesting for developers to create user-perceived high-quality videogames. Future research aims to explore the prediction of the helpfulness score from solely text reviews by using advanced strategies such as the finetuning NLP models.

REFERENCES

Akusok, A., Leal, L. E., Björk, K. M., & Lendasse, A. (2021). Scikit-elm: an extreme learning machine toolbox for dynamic and scalable learning. In *Proceedings of ELM2019* (pp. 69-78). *Springer International Publishing.*

Eberhard, L., Kasper, P., Koncar, P., & Gütl, C. (2018, October). Investigating helpfulness of video game reviews on the steam platform. In 2018 *Fifth International Conference on Social Networks Analysis, Management and Security* (SNAMS) (pp. 43-50). IEEE.

Juniper Research. (May 31, 2021). Global video game market value from 2020 to 2025 (in billion U.S. dollars) [Graph]. In *Statista*. Retrieved February 22, 2023, from https://www-statista-com.skema.idm.oclc.org/statistics/292056/video-game-market-value-worldwide/

Kamel, N. E. L. (2022, January 31). Steam reviews. *Kaggle*. Retrieved March 5, 2023, from https://www.kaggle.com/datasets/nourelkamel/steam-reviews

Kasper, P., Koncar, P., Santos, T., & Gütl, C. (2019, October). On the role of score, genre and text in helpfulness of video game reviews on metacritic. In 2019 *Sixth International Conference on Social Networks Analysis, Management and Security* (SNAMS) (pp. 75-82). IEEE.

Krishnamoorthy, S. (2015). Linguistic features for review helpfulness prediction. *Expert Systems with Applications*, 42(7), 3751-3759.

Lin, D., Bezemer, C. P., Zou, Y., & Hassan, A. E. (2019). An empirical study of game reviews on the Steam platform. *Empirical Software Engineering*, 24, 170-207.

MarketLine. (2021, February). MarketLine Industry Profile: Games Software in the United States. Games *Software Industry Profile: United States*, 1 - 47.

MarketLine. (2015). MarketLine Case Study: *Valve Corporation*. 1 - 23.

Olmedilla, M., Martínez-Torres, M. R., & Toral, S. (2022). Prediction and modelling online reviews helpfulness using 1D Convolutional Neural Networks. Expert Systems with Applications, 198, 116787.

Pedregosa, F., Varoquaux, G., Gramfort, A., Michel, V., Thirion, B., Grisel, O., ... & Duchesnay, E. (2011). Scikit-learn: Machine learning in Python. the Journal of machine Learning research, 12, 2825-2830.

Winter J. (January 6, 2021). SuperData's 2020 Year In Review: Gaming Revenue Up Overall But Fortnite Takes A Tumble, In *Website mmobomb*. Retrieved February 23, 2023, from https://www.mmobomb.com/news/superdatas-2020-year-review-gaming-revenue-overall-fortnite-takes-tumble

Wolf, T., Debut, L., Sanh, V., Chaumond, J., Delangue, C., Moi, A., & Rush, A. M. (2020, October). Transformers: State-of-the-art natural language processing. In *Proceedings of the 2020 conference on empirical methods in natural language processing: system demonstrations* (pp. 38-45).

Wu, J. (2017). Review popularity and review helpfulness: A model for user review effectiveness. *Decision Support Systems*, 97, 92-103.

REAL-TIME TRAFFIC MONITORING SYSTEM BASED ON COMPUTER VISION AND DEEP-LEARNING

Jaime Villa, Arturo de la Escalera and José Mª Armingol
Intelligent Systems Lab (LSI) Research Group, Universidad Carlos III de Madrid
Leganés, Madrid, 28911, Spain

ABSTRACT

In today's busy traffic, effective management and monitoring is crucial, affecting everything from daily transportation routines to environmental sustainability and safety. In this article, we introduce an easy-to-deploy system designed for traffic monitoring through surveillance cameras, utilizing the power of edge computing, computer vision and deep learning technologies. Our system is segmented in to three modules: detection, tracking and analysis. The detection and tracking phases ensure vehicle location, classification, and movement monitoring along the video, while the analysis module provides GPS positioning and speed estimation through visual cues, without the need of traditional GPS location techniques. This approach simplifies the deployment of these systems anywhere, taking the advantage that most worldwide roadways are already video surveilled, providing real-time data crucial for traffic management, safety, and urban planning. With this implementation, the system can easily be scaled up to monitor every corner of a road network, performing all calculations offline without the need for internet connectivity.

KEYWORDS

Deep Learning Algorithms, Computer Vision, Edge Computing, Real-Time Applications, Traffic Monitoring, Intelligent Infrastructure

1. INTRODUCTION

Smart infrastructures represent the present and future of sustainable and secure mobility. Historically, a diverse array of techniques and sensors has been deployed to collect metrics on road traffic. Traditional methodologies primarily utilized sensors embedded in the road surface, such as inductive loops and pneumatic sensors. Additionally, non-intrusive traditional methods have leveraged time-of-flight technologies.

Recent advancements in computer vision and deep learning have elevated the role of cameras to a prominent position for traffic monitoring. These technologies facilitate the use of the extensive network of cameras already deployed along roads worldwide, enabling traffic monitoring without the need for new sensor installation, construction, or traffic interruptions.

Object detection algorithms are crucial for computer vision-based traffic monitoring. The R-CNN and YOLO families represent significant advancements within object detection models, each with distinct advantages. The R-CNN family, starting with R-CNN (Girshick et al., 2013) and evolving through Fast R-CNN (Girshick, 2015) and Faster R-CNN (Ren et al., 2015), introduced a combination of Convolutional Neural Networks (CNNs) with region proposals to accurately localize and classify objects. Redmond et al. (2015) revolutionized the field with the creation of YOLO, which processed images in a single pass, unifying object localization and classification. This approach offered remarkable speed and efficiency for real-time applications.

Subsequent versions proposed by Redmon and Farhadi (2016), Redmon and Farhadi (2018), Bochkovskiy, Wang and Liao (2020), and the most recent such as YOLOv5 (Jocher, 2020) and YOLOv8 (Jocher, Chaurasia, & Qui, 2023) have introduced significant improvements in accuracy, training methodologies, and adaptability to various scenarios. YOLOv5 and YOLOv8 have made a significant impact in the field due to their ease of model training, validation, and deployment, streamlining the development process for real-world applications.

Object tracking algorithms play a pivotal role in video monitoring systems by providing a temporal context for each detected object. Different tracking algorithms can be utilized for this task, but the SORT family stands out among others for its performance. Bewley et al. (2016) proposed the original SORT algorithm and provided a robust foundation for tracking by employing a tracking-by-detection framework. Building upon this, Wojke, Bewley, and Paulus (2017) proposed Deep SORT, which enhances performance by integrating appearance descriptors with deep learning to improve data association, even with long-short term occlusions. BoT-SORT (Aharon, Orfaig, & Bobrovsky, 2022) advances further by incorporating camera motion compensation for dynamic scenes and refining state estimation. A road detection and classification algorithm which tracks several road lanes is presented in (Collado et al., 2006).

Most vision monitoring systems use position estimation techniques based on stereo systems (Jalalat, Nejati & Majidi, 2016), positioning objects in the image employing a disparity map. There are also monocular approaches that use references in the real world to determine the dimensions of strategic objects. Others employ coordinate system transformations that project bidimensional information into a Bird Eye View tridimensional space (Maduro et al., 2008), and even converting image coordinates into GPS information (Blake, 2019), thus allowing distance inference.

This research focuses on the development of a real-time traffic monitoring solution, making significant advances in key areas: (1) the proposed system performs vehicle detection, classification, tracking, and position and speed estimation. (2) The proposed system can run in real-time on an edge computing platform. (3) The proposed system should be easily installed and deployed on the field. (4) This study presents a comparison between speed measurements from an inductive loop road-installed sensor and those from our proposed system.

2. PROPOSED SYSTEM

The proposed system in this research is composed by three main components: an object detection model; based on YOLOv5, a tracker algorithm; based on Deep SORT, and a traffic metrics acquisition module; which allows visual GPS positioning and speed inference.

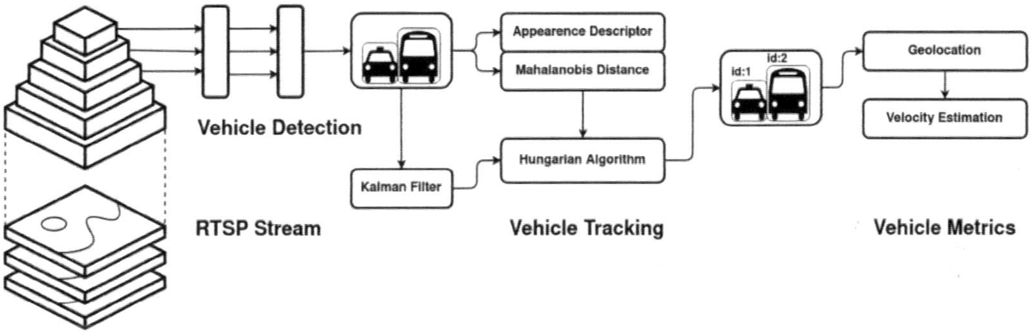

Figure 1. Complete software architecture pipeline. Images are fed into the YOLOv5 object detector, which processes bounding boxes and vehicle classes, then vehicle crops and coordinates are fed to the Deep SORT tracker which assigns identifiers to each detection, finally all the information is processed by the Geolocation and Velocity Estimation modules

Our system pipeline is shown in Figure 1, which explains how information flows along the software parts. In the following sections we will address how each module is built, how the networks where trained and how does the geolocation and velocity inference work.

2.1 Object Detection Module

YOLOv5 is the core of our object detection module and, although we acknowledge the potential of newer versions, these were not available or widely supported during our design and experimental phase. On the other hand, migrating the system to the state-of-the-art YOLO models is straightforward, as the dataset format is generally the same among different YOLO versions, and the detector is designed with a wrapper that post processes the outputs to match the desired format.

The network has been trained on a custom-annotated dataset, composed by images recorded by us in different Spanish locations. The reason behind this decision was to ensure data was aligned with specific camera angles and classification of vehicles into distinct categories: motorcycle, car, van, bus, delivery truck, truck with trailer and articulated truck.

The data was mainly annotated manually, followed by a peer review after each annotation session, which ensured data quality. To assist in the labeling process, the model was trained iteratively, so it could generate new automatic labels that would be later be supervised and adjusted if necessary. With the labelling process completed, data augmentation processes were carried out, which included weather alteration on the raw images, for robust detection in presence of inclement weather conditions like rain or snow. After this whole labeling and augmentation process the resulting dataset contained over 35,000 vehicles distributed over 16,000 labeled images.

$$AP = \sum_{n} (R_n - R_{n-1}) P_n, \qquad mAP = \frac{1}{C} \sum_{c=1}^{C} AP_c, \qquad mAP@IoU = \frac{1}{C} \sum_{c=1}^{C} AP_c @IoU$$

Equation 1. Mean average precision

To evaluate our models, we have employed mAP (mean Average Precision), which is defined in Equation 1, where R_n is the recall and P_n is the precision for the sample n, C is the total number of classes, and IoU is the Intersection over Union threshold that determines the criteria for classifying detections as True Positives (TP), False Positives (FP) and False Negatives (FN). Our final YOLOv5 model achieved an overall 0.954 mAP50 and 0.800 mAP50-95 (Equation 1) on a test set composed by 4,992 images (Table 1). The model training was carried out on a training server equipped with two Nvidia RTX 3090Ti GPUs, which allowed us to train and validate the models as many times necessary in a reasonable amount of time.

Table 1. YOLOv5 final model. Metrics were calculated on a separate test set consisting in 10,616 vehicle instances and 4,992 images across various road traffic scenarios

Class	Images	Instances	P	R	mAP50	mAP50-95
All	4,992	10,616	0.983	0.929	0.954	0.800
Motorbike		1,406	0.989	0.985	0.994	0.762
Car		2,983	0.960	0.849	0.927	0.714
Van		1,771	0.981	0.938	0.962	0.810
Bus		1,046	0.997	0.962	0.976	0.892
Delivery Truck		1,324	0.985	0.985	0.993	0.878
Truck with Trailer		1,561	0.993	0.798	0.835	0.708
Articulated Truck		525	0.978	0.987	0.993	0.837

2.2 Tracking Module

We employed Deep SORT to fulfill the role of tracker in our system, this algorithm builds upon SORT, improving its association metrics. SORT is a tracking-by-detection system that relies on a Kalman Filter to predict object movements and the Hungarian Algorithm to associate the movement predictions with the next object detections. SORT, while effective, struggles with reidentifying objects after occlusions and managing unpredictable movement patterns.

To address SORT's issues, an improved version of the algorithm is employed. Deep SORT introduces a new deep association metric combining the Mahalanobis distance, for motion information, and the Cosine distance, for appearance features. The Mahalanobis distance accounts for the uncertainty in object movement predictions, ensuring consistency between frames. On the other hand, the Cosine distance evaluates similarity between appearance vectors, computed through a CNN (Convolutional Neural Network), which helps the tracker to re-identify objects after long-short term occlusions.

This dual metric provides Deep-SORT with the robustness needed for road surveillance and monitoring tasks, especially in scenarios where larger vehicles like trucks or infrastructure elements might occlude smaller vehicles over several frames.

2.3 Geolocation and Speed Module

In our research, we focused on making the system easy to deploy anywhere. This is the main reason we chose the method proposed by Blake (2019) to infer GPS locations using a monocular vision system, which only requires the calibration of four landmarks, on both the image and the GPS coordinate system.

The geolocation algorithm takes a pair of [x, y] pixel coordinates and turns them into a pair of [lon, lat] GPS coordinates. The perspective transformation is performed using a 3×3 transformation matrix M. This matrix is built using the four landmarks positioned on the road, and solving an equation system, which can be easily computed using a built-in OpenCV function.

$$\begin{bmatrix} \widehat{lon} \\ \widehat{lat} \\ \alpha \end{bmatrix} = M \begin{bmatrix} x \\ y \\ 1 \end{bmatrix}, \qquad lon = \frac{\widehat{lon}}{\alpha}, \qquad lat = \frac{\widehat{lat}}{\alpha}$$

Equation 2. Perspective transformation

After obtaining M, the perspective transformation is straightforward as shown in Equation 2 (where α is a scaling factor), this allows locating objects that appear in the camera angle in GPS format. This method is very effective when the road is flat but has difficulties if the surface slope changes drastically.

Speed inference is not straightforward when using GPS coordinates, this is where the Haversine formula takes the center stage. This formula uses the Earth's radius to find the spherical arc that must be traveled from one pair of coordinates to another. With this set of tools, the module can infer the traveled distance of any vehicle along two video frames, which can lead to average speed estimation considering the time elapsed between each pair of frames.

We developed a setup to validate our velocity estimation system, which consists of a security camera mounted on a gantry over the road, which already had a set of paired inductive loops installed under the asphalt, to detect vehicle presence and measure their speed (Figure 2).

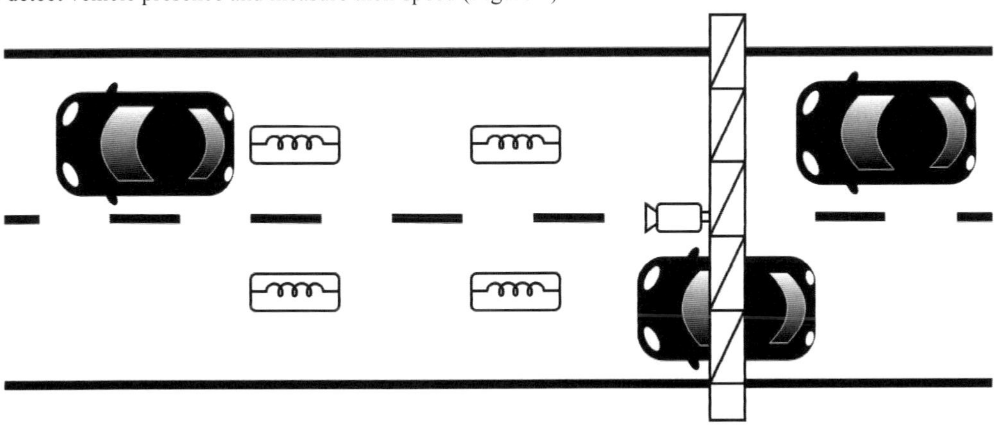

Figure 2. Speed estimation validation setup. The camera is situated on a gantry above the center of the road, while the inductive loops are embedded into the asphalt

2.4 Containerization and Deployment

Software deployment is carried out on edge computing hardware. A platform with great support and widespread use in robotics and deep learning at the edge is the Nvidia Jetson Orin AGX developer kit 64GB, which offers sufficient computing power to run such applications in real time.

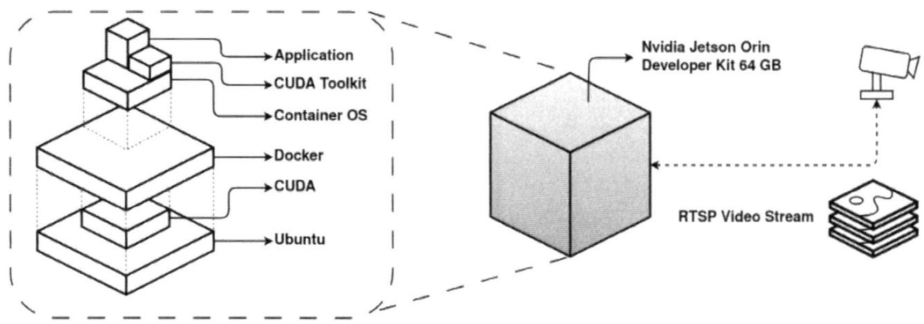

Figure 3. Software containerization using Docker. Employing containers reduces installation problematics and facilitates deployment

Development and deployment are greatly facilitated using containers, which allows for the installation of multiple units with very little effort, making the proposed architecture a scalable solution. As illustrated in Figure 3, in the final system, the container runs on the Nvidia card, executing the application. To feed the video stream, the RTSP protocol is used with correct port forwarding.

3. RESULTS

In this section, we show the findings of our study, the results are detailed in accordance with the experimental setup described previously. As mentioned before, the model responsible for vehicle detection has been trained iteratively as the dataset grew larger. This has allowed for progressive study of its performance improvement, for instance, going from mAP50-95 (Equation 1) value of 0.479 in its first version to 0.800 in the final version, whose metrics are represented in Table 1.

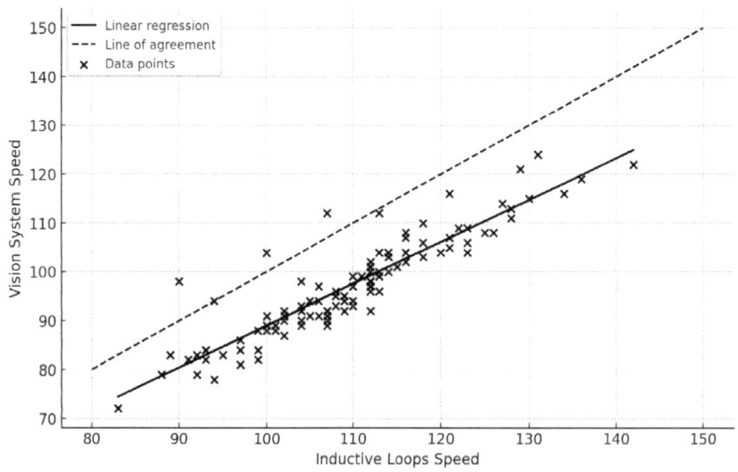

Figure 4. Speed measurement validation between vision and inductive loop systems (kilometers per hour), the dashed line represents the perfect agreement between both systems, while the regression line represents the real relationship between them

Figure 5. Complete deployment. The system performs vehicle detection, classification, tracking, speed estimation and direction analysis

To validate the speed estimation system, the setup from Figure 2 was used, resulting in the data illustrated in Figure 4. Analysis of these samples reveals that the computer vision-based system exhibits an average error rate of 12.68km/h, predominantly underestimating vehicle speeds. Despite the strong positive correlation between both methods, the variation of data becomes more pronounced at higher speeds. It is essential to acknowledge that loop detection systems, serving as a benchmark in this context, are not infallible. Their inherent inaccuracies must be factored into the interpretation of these experiments.

In Figure 5, the proposed system is illustrated in operation, showcasing the detection of vehicles delimited by bounding boxes, each annotated with its respective class and tracking id. Additionally, the velocity of each vehicle, expressed in kilometers per hour (km/h), is displayed, derived from analyzing the estimated GPS position of each vehicle using the techniques previously discussed. Finally, a directional arrow, positioned at the center of each vehicle's bounding box, indicates their movement direction.

Our solution, embedded in the specified Jetson device as outlined in Figure 3, performs all detections and computes traffic metrics in real time. The system's efficiency varies mainly with traffic density; for instance, in scenarios similar to the one depicted in Figure 5, across a 1-hour video, an average frame rate of 23.39 fps is achieved, as detailed in Figure 6. Although the system does not achieve 25 fps, its performance demonstrates robustness and stability suitable for real-time applications.

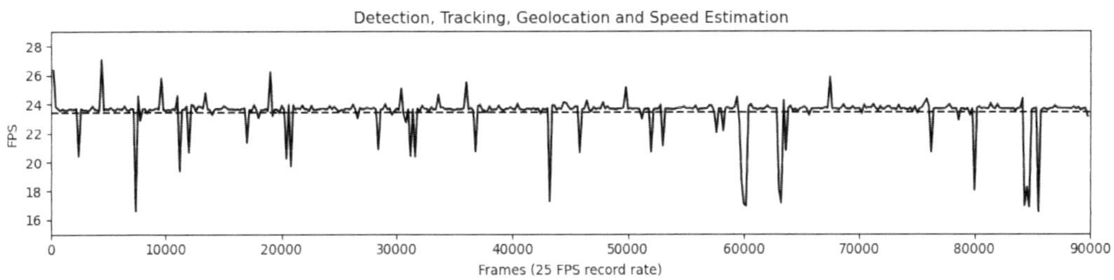

Figure 6. System performance (frame rate) running on a 1-hour video with medium traffic, performing detection, tracking, and analysis of vehicles

4. CONCLUSION

In conclusion, we have developed a real-time traffic monitoring system on an edge computing device. This system is composed of three main components: YOLOv5; for vehicle detection, Deep SORT; for tracking, and a module for traffic analysis. Our validation of the speed estimation method shows promise, despite a noticeable underestimation error compared to the validation system, which remains consistent across different speed ranges showing a strong correlation.

For future development, work could focus on implementing more efficient detection models like YOLOv8 and subsequent versions, along with improved tracking algorithms that can perform more robustly and faster. Moreover, optimization of detection, classification and tracking techniques through highly parallelized implementation could improve system efficiency, GPU utilization, and batch processing for multi-steam analysis in real-time.

In the proposed system, images are gathered by a thread and processed on a separate one, this allows the system to access updated information. If the system is not fast enough, some information might be lost. Scheduling in image computation could be improved, allowing for first-in-first-out frame processing when the traffic is high and synchronous processing when traffic is low.

Refining the geolocation system calibration techniques could lead to improvements in accuracy. Speed estimation could benefit from more precise calibration by directly measuring distances in the actual location and introducing those parameters into the transformation matrix method, rather than relying on the GPS landmarks, thereby simplifying the transformation problem. Although this approach will probably predict more accurate vehicle speeds, the geolocation feature would be indirectly removed. Finally, designing a monocular depth estimation model, alongside this approach, could automate the setup of landmarks reducing deployment efforts.

ACKNOWLEDGEMENT

Grant PID2021-124335OB-C21, PID2022-140554OB-C32 and PDC2022-133684-C31 funded by MCIN/AEI/10.13039/5011000 11033.

REFERENCES

Aharon, N., Orfaig, R., & Bobrovsky, B., (2022). Bot-SORT: Robust Associations Multi-Pedestrian Tracking. *In Arxiv*, abs/2206.14651.

Bewley, A. et al., (2016). Simple Online Realtime Tracking. *In CoRR*, abs/1602.00763.

Blake, S., (2019). How to Track Objects in the Real World with Tensorflow, SORT and OpenCV. *In Medium*, HAL24k Techblog.

Bochkovskiy, A., Wang, C. & Liao, H. W., (2020). YOLOv4: Optimal Speed and Accuracy of Object Detection. *In CoRR*, abs/2004.10934.

Collado, J. et al., (2006). Adaptative Road Line Detection and Classification. *Advanced Concepts for Intelligent Vision Systems*, Antwerp, Belgium, pp1151-1162.

Girshick, R. B., (2015). Fast R-CNN. *In CoRR*, abs/1504.08083.

Girshick, R. B. et al., (2013). Rich Feature Hieraarchies for Accurate Object Detection and Semantic Segmentation. *In CoRR*, abs/1311.2513.

Jalalat, M., Nejati, M., & Majidi, A., (2016). Vehicle Detection and Speed Estimation using a Cascade classifier and Sub-Pixel Stereo Matching. *2016 2nd International Conference of Signal Processing and Intelligent Systems*, Tehran, Iran, pp1-5.

Jocher, G., (2020). YOLOv5. *Ultralytics.*

Jocher, G., Chaurasia, A. & Qui, J., (2023). YOLOv8. *Ultralytics.*

Maduro, C., et al., (2008). Estimation of Vehicle Velocity and Traffic Intensity Using Rectified Images. *In ICIP*, pp791-795.

Redmon, J. et al., (2015). You Only Look Once: Unified, Real Time Object Detection. *In CoRR*, abs/1506.02640.

Redmon, J. & Farhadi, A., (2016). YOLO9000: Better, Faster, Stronger. *In CoRR*, abs/1612.08242.

Redmon, J. & Farhadi, A., (2018). YOLOv3: An Incremental Improvement. *In CoRR*, abs/1804.02767.

Ren, S., et al., (2015). Faster R-CNN: Towards Real-Time Object Detection with Region Proposal Networks. *In CoRR*, abs/1506.01497.

Wojke, N., Bewley, A. & Paulus, D., (2017). Simple Online and Realtime Tracking with a Deep Association Metric. *In CoRR*, abs/1703.07402.

TRANSFORMING ROME: STRATEGIC INITIATIVES FOR DEVELOPING A SMART CITY

Vincenzo Gattulli and Spartaco Paris
Department of Structural and Geotechnical Engineering, Sapienza University of Rome
Via Eudossiana 18, 00193, Rome, Italy

ABSTRACT

Several activities have characterized in the last years the Capitoline municipality to transform Rome in a smart city. The work opens a multidisciplinary discussion on the potentiality of the digital transformation for an effective enhancement of city life in Rome, in view of 2025 Jubilee. The results of two initiatives are presented related to the architecture of urban pavements designed by a multi-scale digital method and to the use of the Digital Twin of Marcus Aurelius' Hall in the management of a complex museum. In the first case, themes related to innovative design guidelines are introduced, while the second case permits a discussion on the technology effectiveness of embedding IoT technology directly in the BIM of an important architectural realization immersed in a cultural heritage site, dense of historic masterpiece. The interaction between different sources of data coming by advanced monitoring services and management issues are discussed to promote a useful utilization of the developed technology within the concept of Rome as smart city.

KEYWORDS

Digital Twin, Structural Monitoring, IoT, MEM Sensors, Urban Quality, Cultural Heritage

1. INTRODUCTION

A definition of smart city is a hard task as already mentioned in the book of Renata Paola Dameri on 2017 which in chapter 1 discusses a series of classical definition. The debate regards the main components of a smart city: for academicians, smart city is especially based on components such as intellectual and social capital of the city, city governance and smart city strategic planning; in the empirical vision expressed by large companies such as IBM, Cisco and so on, the main component is the technology.

Compared to the initial introduction of the concept of a smart city (Albino et al., 2015) within the international debate, urban policies for city governance have begun a slow transformation process, not only from an instrumental point of view but also from a strategic perspective. The term, although inevitably overused in political communication, has gradually acquired more precise connotations.

In Europe, the area of planning has experienced a contraction both in experimentation and application opportunities according to a growth line - in recent years we have witnessed the dominance of zero-growth. However, the area of urban management is expanding rapidly and seeking new tools and methodologies to become more efficient.

It has been appropriately observed (Zamboni, 2014) that the concept of a smart city itself is oriented towards defining management systems for services - both material and immaterial - for urban life. It represents a new 'layer' of an apparently immaterial nature, characterizing the new modes of interaction between people - city users, no longer - perhaps - citizens, and the city. It has been rightly noted that what is being defined primarily concerns the field of services and, to a lesser extent - though not insignificantly - the transformation of the physical configuration of our cities.

To the layer of the city made of stone, concrete, and glass that has stratified over centuries, with its urban framework of streets, squares, and the articulation between public and private spaces that constitute our real cities, the "smart city" is adding a layer of material and immaterial networks and interfaces for data transmission. Yet, this new layer takes shape in various interfaces that, although minimal in size and form, allow new interactions between humans and the physical artifacts that make up the city: infrastructures, transportation systems, dwelling, offices, industries, settlements for water, waste, and energy, supply and management, green infrastructures.

It has been widely outlined that "The Smart City", in addition to being a concept, has become a slogan – somewhat like sustainability – to qualify the transformation and management of today's cities. When dealing with this slogan, it is necessary to consider the structure of the city's decision-making frameworks, in terms of efficiency and effectiveness regarding who makes the decisions, and in terms of the breadth of inclusivity regarding the beneficiaries. (Rison, 2014).

The smart city, as well as being a concept, also seems to be a slogan – as sustainability is – used to describe the transformation and the management of the contemporary city. But when dealing with the slogan, we still have to consider the basic elements of urban decision-making structures in terms of efficiency and effectiveness – in terms of who decides, and how we can ensure broad inclusiveness among the beneficiaries.

According to the views of the doyen of public-space design, as Richard Rogers define Jan Gehl in the preface of his marvelous book (Gehl J., 2010), a city can be recognized as smart not only if it improves people's quality of life, but if it places people, rather than the increasingly sophisticated technologies capable of providing massive amounts of data, at the center of its focus.

Smart Cities that prioritize people are equipped with superior tools to address the needs of their citizens, offer access to information, and facilitate meaningful dialogue and interaction between residents and administrators. This enables a Smart City to adapt more effectively to the fast-paced changes in urban culture.

Unfortunately, when the "smart city" concept is branded, it often becomes associated with specific technologies or a one-size-fits-all approach. What remains true for cities both in the West and the East is that trust is essential for unlocking the potential of a place and its people. Smart cities should aim to build and reinforce this trust, though the methods to achieve it will vary greatly between different cities. Additionally, smart cities must ensure the efficient use of our most valuable resources: we need to enhance how we manage space, time, and people. The challenge for a proper smart city could be in the intersection of the built environment and quality of life. Put another way, this is the intersection of the material and the immaterial (Rison, 2014).

Undoubtedly the recent advance in several technologies have permitted novel vision for cities. For example, the integration of augmented reality and drones allows past and future landscapes to be visualized from an aerial perspective (Kikuchi et al. 2022). By creating highly detailed 3D models of cities, drones can provide planners with valuable insights and data to inform a wide range of initiatives. From smart city planning to historic preservation efforts and disaster response and management, 3D drones modeling could prove to be an indispensable tool for visualizing and analyzing cities (Hu and Minner, 2023).

The use of different data set is particularly considered for cities dense of cultural heritage. Indeed, highlighting the potential impact of innovative data integration, methods, and workflows on architectural renovation and energy upgrades, the integration of conservation information for heritage buildings and large-scale environmental analysis is possible for historic clusters in modern cities (Artopoulos et al 2024).

2. ACTIONS FOR ROME AS SMART CITY

Rome is actively undergoing a transformation into a smart city, with numerous initiatives aimed at enhancing its infrastructure, sustainability, and overall quality of life for residents and visitors. This effort involves integrating advanced technologies and innovative solutions across various sectors. In this respect it is not easy to follow the historical development of all of them.

Starting from the results of the study "Efficient cities Siemens 2012", cited in (Bellone and Geropanta, 2019), the authors presents a review on the smart city initiatives for ameliorating the mobility of Rome, launched by the Municipality of Rome the last 6 years, and makes a reflection about their effectiveness to bring about regeneration in the city. The first initiative concerns the "Sustainable Mobility Plan" of Rome (SUMP) with the objective of "reconnect the city, reduce the use of private vehicles (cars and motorbikes, recover and redistribute public spaces and improve the environmental status".

SUMP was announced in 2012 and works with three objectives. The first one is the systemization of the transport system, by expanding two metro lines (B and C) and ameliorates the tramline. The second regards a number of e-mobility devices, car sharing and car-pooling for the city. The last one regards the use of ICT that could have the ability, firstly control several of the traffic systems and then to allow citizens' participation.

However, the web sources of information are not more available for this project and the authors conclude that more time for feedback and a considerable increase in the level of technical competencies is needed among the pillars of planning disciplines.

In 2017 a key-player in Rome ACEA announce a serious of needs to transform Rome in a Smart City (Acea, 2017) among several items, structural monitoring has been mentioned.

On 5 September 2023, with act n.144 (Act 144, 2023), the municipality of Rome approves the "Establishment of the Roma Smart City Lab Council and approval of its Regulations" in order to accomplish the Digital Transition which is indicated, in the 2021-2026 Programmatic Guidelines for the government of Rome Capital. In such guidelines among other goals, it is mentioned the necessity to "build a model based on the three tools of CIM (city information modeling), the BIM (building information modeling) and GIS (geographic information system) for implement a urban maintenance system and ensure accessibility to all places by all. Within this context a series of initiatives has started and mentioned in the following.

In the field of *5G Infrastructure and Connectivity, t*he #Roma5G project, a collaboration between Roma Capitale and Boldyn Networks, is central to Rome's smart city development. This project involves building a comprehensive 5G infrastructure, covering metro lines, installing IoT sensors, and deploying thousands of 5G CCTV cameras. The initiative aims to enhance public safety, provide extensive connectivity, and support various smart services throughout the city (Boldyn, 2023).

In the field of *Digital Transformation and Data Management* the quantities of real-time data available in a modern city is tremendous, consequently a series of project have been proposed and developed with several difficulties which are pushing towards the adoption of data management frameworks (Bilal et al., 2020).

Rome has been leveraging modern data platforms to improve its urban management and services. In collaboration with MongoDB, Roma Capitale has developed a centralized data management system that facilitates real-time insights and efficient service delivery. This platform supports the city's goal of being user-centric, sustainable, and data-driven (MongoDB, 2024). Rome is enhancing its transportation systems through intelligent traffic management, smart parking solutions, and the promotion of electric and shared mobility options. These measures aim to reduce traffic congestion, lower emissions, and improve the overall efficiency of urban transport. The city is expanding its public Wi-Fi network, particularly in high-tourist areas, to provide free internet access. Additionally, the deployment of IoT sensors across Rome allows for better environmental monitoring, resource management, and data collection, which are crucial for informed decision-making and urban planning. Rome is focusing on enhancing transparency and citizen engagement through digital platforms. These platforms enable residents to participate in governance processes, access public services online, and stay informed about city initiatives. The integration of big data analytics helps the city administration to make more informed and effective policy decisions. Efforts to improve energy efficiency and promote renewable energy sources are also underway. Rome is retrofitting buildings with smart meters and energy management systems, and expanding the use of solar and wind energy. Environmental monitoring systems are being implemented to track air quality, noise levels, and water quality, contributing to a healthier urban environment.

Rome's journey towards becoming a smart city involves a comprehensive approach that integrates advanced technology, sustainable practices, and active citizen engagement. By focusing on these key areas, Rome is set to improve its urban infrastructure, enhance the quality of life for its inhabitants, and maintain its historical charm while embracing modernity.

These initiatives position Rome as a leading example of smart city development in Europe, reflecting a commitment to innovation and sustainability. The city's ongoing projects and strategic partnerships illustrate its dedication to transforming into a connected, efficient, and resilient urban center (Juniper Resarch, 2023).

2.1 Rome and its Urban Streets Pavements

Recent applied research in the city of Rome is here described, which heavily relies on the use of new digital technologies applied to the road infrastructure of Rome, particularly to the pavement surfaces. Indeed, a more efficient and effective management of maintenance activities requires an acceleration of digital mapping processes of the Roman road system through GIS tools and parametric models aimed at management by specialized competent offices. As an example, one can look at the digital mapping model used by the Paris administration for managing interventions on the road network, through the continuous updating of a thematic digital mapping managed by a dedicated agency (APUR).

Mappatura urbana dei siti gestiti dal Dipartimento Coordinamento Sviluppo Infrastrutture e Manutenzione Urbana (S.I.M.U.) del Comune di Roma

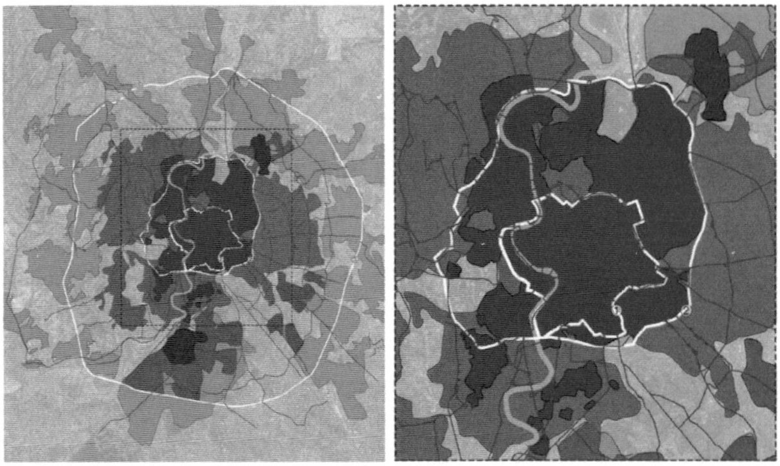

Figure 1. The mapping of the 599 sites managed by the S.I.M.U. shows how the network represented in the table is complex and made up of different types of spaces such as: streets, squares, wide open spaces, bridges and monumental gates, each morphologically and structurally different from the other. The coordination of management and maintenance activities, including all its facets, is a highly complex operation. It is in this context that the 'Map of the urban pavements of the City of Rome' produced by the CITERA study (2022) fits in, which provides the Administration a simple, good digital governance tool that, however, is still lacking in Rome

In 2022, Citera (Interdepartment Centre for Research Territory, Building, Restoration, Environment of Sapienza University) was commissioned by the Department of Infrastructure Development and Urban Maintenance Coordination of the Public Works and Infrastructure Department of Rome Capital to develop the study "Guidelines and operational tools for improving the urban quality of drivable, cycleable, and pedestrian pavements in public spaces of Rome Capital". The activity, while delimited, is in progress.

However, while focusing on a specific aspect strictly related to the built environment, such as the quality of urban pavements, the study must engage with the broader framework of material and immaterial components contributing to the quality and decorum of the city's public spaces.

Streets amount to roughly 20-30 per cent of the space in cities, typically 80 per cent of the open space. This means that a third of the urban space is being underutilized and plays a significant role for the urban life quality.

Due to Rome's complex topography, urban morphology, and the plurality of characteristics defining its parts, it serves as a laboratory where multiple planning and management tools have evolved over time to achieve urban quality. Significant progress lies in defining tools aimed at guiding interventions on the built environment of the city's open spaces, both historic and contemporary.

The objectives agreed upon in the agreement between CSIMU and the Centro CITERA de La Sapienza, consistent with the specific competencies of the Capitol Department, have led to circumscribing the scope of analysis and experimentation activities to the road pavements of the city of Rome (figure 1).

Currently, the capital is traversed by over 8,000 kilometers of roads, of which 800 kilometers fall under the management jurisdiction of the SIMU Department - approximately 10%. While this represents a relatively small portion, it is extremely significant as it includes the most relevant roads in the city. Consequently, it can guide strategic choices and operational programming and management activities of other involved institutional bodies. The research defined a framework to enhance the quality of roads in Rome, including analyses of innovative models, existing tools, and a historical overview of urban sidewalk evolution. Best practices and innovative solutions were identified for various types of pavement, aligning with sustainability goals (figure A strategic subdivision of Rome into four main intervention environments was proposed.

The articulation of Rome through its "fabrics" was acknowledged as crucial for valorizing the city's complexity and diverse identities. Lastly, a classification and coding system were developed to identify key pavement types and nodes in the BIM environment and with a digital framework (Paris at al., 2023).

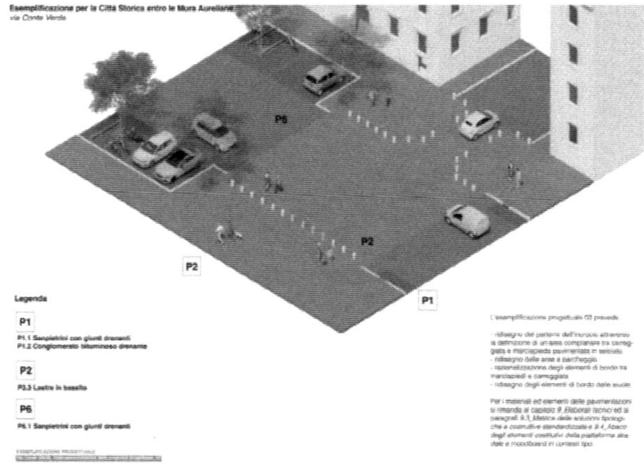

Figure 2. Design Example, developed through BIM methodology, envisages a pavement redevelopment operation with the following actions: - redesign of the parking areas; - rationalisation of the edge elements between the pavement and the roadway; - redesign of the edge elements of the flowerbeds. 3_Matrix of standardised typological and constructive solutions and 9.4_Application of the constituent elements of the road platform and moodboard in typical contexts

The main result achieved by the research is the transfer of data from the analogue to the digital sphere; until today the Department of the municipality of Rome manages ordinary and extraordinary maintenance works on urban pavements through traditional methods, that cause redundancies, repetitions, cost increases and uncertain time management. Therefore, a set of 11 recurring types of pavements and building elements - which combined define the city's pavement surfaces in different contexts - has been created in a BIM library developed 'ad hoc' and integrated with the standard specifications and related costs for carrying out interventions. This tool, in addition to updating the skills of the administration's technicians, makes it possible to optimize the time and costs for standard interventions, estimating at least 30% time savings in control and management and the reduction of maintenance interventions by 'emergency' (there is a lack of experimentation on this data to quantitatively assess the benefit). The further steps of the ongoing research – according to Municipality, should focuses the definition of a digital mapping in GIS format of the roads under the jurisdiction of CSIMU, serving as a foundation for experimenting with the management of development interventions and maintenance across the entire road network and public spaces of the city. This involves creating a map of the roads under CSIMU's jurisdiction (approximately 600 roads), in georeferenced digital format, indicating for each the length, width, and articulation between the roadway and pedestrian areas. Such a tool will enable the Administration to update planning, implementation, and management activities within a dynamic digital framework.

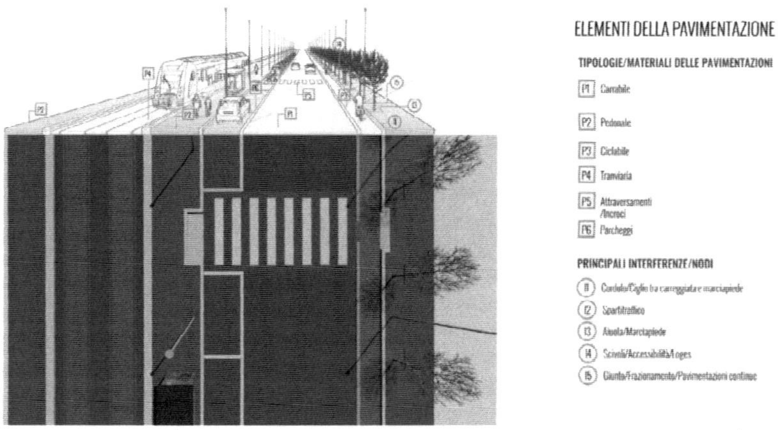

Figure 3. Set of elements and components which define the urban pavement platform of the city. These 11elements has been modeled with BIM methodology and linked with specific data sheets that describe requirements, features, and costs, in order to upgrade the management of repair interventions. (Source. Citera/ Municipality of Rome)

2.2 Digital Twins in Smart City

Digital twins, which are virtual replicas of physical entities, systems, or processes, are revolutionizing the way cities are planned, developed, and managed. By integrating real-time data and simulations, digital twins provide a dynamic and interactive model of urban environments.

Very recently a series of studies have evidenced specific uses of Digital Twins to enhance specific functions in Smart Cities. The transforming role of Digital Twins in Smart City development has been clearly evidenced by the state-of-art article of Peldon et al. 2024 in which the integration of DTs in smart city frameworks is presented as transformative potential for urban development. The work evidenced that nnumerous cities are adopting DT idea to facilitate urban administration, development, and decision-making processes even if the complete realization of a city-scale DT has not yet been achieved, probably because, significant barriers such as data management complexities, interoperability issues, and cybersecurity should be addressed in future developments. The study evidences also as future research should specifically address the integration of social community dynamics and stakeholders should collaborate to ensure these systems are not only technologically robust but also socially attuned, facilitating the creation of resilient, sustainable, and truly smart urban ecosystems. In (Pan et al., 2024) the concept of Smart City Digital Twins (SCDTs) is introduced as an emerging approach to understanding and addressing urban challenges, such as unhealthy built environments, high energy consumption, and disaster responses. The method proposed by the authors utilize meteorological sensors and street video streams to continuously collect data in real-time and to take action to mitigate the negative impacts of heat exposure on city residents.

Moreover, the growing technical capacity to easy deploy digitalization efforts has induced widespread international interests regarding their potential implications for the preservation, presentation, and management of cultural heritage assets. The classical collection through manuscript recordings and manual measurements, for esthetic and technical information of cultural heritage, can be enhanced by the use of digital technology in fully reproducing construction details, material properties, and eventually considering permanent monitoring measurements.

In this respect the concept of digital twin, has been very recently introduced for its potential effectiveness in the collection, processing, and utilization of graphical, spatial and other attribute data for safety diagnose, virtual reproduction and exhibition, interventions and management of cultural heritage (Dang et al., 2023).

Digital twin targets to virtually reproduce the physical entity based on static and real-time data (Eleftheriou and Anagnostopoulos, 2022). In the framework of structural monitoring, an automated long-term monitoring system could involve the continuous monitoring of a building's structural integrity using sensors such as accelerometers. Since 2021 the idea to construct a digital representation of an important area of the Capitoline Museum of Rome has been supported by different research projects. The case study is the Marcus Aurelius Exedra Hall within the complex of the Capitoline Museum of Rome (see Figure 4).

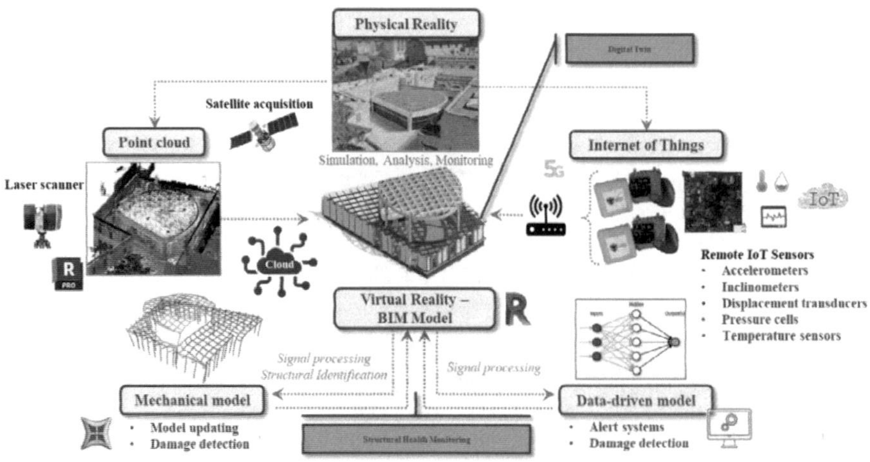

Figure 4. Conceptualization of the Exedra Digital Twin in Rome

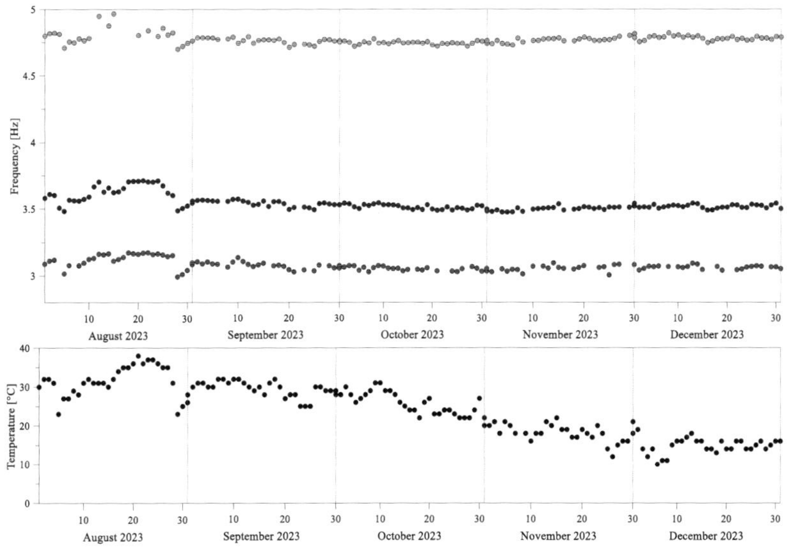

Figure 5. Natural frequencies tracking along five months and temperature evolution

The steel and glass structure is the new grand glass-roofed hall designed by the architect Carlo Aymonino, and it has a semi-elliptical plan, supported by six 75 cm-diameter steel pillars. The six columns support the massive box section girders at two different altitudes that hold up the perimetrical glass enclosures: the horizontal elliptical roof, and the lateral surfaces. The steel and glass structure is the new grand glass-roofed hall designed by the architect Carlo Aymonino, and it has a semi-elliptical plan, supported by six 75 cm-diameter steel pillars.

The six columns support the massive box section girders at two different altitudes that hold up the perimetrical glass enclosures: the horizontal elliptical roof, and the lateral surfaces. The proposed monitoring system is made up of a gateway for SHM, six wired sensor nodes measuring acceleration in three orthogonal directions, a Switch PoE, and a Router.

The implementation of the system requires the preliminary realization of a star network topology using Ethernet cabling. The circuit has been designed to perform measurements, process data, and send results to the gateway (Crognale et al. 2024). The implemented system furnishes information to a BIM model which interact with a FEM model according to the diagram depicted in Figure 4. The data are treated by hybrid approach, physical and data-driven to furnish alert and information on damage detection. In Figure 5 are depicted the natural frequencies extracted by the acquired data through operational modal analysis and specific data clustering techniques. The frequencies are directly compared with information coming from temperature and it demonstrate how high levels of temperature experienced in August 2023 in Rome have specific effects on the steel structure of the Exedra. The information is extremely important to start to analyse how large increment of temperature in the cities may determine increases of thermal stresses and potential safety problems.

The management of data permits to obtain valuable information also by satellite measurements (De Iulis et al. 2023). Digitalization and modelling of Marco Aurelio monuments poses a basis for the integration of different objects at different scale.

3. CONCLUSION

The paper discusses two recent experiences in the use of digital technology in the management of pavement renovation and structural monitoring developed in the city of Rome. The applications show how the use of digital modelling in some cases integrated by on-line measurement permits to enhance the knowledge of the studied system. Consequently, an enhanced management of actions in a complex city is a chance which cannot be lost for the transformation of Rome in a Smart City. The perspective on the use of Digital Twins of the cultural heritage sites open the future scenario to a tremendous set of different opportunities from monitoring the effect of climate change on monuments to the control of the increasing number of tourists visiting the Forum area in Rome which is continuously renovating the pavement often still constructed with the famous "*sanpietrini*".

ACKNOWLEDGEMENT

The research leading to these results has received funding in part sponsored by IRIS Project, NATO Science for Peace and Security Programme under grant id. G5924 https://irisnatoproject.eu/

REFERENCES

Acea: Smart City a Roma: Progetti per la Mobilità Sostenibile. (2017). Retrieved from https://romamobilita.it/sites/default/files/pdf/mobilityweek2017/PPT_Acea_DEF. pdf

Act 144, Municipality of Rome. (2023). Retrieved from Laboratorio Smart City Roma Capitale (labsmartcityroma.it)

Albino, V., et al. (2015). Smart cities: Definitions, dimensions, performance, and initiatives. *Journal of Urban Technology,* Vol. 22(1), pp. 1-19.

Artopoulos, G., Fokaides, P., Lysandrou, V., Deligiorgi, M., Sabatakos, P., Agapiou, A. (2023). Data-Driven Multi-Scale Study of Historic Urban Environments by Accessing Earth Observation and Non-Destructive Testing Information via an HBIM-Supported Platform. *International Journal of Architectural Heritage*, vol. 18(6), pp. 920–939.

Bellone, C., Geropanta, V. (2019). *The 'Smart' as a Project for the City Smart Technologies for Territorial Management Planning Strategies*. In: Vasant, P., Zelinka, I., Weber, GW. (eds) Intelligent Computing & Optimization. ICO 2018. Advances in Intelligent Systems and Computing, vol 866. Springer, Cham., pp. 66-75

Bilal, M. et al. (2020). Smart Cities Data: Framework, Applications, and Challenges. In: Augusto, J. (eds) Handbook of Smart Cities. Springer, Cham., pp. 1-29

Boldyn (2023). Roma Capitale and Boldyn Networks kick off #Roma5G, Rome 12 Dec. Retrieved from Boldyn.

Crognale, M. et al. (2024). Developing and Testing High-Performance SHM Sensors Mounting Low-Noise MEMS Accelerometers. Sensors, 24, 2435.

Dameri R.P. (2017). *Smart City Implementation: Creating Economic and Public Value in Innovative Urban Systems.* Springer Cham.

Dang, X. et al. (2023). Digital twin applications on cultural world heritage sites in China: A state-of-the-art overview. *Journal of Cultural Heritage*, Vol. 64, pp 228-243.

De Iuliis, M., et al. (2024). On the combined use of satellite and on-site information for monitoring anomalous trends in structures within cultural heritage sites. *Journal of Civil Structural Health Monitoring*. On-line Open Access.

Eleftheriou O.T., Anagnostopoulos C.N. (2022). Digital twins: A brief overview of applications, challenges and enabling technologies in the last decade [version 1; peer review: 2 not approved]. *Digital Twin*, 2:2

Gehl J. (2010). *Cities for People.* Island Press, Washington, Covelo, London.

Juniper Research. (2023). Berlin Ranked Number 1 Smart City in Europe in 2023. *Sustainability & Smart Cities.* Vol. July.

Hu D, Minner J. (2023). UAVs and 3D City Modeling to Aid Urban Planning and Historic Preservation: A Systematic Review. *Remote Sensing*. vol. 15(23), 5507.

Kikuchi, N., Fukuda, T., Yabuki, N. (2022). Future landscape visualization using a city digital twin: integration of augmented reality and drones with implementation of 3D model-based occlusion handling, *Journal of Computational Design and Engineering*, vol. 9, Issue 2, April 2022, pp. 837–856,

MongoDB. (2024). The Eternal Smart City Powered by Roma Capitale and MongoDB. Retrieved from MongoDB

Pan, X, Mavrokapnidis, D., Hoang T. L., Mohammadi N., Taylor J. E. (2024). Assessing and forecasting collective urban heat exposure with smart city digital twins. *Scientific Reports,* Vol 14, 9653

Paris, S. et al. (2023). Architecture of urban pavements. Multi-scale and digital methods for an ecological transition. *AGATHÓN | International Journal of Architecture, Art and Design*, 13, pp. 217–226.

Peldon, D. Banihashemi, S. LeNguyen, K. Derrible, S. (2024). Navigating urban complexity: The transformative role of digital twins in smart city development, *Sustainable Cities and Society*, Vol 111, 105583, ISSN 2210-6707

Rinaldi, C. et al, 2024. Long-term vibrational monitoring of the Exedra of Marcus Aurelius' Hall. *Proceedings of the 10th International Operational Modal Analysis Conference (IOMAC 2024).* Napoli, Italy.

Rison J. (2014). Un'urbanistica per le persone, *Domus*, Supplement Smart City, n. 985, November, pp. 10-13

Zamboni, A., ed. (2015). Smart City. Domus supplement, Domus n. 985, nov. 2014, Editoriale Domus, Rozzano, pp. 2-3.

IS TOYOTA'S WOVEN CITY A SMART CITY?

Hiroshi Ito and Kazuhiko Kato
Graduate School of Management, Nagoya University of Commerce and Business
1-3-1 Nishiki Naka, Nagoya, Aichi, Japan 460-0003

ABSTRACT

The concept and implementation of corporate smart cities have been gaining widespread recognition on a global scale. Notable examples of such initiatives include Songdo, Masdar, Fujisawa and Quayside (although the latter was abandoned in 2020). In line with this growing trend, the Toyota Motor Corporation has embarked on the planning and development of "Woven City", a smart city being constructed on the former site of Toyota's East Fuji Plant in the city of Susono, Shizuoka Prefecture, Japan. Woven City is a remarkable case study in several aspects. Although it is physically located within Susono, it stands out as a "green field" development, built from scratch and intends to be populated exclusively by individuals from outside. Moreover, Woven City holds the distinction of being the first smart city project led by an automobile manufacturer. However, comprehensive and publicly accessible information about the project is limited. Therefore, the present study examines the nature and characteristics of Woven City in comparison to other corporate smart cities. To achieve this, we conducted document analysis and interviews with Toyota personnel engaged in the project, a Susono city government official and a Shizuoka University professor. The findings are expected to offer insights for city planners and policymakers involved in the development of corporate smart city initiatives.

KEYWORDS

Corporate Smart City, City Planning, Living Labs, Woven City, Toyota Motor Corporation

1. INTRODUCTION

The advent of "smart cities" has been a pivotal development in urban planning since the term was first coined in the 1980s (Holguín Rengifo, Herrera Vargas, and Valencia-Arias, 2023). These digitally interconnected urban centers are designed to foster sustainable living and user-driven innovation, a concept well-exemplified by Amsterdam's Digital City, established in 1994 and often regarded as the world's first smart city (Dameri, 2017; Matsumoto, Esaka, and Izumiyama, 2023). Within this framework, the corporate smart city has emerged, characterized by corporate entities implementing technologies and services to create a new urban environment (Kummitha, 2019). This study focuses on Toyota's Woven City, located in Susono, Shizuoka Prefecture, Japan, and analyzes its potential as a corporate smart city. Despite limited publicly available information, this research draws on document analysis and interviews with key stakeholders to examine Woven City's characteristics and differentiation from other examples such as Masdar, Songdo, Fujisawa, and Quayside. The findings will contribute to the discourse on the integration of corporate strategies within urban development (Cardullo and Kitchin, 2019) and the role of car-mobility companies in this innovative domain.

2. SUMMARY OF LITERATURE REVIEW

Corporate smart city projects offer promising solutions for sustainable urban development, but their success hinges on addressing environmental and climate challenges, fostering collaboration, ensuring inclusivity and prioritising ethical technology use. The development of corporate smart cities such as Masdar, Songdo, Fujisawa, and Quayside have implications pertinent to urban planners and practitioners managing contemporary and forthcoming corporate smart city initiatives. The following implications were selected as focal points for analysis because they emerge as common characteristics among the urban development projects of Masdar City, Songdo City, Fujisawa Sustainable Smart Town, and Quayside. Each of these projects demonstrates the intricate balance and interplay between ambitious technological aspirations and the pragmatic pace of actual urban development. By examining these implications, the research can offer insights into the fundamental challenges and opportunities that define corporate smart cities.

2.1 Gradual Progress

Corporate smart city projects may take longer than initially anticipated to fully develop and become operational. Factors like budget constraints, bureaucratic processes and the need to integrate multiple technologies can slow down progress. It is essential to manage expectations and communicate effectively with the public about the timeline and milestones (Angelidou, 2014; Cugurullo, 2018; Randeree and Ahmed, 2019).

2.2 Public Scepticism

The stagnation or discontinuation of some corporate smart city projects, such as the example of Quayside in Toronto, can lead to public scepticism about corporate-led initiatives. Concerns about privacy, transparency and the influence of private interests may undermine public trust, smart city initiatives should prioritise inclusive public engagement, democratic decision-making and transparent planning processes to gain public acceptance (Cugurullo, 2018; Flynn and Valverde, 2019; Filion, Moos and Sands, 2023).

2.3 Technology Innovation

Corporate smart city projects serve as living labs and testbeds for various technological innovations. The projects allow companies to develop and commercialise new technologies, ranging from renewable energy solutions and smart infrastructure to surveillance and data analytics. However, ensuring responsible and ethical use of these technologies is essential to address concerns about privacy and social inclusion (Randeree and Ahmed, 2019; Sgouridis and Kennedy, 2010; Tokoro, 2018).

2.4 Sustainable Urban Development

The focus on sustainability in smart city projects reflects the growing awareness of environmental concerns. Projects aim to reduce carbon emissions, promote renewable energy sources, and adopt eco-friendly practices. However, striking a balance between environmental goals and economic viability remains crucial to achieving long-term sustainability (Barrett, DeWit and Yarime, 2021; Cugurullo, 2018; Griffiths and Sovacool, 2020).

2.5 Energy and Climate Insights

The dawn of smart cities has brought to the forefront projects, each grappling with the nuances of integrating technology with sustainability and climate objectives (Griffiths and Sovacool, 2020; Cugurullo, 2018; Randeree and Ahmed, 2019; Kuecker and Hartley, 2019). For instance, Masdar City's recalibrated carbon-neutrality goals and energy challenges (Sgouridis and Kennedy, 2010) and FSST's smart energy systems within Panasonic's vision (FSST Management, 2023), reflect smart city aspirations with the realities of energy and climate imperatives. The experiences of these smart cities reinforce the necessity of a multi-faceted approach to energy and climate issues. By acknowledging the challenges and embracing the dynamic nature of smart city development, future projects can better navigate the intricacies of sustainable urbanisation.

Corporate smart city initiatives like Masdar, Songdo, Fujisawa, and Quayside exemplify the complex integration of technology with sustainable urban development, aiming to address climate challenges and promote environmental sustainability. These projects serve as testbeds for technological innovations, including renewable energy and smart infrastructure, while grappling with the need for responsible and ethical technology use to ensure privacy and social inclusion. However, the development of such cities often faces hurdles like budget constraints, bureaucratic delays, and public scepticism due to concerns over privacy and the influence of private interests, potentially undermining trust and acceptance. The balance between environmental goals and economic viability is crucial, as these cities strive to reduce carbon emissions and promote eco-friendly practices, but must also manage expectations regarding their progress and operational timelines. The experiences from these cities highlight the importance of inclusive public engagement and transparent planning to navigate the complexities of sustainable urbanisation effectively.

3. RESULTS

Our investigation into Woven City reveals a forward-looking enterprise where Toyota reshapes urban living with advanced mobility and technology. However, this transition is not without its complexities, as the project navigates between corporate governance and community engagement. As a microcosm of innovation, Woven City embodies a corporate determination to harness technology for societal benefit, yet it must reconcile the scope of public involvement with its top-down implementation approach.

3.1 Findings from Document Analysis

The advancement of smart cities requires a convergence of city planning, technological innovation, and corporate willingness to create futuristic, efficient, and sustainable habitats. The findings, drawn from the document analysis, shed light on the origins, objectives, structure, and operational systems of Woven City. Woven City is an ambitious initiative that was first announced by Toyota's former president, Akio Toyoda (2009-2023), at the Consumer Electronics Show (CES) 2020. The project aims to create a smart city where various technologies are integrated, such as automated driving, Mobility as a Service (MaaS), smart homes and artificial intelligence (Sakuma and Kato, 2020).

The Woven layout features a network of three types of streets with distinct functions: one dedicated to self-driving and eco-friendly vehicles; another thoughtfully configured to accommodate a blend of slower modes of mobility alongside pedestrian traffic and finally, broad thoroughfares exclusively reserved for pedestrian use. The nomenclature of this city, Woven, is inspired by the fabric of its streets, artfully intertwined to create a harmonious urban landscape (Sakuma and Kato, 2020).

Woven City has been planned and developed with WbyT, which is the mobility technology subsidiary of Toyota Motor Corporation. Its mission is to deliver safe, intelligent, human-centred mobility for all, through Woven City – a test course for advanced mobility (Toyota, 2023a). WbyT's vision focuses on extending the value of the car, expanding mobility into new realms and integrating mobility with social systems. With this framework, WbyT helps Toyota to develop next-generation cars and to realise a mobility society in which everyone can move freely, happily, and safely (Woven by Toyota, 2023).

WbyT was renamed in April 2023, from Woven Planet Holdings, which comprised former Woven Core and Woven Alpha (effective from January 2021), while Woven Capital remains as an investment arm of WbyT. Woven Core was responsible for the development of automated driving technology and advanced driver assistance systems. Woven Alpha aimed to expand its business through innovative projects such as Woven City. Woven Capital was a global investment fund, which invested in growing companies that are developing innovative technologies. Woven Planet Holdings originated as the Toyota Research Institute-Advanced Development (TRI-AD) established in Tokyo in March 2018. TRI-AD was formed to provide fully integrated, production-quality software for automated driving. Its mission was to create world-class technology and build safe mobility products for the world (Toyota, 2023b). Daisuke Toyoda, the son of Akio Toyoda, serves the senior vice president of these organisations.

Akio Toyoda, the former president of Toyota, claimed, "we will work with the local community, step by step, towards the future" and "cooperate in the development of the Iwanami Station [the nearest station to Woven City)" (*Toyota Times*, 2021). However, Woven City will be a corporate smart city, where the design and implementation of the technological infrastructure are controlled by Toyota. Therefore, local citizen participation in the decision-making process has been limited (Kitayama, 2021).

In *Toyota Times* (2021), an online platform and publication established by Toyota to serve as a source of news, insights and stories related to Toyota's innovations and corporate activities, Akio Toyoda claimed:

> Woven City is a test course for Toyota to become a mobility company. This experiment can only be conducted in the framework of building of a "city". However, this city will never be completed, as the goal is not to bring a product to completion. To be honest, there is no need to explain every single detail about Woven City. We have no intention to make the process democratic. I want to

take the first step in a closed forum to be dictated by Toyota Motor Corporation, using its own land and own funds without using any tax money of the city of ... It should be a top-down decision-making process in the initial phases.

The website of Woven City (2023a) further elaborates that in its system, developers from various companies including Toyota, start-ups, entrepreneurs, and residents, work together to mass-produce future happiness for others – specifically, for someone other than themselves. Sakichi Toyoda, the founder of Toyota, invented the automatic loom with a single thought in mind: to lighten the burden of weaving for his mother. This principle – make someone happy – is the starting point of Toyota, and it remains unchanged even today. It is also the inspiration for the name Woven, which references weaving. Toyota has transitioned from being an automobile manufacturer to a mobility company, guided by the vision Mobility for All, with an aim to transcend the conventional concept of mobility to bring freedom and enjoyment of mobility to people of all backgrounds. Toyota, along with its various partners, will participate in real-life scenarios and collaborate to shape the future (Woven City, 2023b; 2023c).

Woven City provides a testing ground for such experiments, as follows: for instance, if one has developed a mobility system that makes it easier to get around, but finds it difficult to find a place to test it; one has developed a new app but is unable to gather user feedback effectively; or one has developed technology to reduce traffic accidents and needs to test it (Woven City, 2023c).

In the evolving landscape of urban development, Woven City distinguishes itself as an example of innovation, sustainability and a community-centric ethos (despite the ethos, however, the involvement of the local community in actual governance has been limited). By interweaving insights from key stakeholders, including Toyota Group insiders, a government official and a university professor, we aim to present a comprehensive picture of Woven City, shedding light on the nuances that define this intriguing urban experiment.

3.2 How Did the Project Get The Name Woven?

Upon asking about the origin of the term "Woven" in Woven City, one respondent from Toyota Group said, "Woven can refer to the various types of streets that are folded together in Woven City" (T-5). However, he continued:

> Rather, the name Woven originates from the automatic loom (weaving machine) that Sakichi Toyoda invented to help his mother. He then went on to establish Toyota Industries Corporation, the first Toyota Group entity. The core philosophy of the company since then is "for someone other than oneself". (T-5)

Thus, Woven City finds its roots in the innovative history of the Toyota Group: it harks back to the automatic weaving machine invented by Sakichi Toyoda, the founder of the Toyota Group.

3.3 How Did the Concept Of Woven City Originate?

Subsequently, we delved into the origins of the Woven City concept, seeking to unravel the narrative behind its inception. One respondent from Toyota Group reported that the concept was proposed when the decision was taken to close Toyota's East Fuji Plant in the city of Susono in 2018:

> after the Great East Japan Earthquake in March 2011, there was talk of relocating car production plants to Tohoku (the region where the earthquake struck), and the decision was made to close Toyota's East Fuji Plant in Susono in 2018. The then-president Toyoda had a talk with the East Fuji Plant workers. The atmosphere was bleak. One of the workers asked him, "What will happen to our jobs after the plant closes?" He continued, "I can go to Tohoku, but some of my colleagues cannot". After a few moments of silence, Toyoda replied, "We will establish a testing ground here for inventions that will benefit people". (T-5)

He explained how the concept of Woven City emerged as a testing ground for the comprehensive exploration of mobility solutions:

> in 2019, Toyota's transition to a mobility company was announced. Just as the loom became an automobile company, Toyota will now become a mobility company. Woven City is a test course for mobility. The word "Move" in "Mobility" refers not only to the movement of people, goods and information, but also to emotions. (T-5)

While the narrative presented so far serves as the "official explanation" for the genesis and realisation of the concept of Woven City, a Susono city government official offered additional insights:

> about 10 years ago, after the Great East Japan Earthquake, there was talk of Toyota moving its plant to Tohoku, as part of its contribution to the region, and establishing a smart city that would utilise hydrogen and solar power and efficiently run its energy infrastructure, with the plant at its core. I think the idea was to make individual lives smarter, but that story was cut short and the talk turned to creating a "Woven City" in the city of Susono. (G-1)

The concept of Woven City originated in response to the closure of Toyota's East Fuji Plant in Susono in 2018, which followed discussions about relocating car production plants post the Great East Japan Earthquake in 2011. However, an alternative perspective suggests that discussions about Toyota relocating its plant and establishing a smart city with sustainable energy practices to enhance individual lives with smart technology has been ongoing for a decade. Over time, this vision may have transformed into the concept of Woven City in Susono, demonstrating the complex evolution of city planning and development.

3.4 Can Woven City Be Considered a Corporate Smart City?

Our subsequent inquiry delved into the perception of Woven City as a corporate smart city. We were eager to discern whether those engaged in the project viewed it primarily as a corporate-driven smart city or if there were alternative perspectives that portrayed it differently. One respondent from the Toyota Group noted that Woven City is destined to serve as an experimental smart city: "Woven City will be an experimental smart city with a large early adopter population" (T-2). Another respondent from Toyota-Group referred to Woven City as a visionary corporate smart city:

> Woven is a smart city to be built by Toyota, and is designed as a next-generation urban form. This is an advanced experimental approach to intelligent energy, mobility and lifestyle. Woven is a place with no restrictions on personal information. It will be a long time before the data obtained from within is made available to the outside world. Most of the residents will be researchers from abroad or Toyota employees. (T-1)

However, one respondent from the Toyota Group provided a perspective that Woven City is a test course: "inside WbyT, Woven City is perceived as a test course rather than a city. It is not that we want to create a city, but rather we want to know what residents will do there" (T-3). Another respondent from the Toyota Group emphasised that the term "smart city" should not be used within WbyT because Woven City, in their view, is not a typical smart city:

> we are not supposed to use the term "smart city" inside WbyT because Woven City is a test course to accelerate innovation through experiments. A smart city is associated with the image of a convenient city with the use of ICT, but Woven City is not one (a convenient city). Toyota is huge, so there may be some misunderstanding about it even among Toyota employees. (T-4)

Another respondent from WbyT also claimed that Woven City is not classified as a conventional smart city; instead, it is best considered a test course designed to enhance people's quality of life:

> Woven City is definitely not a smart city: it is a test course for people to live better. It does contain elements of a smart city, and it is a city in the sense that people actually live in it. For us, however, it is a test course in the form of a city, and city development is not our primary objective. Essentially, we want inventors or people with an inventor mindset to live there. We aim to attract people who are eager to contribute to the betterment of others. Our ultimate goal is the mass production of happiness. (T-5)

The respondent from academia expressed the belief that Woven City is a smart city but highlights a difference in its approach compared to typical smart cities:

> I consider Woven City to be a smart city. However, smart cities typically prioritise addressing specific issues, such as the healthcare needs of the elderly, which can have a global appeal. In contrast, Woven City does not appear to focus on solving specific problems. (A-1)

These findings suggest that Woven City is multifaceted and may encompass elements of a smart city, but its primary purpose is experimental innovation and improving the well-being of its residents, potentially differentiating it from traditional smart city concepts. The lack of a unanimous definition underscores the evolving nature of this unique project.

4. DISCUSSION: CAN WOVEN CITY BE CONSIDERED A CORPORATE SMART CITY?

From the authors' perspectives, based on the analyses of the findings from the literature review, document analysis, and interview data, we suggest considering Woven City a type of corporate smart city, as it shares many characteristics with other corporate smart cities in the literature review. These similarities include corporate leadership and funding, technological integration and innovation hubs.

Corporate smart cities such as Woven City, Masdar, Songdo, Fujisawa, and Quayside, are spearheaded, funded, and managed by large corporations. Corporate smart cities often represent the integration of cutting-edge technologies, including AI, automation, smart homes, and advanced mobility such as MaaS (Griffiths and Sovacool, 2020; Tokoro, 2018) – traits also shared by Woven City. All these cities serve as testing grounds or living labs for corporations to experiment with new technologies, solutions, and ideas in a real-world urban setting (Kummitha, 2019; Randeree and Ahmed, 2019; Strickland, 2011; Toyota, 2023a).

However, the interviews revealed that some within Toyota classify Woven City as a conventional smart city (T-2, 3, 4), while others do not (T-1, 5). It may differ from other corporate smart cities in various respects, notably Toyota's complete control and ownership, emphasis on experimental innovation, focus on mobility, unique identity and potential inconvenience (which will be discussed shortly in this section). Unlike some other corporate smart cities that collaborate with external partners and investors, Woven City is fully controlled by Toyota: the design and implementation of technological infrastructure are under Toyota's authority, limiting local stakeholder participation in decision-making (Kitayama, 2021).

The Woven project is being conducted on Toyota's privately own land, funded by the corporation itself, without reliance on tax money. In contrast, the ownership structure, funding mechanisms and involvement of public funds vary significantly among other corporate smart cities. Toyota's robust ownership is also relevant to the control of privacy or protection of personal data within Woven City. This is because the project is owned by Toyota and only those who provide informed consent to be part of the experiment can enter the city, as some interviewees reported (T-1-4).

It is also noteworthy that although most corporate smart cities prioritise technology and efficiency, Woven City has a particular emphasis on mobility with a particular focus on automated driving and artificial intelligence, as it is backed by Toyota. Although other smart cities such as Masdar and FSST also incorporate mobility as a significant element, Toyota's focus on mobility distinguishes Woven City from other corporate smart cities that may prioritise different aspects of urban living.

Another notable difference is that Toyota seems to make a deliberate attempt to distance Woven City from the smart city label; The emphasis is on it being a test course in the form of a city. This is what lends it a unique identity, unlike other corporate smart cities that embrace the smart city label, emphasising technological integration as their primary identity.

Woven City is not a typical smart city in that its principal feature is to provide a test course for experiments and this can lead to residents or visitors having to face uncomfortable experiences in the processes towards mass-production of happiness. As one respondent from the Toyota Group (T-4) said, "Woven is not a smart city, which is typically associated with convenience". To reiterate the authors' perspectives, Woven City aims to be a convenient city – whether consciously or subconsciously – as a means to improve quality of life because convenience is a fundamental aspect of the city's mission – the mass production of happiness – albeit not its sole focus.

In the ambit of sustainability, Woven City exemplifies a deliberate effort to synergise technological progress with environmental stewardship. Toyota's investment in automated driving and AI is complemented by the adoption of renewable energy systems and eco-friendly materials, positioning the project in line with broader global objectives of energy efficiency and climate action. The city's design anticipates a low-carbon future, leveraging Toyota's expertise in hydrogen fuel cells and renewable energy sources to reinforce its role as a paradigm of eco-conscious urbanism. By prioritising environmental considerations alongside technological innovation, Woven City sets out to redefine corporate smart cities, aspiring not only innovation hubs but vanguards of sustainable urban life.

5. CONCLUSION

Woven City deviates from the standard smart city concept, embracing an experimental ethos centered on enhancing life quality within a sustainable framework. It seeks to set new benchmarks for urban living by focusing on technology while simultaneously striving to minimize carbon emissions, advance clean mobility, and support a shift to a low-carbon economy. The Woven City narrative champions transformative urbanism aligned with the urgent global need for cities that are innovative and responsive to climate challenges, positioning Toyota as a leader in the development of smart, sustainable urban futures This study contributes to the discourse on contemporary smart city development showcasing Toyota's Woven City project as a case study that integrates corporate innovation with a steadfast commitment to sustainability, energy efficiency, and climate responsiveness. Diverging from traditional smart city models, Toyota's Woven City exemplifies a novel approach by merging its mobility prowess with environmental foresight. This alignment results in a unique urban fabric where top-down decision-making and technological innovation contribute to the shaping of the city's development in harmony with sustainability.

While Woven City may diverge from the conventional smart city label, opting instead for an experimental ethos, its essence lies in fostering an improved quality of life within a sustainable framework. It aspires to establish new standards for urban living that prioritise not only technological innovation but also a concerted effort to reduce carbon emissions, promote clean mobility solutions, and support a transition towards a low-carbon economy. Ultimately, Woven City's narrative is one of transformative urbanism that resonates with the global imperative of crafting cities that are both innovative and attuned to the exigencies of our climate, positioning Toyota at the vanguard of shaping smart, sustainable futures.

REFERENCES

Angelidou, M. 2014. Smart city policies: A spatial approach, *Cities*, Vol. 41, pp. 3–11.

Barrett, B.F.D., DeWit, A. and Yarime, M. 2021 Japanese smart cities and communities: integrating technological and institutional innovation for Society 5.0, in Kim, H.M., Sabri, S. and Kent, A. (eds.) *Smart cities for technological and social innovation: case studies, current trends, and future steps*. Cambridge, MA: Academic Press, pp. 73–94.

Cugurullo, F. 2018. Exposing smart cities and eco-cities: Frankenstein urbanism and the sustainability challenges of the experimental city, *Environment and Planning A: Economy and Space*, Vol. 50, No. 1, pp. 73–92.

Dameri, R. P. 2017. *Smart city implementation*. Cham: Springer.

Filion, P., Moos, M. and Sands, G. 2023. Urban neoliberalism, smart city, and Big Tech: the aborted Sidewalk Labs Toronto experiment, *Journal of Urban Affairs*, Vol. 45, No. 9, pp. 1625–1643.

Flynn, A. and Valverde, M. 2019. Planning on the waterfront: setting the agenda for Toronto's 'smart city' project, *Planning Theory & Practice*, Vol. 20, No. 5, pp. 769–775.

FSST Management. 2023. *Fujisawa SST.* Fujisawa SST Council.

Griffiths, S. and Sovacool, B.K. 2020. Rethinking the future low-carbon city: carbon neutrality, green design, and sustainability tensions in the making of Masdar City, *Energy Research & Social Science*, Vol. 62, pp. 1–9.

Holguín Rengifo, Y. X., Herrera Vargas, J. F. and Valencia-Arias, A. 2023. 'Proposal for a comprehensive tool to measure smart cities under the tiple-helix model: Capacities learning, research, and development'. *Sustainability*, Vol. 15, 1-21.

Kitayama, S. 2021. *Automotive-driven city of smartness: study of compatibility between Toyota's Woven City and its current business model*, Master thesis, Keio Business School, Japan.

Kuecker, G.D. and Hartley, K. 2019. How smart cities became the urban norm: Power and knowledge in New Songdo city, *Annals of the American Association of Geographers*, Vol. 110, pp. 516–524.

Kummitha, R.R.K. 2019. Smart cities and entrepreneurship: an agenda for future research, *Technological Forecasting & Social Change*, Vol. 149, pp. 1–10.

Randeree, K. and Ahmed, N. (2019) The social imperative in sustainable urban development: the case of Masdar City in the United Arab Emirates, *Smart and Sustainable Built Environment*, Vol. 8, No. 2, pp. 138–149.

Sakuma, D. and Kato, K. 2020. The background of the automotive industry's smart city strategy and the pros and cons of winner takes all (WTA): Insights from the case of Toyota's Woven City, *Development Engineering*, Vol. 40, No. 1, pp. 37-40.

Sgouridis, S. and Kennedy, S. 2010. Tangible and fungible energy: Hybrid energy market and currency system for total energy management: A Masdar City case study, *Energy Policy*, Vol. 38, No. 4, 1749-1758.

Strickland, E. (2011). Cisco bets on South Korean smart city, *IEEE Spectrum*, Vol. 48, No. 8, pp. 11–12.

Tokoro, N. 2018. Realization of a health support ecosystem through a smart city concept: a collaborative dynamic capabilities perspective, in Kodama, M. (ed.) *Collaborative dynamic capabilities for service innovation: creating a new healthcare ecosystem*. Cham, Switzerland: Palgrave Macmillan, pp. 135–151.

Toyota Times. 2021. Woven City started'. https://toyotatimes.jp/toyota_news/construction_wovencity/120.html

Toyota. 2023a. *Woven by Toyota to accelerate Toyota's vision for mobility*. https://global.toyota/en/newsroom/corporate/39070846.html

Toyota. 2023b. *Toyota Research Institute - Advanced Development Inc. (TRI-AD) announces it will expand and its operations by forming Woven Planet Holdings and two new operating companies, Woven Core and Woven Alpha*. https://global.toyota/en/newsroom/corporate/33786527.html

Woven City. 2023a. *Above-ground and underground mobility*. https://www.woven-city.global/

Woven City. 2023b. *Woven City Press. Vol. 1*. https://www.woven-city.global/

Woven City. 2023c. *Woven City Press. Vol. 2*. https://www.woven-city.global/

Woven by Toyota. 2023. *What we build*. https://woven.toyota/en/what-we-build/

A REVIEW ON DATA PROTECTION AND REGULATION IN SMART CITY CONCEPT

Andrej Ignjatić
University of Zagreb
Faculty of Organization and Informatics
Pavlinska 2, Varaždin, Croatia

ABSTRACT

This review paper provides an overview of the development of defining the concept of a Smart City and the similarities and differences in the definitions by various authors of papers. Furthermore, it presents data protection regulations in the European Union, the United States of America, and China. Finally, the review concludes with an overview of data protection in Smart Cities, linking data protection to the development of Smart Cities. Five citation databases were utilized to conduct the literature review: Web of Science, Scopus, ScienceDirect, ResearchGate, and IEEE Xplore. In total, 21 articles were analysed. Furthermore, differences between the General Data Protection Regulation, the California Consumer Privacy Act of 2018 and the Personal Information Protection Law of the People's Republic of China are shown.

KEYWORDS

Smart City, Regulation, Data Protection, Privacy

1. INTRODUCTION

The concept of a Smart City is closely associated with data and its security, or protection, a matter of utmost importance in our digital age. Cities are adopting technologies through their smart services which process "personal data rapidly (Vandercruysse et al., 2022). National and international regulators are making significant efforts to govern data security and protection in various environments using different technologies, thus aiming to regulate data security and protection within Smart Cities. The research question this review paper seeks to address is: How does the enactment of data protection regulations, a critical aspect of Smart City development, affect the development of Smart Cities?

The literature search utilised five citation databases: Web of Science, Scopus, ScienceDirect, ResearchGate, and IEEE Xplore. During the paper search, the following terms were used: 'Regulation AND Smart City', 'Legislation AND Smart City', and 'Data protection AND Smart City'. Additionally, papers referenced in the findings from these databases were also searched. From the analysis, papers published before 2017 were excluded. English was defined as the language of the paper.

The rest of the paper is structured as follows: In section 2, the concept of a Smart City is tried to be defined. Section 3 provides an overview of data protection regulations in the European Union, the United States of America, and China. In section 4, the need for data privacy and protection regulations in Smart Cities is shown. At the end of this paper, there is a discussion and conclusion.

2. DEFINING THE TERM OR CONCEPT OF A SMART CITY

All papers that attempt to define the term 'Smart City' agree that there is no universally accepted definition as such. Founoun, Hayar, and Haqiq (2019) outline two primary academic viewpoints regarding the definition of Smart City:

- Smart City is a mean of sustainable urbanisation
- A Smart City is a technological advancement that should be integrated into everyday life.

Vojković (2018) remarked that a city could be defined as smart when information and communication technologies and the concept of the Internet of Things integrate its various systems of the city. This integration necessitates a discerning approach to data collection, access and data management.

Herath and Mittal (2022) argue that the Internet of Things is a key component of a Smart City because it produces substantial volumes of data. Equivalently, Taeihagh and Ying Tan (2022) have observed that all definitions typically emphasize the use of information and communication technologies to improve the quality of life of citizens and/or competitiveness and operational efficiency of urban systems.

In their paper, Srebalova and Peraček (2022) refer to the definition of a Smart City by the European Parliament. According to this definition, the European Parliament considers any city that attempts to solve public issues using information and communication solutions as a Smart City. To solve public issues, public administration collects a large amount of data (Breuer and Pierson, 2020) by using sensors often in public spaces (Vandercruysse et al., 2022) or by using mobile devices (Daoudagh et al., 2021), by letting the question of data protection to remain on margin (Vaniš et al., 2023), especially question of user location (geolocation) and users' identity and their interaction behaviour (Daoudagh et al., 2021; Fabrègue and Bogoni, 2023).

In their research paper, Ehwi, Holmes, Maslova, and Burgess (2022) refer to information and communication technologies as "digital technologies." They define Smart Cities through the use of large amounts of data.

Unlike the previously mentioned authors, Founoun and Hayar (2018) did not mention information and communication technologies or data types in their definitions. They believe that a city should be called "intelligent" if it can gather resources and achieve goals by meeting the needs of its citizens to a certain degree.

Franke and Gailhofer (2021) emphasise that cities should be smart and sustainable in the future. Moore, Lightcap, and Buttler (2024) noticed this. They defined Smart Cities as using technology to "create energy-efficient and sustainable urban centres."

Golubchikov and Thornbush (2022) deepen the definition of a Smart City by stating that a Smart City is divided into three levels: cybernetic, physical, and social. Some authors expand the definition of smart cities by suggesting that the functionalities of Smart Cities can be categorised into one of the following six concepts (Founoun and Hayar, 2018; Founoun, Hayar, and Haqiq, 2019; Luckey et al., 2020):

- Smart Living
- Smart Economy
- Smart Citizens
- Smart Governance
- Smart Mobility
- Smart Environment.

Specific papers attempted to define a Smart City and mentioned terms like 'artificial intelligence' and 'big data'. Golubchikov and Thornbush (2022) note that transparency in city management is based on artificial intelligence algorithms and big data, while Founoun, Hayar, and Haqiq (2019) believe that the application of artificial intelligence algorithms can be implemented in the development of Smart Cities themselves. Luckey et al. (2020) state that artificial intelligence algorithms provide the ability to process large amounts of data and discover patterns that are not detectable through traditional approaches.

3. OVERVIEW OF DATA PROTECTION REGULATION

Data protection acts, which are going to be mentioned in the rest of this paper, are graphically shown in Figure 1.

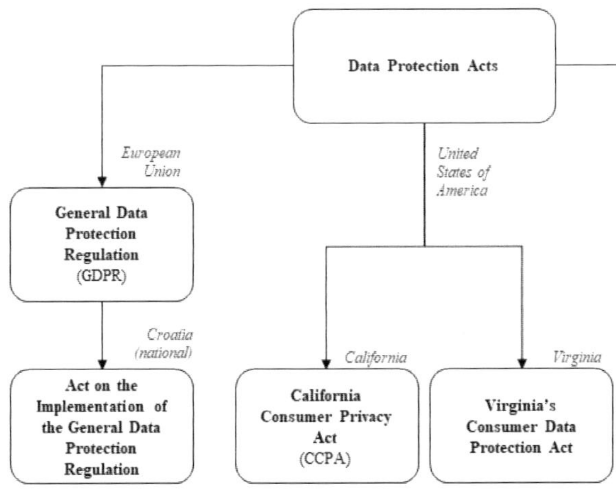

Figure 1. Data protection acts in the European Union, USA (California), and China

The most well-known legal regulation governing personal data protection within the European Union is the General Data Protection Regulation EU 2016/679, globally recognised by the GDPR. The purpose of GDPR is to protect privacy and guard against potential data breaches for all citizens of the European Union in a data-driven world. Additionally, the GDPR defines personal data as any data that can be used to identify an individual or assist in identifying an individual (Vojković, 2018).

The GDPR is based on several fundamental principles graphically shown in Figure 2. (Vojković, 2018; Lučić, Boban, Mileta, 2018; El Majdoubi, El Bakkali and Sadki, 2020).

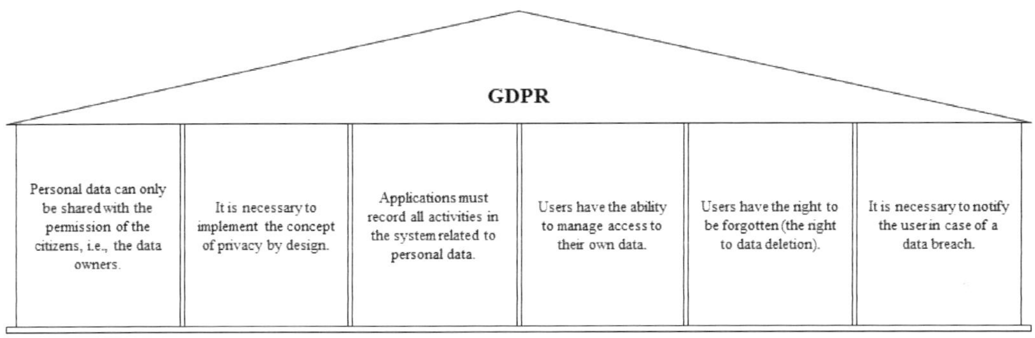

Figure 2. Fundamentals principles of GDPR

According to El Majdoubi, El Bakkali, and Sadki (2020), the fundamental data protection principles recognized in the GDPR can only be achieved through decentralized data management systems. Blockchain technology can "ensure protection, transparency, and decentralised trust" in the system.

Apart from the GDPR, Srebalova and Paraček (2022) cite the Directive of the European Parliament and of the Council on measures for a high common level of security of network and information systems across the EU 2016/1148 (referred to as 'Directive 2016/1148'). Directive 2016/1148 states that "network and information systems and services play a pivotal role" in the information society (2016). In 2022, Directive 2016/1148 was repealed when Directive (EU) 2022/2555 of the European Parliament and of the Council of 14 December 2022 on measures for a high common level of cybersecurity across the Union, amending Regulation (EU) No 910/2014 and Directive (EU) 2018/1972, and repealing Directive (EU) 2016/1148 (NIS 2 Directive) (2022) took effect.

In addition to the GDPR, another regulation defined in protecting citizens' data within the Union is the Regulation of the European Parliament and of the Council on European data governance (2020). This regulation focuses on data generated from Internet of Things solutions and relevant services while excluding major tech companies, digital platforms, and telecom operators, considered the most significant data producers and holders (Calzada, 2022).

The Republic of Croatia's GDPR implementation will be mentioned. The Republic of Croatia has implemented the General Data Protection Regulation (GDPR) and the Directive of the European Parliament and of the Council on measures for a high common level of security of network and information systems across the Union into its national legislation. Accordingly, Directive 2022/2555 has been transposed into the Cybersecurity Act (2024), and the GDPR has been incorporated into the Act on the Implementation of the General Data Protection Regulation (2018).

In addition to the GDPR, Calzada (2022) recognises two other regulations for protecting citizens' data – in China and California. The Personal Information Protection Law of the People's Republic of China (PIPL) applies in China. Like the GDPR, the CCPA defines privacy norms and illustrates the rights of organisations and citizens, setting limits on the issue rather than describing it. Although the core of PIPL is defining data privacy, PIPL also outlines a framework for regulating information protection and cybersecurity. In the United States, specifically in California, the California Consumer Privacy Act of 2018 (CCPA) is in effect, and in Virginia - Virginia's Consumer Data Protection Act. These two acts gave consumers the right to access, delete, and control the sale of any personal data (opt-out). Furthermore, CCPA requires organisations to disclose the categories of personal data collected and the purpose and collection of use and inform users to whom the data will be shared (Moore, Lightcap and Buttler, 2024). Differences between the General Data Protection Regulation (2016), the California Consumer Privacy Act of 2018 (2024) and the Personal Information Protection Law of the People's Republic of China (2021) are shown in Table 1.

Table 1. Differences between GDPR, CCPA and PIPL

Characteristic	**General Data Protection Regulation (GDPR)**	**California Consumer Privacy Act (CCPA)**	**Personal Information Protection Law (PIPL)**
Effective Date	25 May 2018	01 January 2020	01 November 2021
Geo Effectiveness	All entities in the EU and those processing data of EU persons (Article 3)	All entities doing business in California or processing data of California residents (Section 1798.140)	All entities processing data of Chinese persons, regardless of location (Article 3)
Definition of Personal Data	Any information related to an identified or identifiable natural person (Article 4)	Any information that identifies relates to, describes, is capable of being associated with, or could reasonably be linked, directly or indirectly, with a particular consumer or household (Section 1798.140)	Any information related to an identified or identifiable natural person (Article 4)
Person Rights	Right to access, rectify, erase, restrict processing, data portability, object, and right not to be subject to automated decision-making (Articles 15-22)	Right to access, delete, and right to opt-out of the sale of personal data (Sections 1798.100-1798.125)	Right to know, restrict or refuse, consult or copy, make correction, delete (Articles 44-47)
Children's Privacy	Processing personal data of children under 16 requires parental consent (Article 8)	Processing data of children under 13 requires parental consent (Section 1798.185)	Processing personal data of children under 14 requires parental consent (Article 28)
Data Breach Notification	Without undue delay, within 72 hours if there is a risk to individuals' rights and freedoms (Article 33)	Within 30 days (Section 1798.150)	Without undue delay (Article 57)
Outside Data Transfer	Allowed only to countries with adequate data protection or under appropriate safeguards (Articles 44-50)	-	Subject to strict conditions and security assessments (Articles 38-43)

Table 1. *continued*

Highest Non-Compliance Penalties	Up to 20 EUR million or 4% of annual global turnover, whichever is greater (Article 83)	Up to 7,500 USD per violation (Section 1798.155)	Up to 50 million CNY or 5% of annual revenue, whichever is greater (Article 66)
[Penalties in other currencies (ECB, 2024)]	[21,498,000 USD] [156,010,000 CNY]	[54,427.16 CNY] [6.977,39 EUR]	[6,409,845.52 EUR] [6,889,942.95 USD]
Smart City Impact	Data processing must be lawful, and transparent, and ensure data minimisation and privacy by design (Articles 5, 25)	Must provide notice of data collection and allow opt-out options; security measures must protect collected data (Sections 1798.100, 1798.110)	Must ensure data security, obtain explicit consent for data collection, and conduct security assessments for large-scale data processing (Articles 13, 14, 40, 51)

4. DATA PROTECTION IN SMART CITIES

In recent years, there have been global discussions about the privacy of citizens' data. One recurring theme in these discussions is the issue of privacy in Smart Cities and the introduction of various regulatory frameworks. Besides the need to define the means of data protection itself, it is necessary to regulate data ownership, data access rights, the transparency of artificial intelligence operations, algorithm automation, and more (Calzada, 2022).

The task of regulating is twofold. On one hand, regulation should establish market structures and enable all stakeholders to compete effectively in the market. On the other hand, regulation aims to preserve the public interest associated with providing essential services, including data protection (Calzada, 2022). Furthermore, this is important because for public administration, processing personal data is a standard task, which they perform without consulting with other stakeholders (Breuer and Pierson, 2020).

Ehwi, Holmes, Maslova, and Burgess (2022) also link data protection to ethical considerations. They question the existence of ethical issues given regulatory norms and explicitly cite the General Data Protection Regulation in their paper. They argue that regulation alone "is insufficient to guide society in the right direction but merely to define what is lawful and what is not."

Taeihagh and Ying Tan (2022) conducted a study identifying factors crucial for developing smart cities. In their paper, they used the Delphi method to gather expert opinions. The experts did not recognise data protection regulation as critical for developing smart cities. However, Taeihagh and Ying Tan acknowledged the connection between data protection and technological development, citing examples like the General Data Protection Regulation and Singapore's Personal Data Protection Act. However, they recognised the need for regulation in other ways, such as bridging the gap between regulation and actual development, creating equal rights for all startup organisations, and encouraging public participation.

Founoun and Hayar (2018) do not recognise the need to regulate personal data protection in Smart Cities. However, they acknowledge the need for and impact of regulation in the conceptual model of a Smart City.

Calzada (2022) notes that given the Internet of Things' importance for the development of Smart Cities, the European Parliament and Council's Regulation on European data governance could significantly affect city authorities' ability to access data to make better management and development policies and create urban data markets.

Vojković (2018) argues that the "General Data Protection Regulation (GDPR) should not be seen as a threat to Smart Cities, but rather as a set of conditions that can aid in their development." He explains that citizens are likelier to embrace a Smart City if it does not remind them of the concept of "Big Brother or some totalitarian regime that controls personal freedoms." He concludes that the GDPR helps develop Smart Cities by increasing data protection levels, thereby reducing fears of potential misuse of the cities' control functions.

In contrast, Lučić, Boban, and Mileta (2018) believe that the GDPR could negatively impact development due to its requirements for pseudonymisation. They also mention the need for a national regulatory body to work alongside existing national regulations to reduce citizens' concerns about potential human rights violations or privacy breaches.

In their paper, Franke and Gailhofer (2021) state that heavily regulated data access protection can negatively impact development. They argue that regulation can prevent significant innovations and, consequently, the prosperity of individuals. They note that citizens will not accept data management strategies by organisations, using the "Sidewalk Toronto" project as an example, where such data management grants organisations certain powers by placing them in a controlling position.

Luckey et al. (2020) state that implementing European data protection regulations necessitates transparent decision-making in artificial intelligence, which limits the distribution and implementation of machine learning algorithms.

Herath and Mittal (2022) recognise that while artificial intelligence in Smart Cities has various advantages, it also opens up specific regulatory issues, such as data privacy concerns.

From a technical perspective, papers (Qian et al., 2018; El Majdoubi, El Bakkali, and Sadki, 2020; Makhdoom et al., 2020; Calzada, 2022) suggest using Blockchain technology for data protection.

Due to artificial intelligence, the European Union recognized the possibility of collecting a large amount of data. Regarding that, the European Union introduced a Proposal for a Regulation laying down harmonised rules on artificial intelligence (Artificial Intelligence Act) (2021), which defines that non-biometric user data that do not constitute personal data or a special category of personal data can be collected.

5. DISCUSSION AND CONCLUSION

Data protection in Smart Cities must be enforced to preserve user privacy, that is, citizens' privacy. Society tries to define data protection methods through regulation, including the General Data Protection Regulation (GDPR).

A fundamental principle of the GDPR is that data sharing is allowed only with the consent of the data owner, who, in the case of Smart Cities, is the user of the service (Vojković, 2018; El Majdoubi, El Bakkali and Sadki, 2020). This enables users to influence the development of the service directly and, consequently, the evolution of the Smart City since data is needed for purposes such as artificial intelligence learning or other functions.

Additionally, the GDPR has replaced lengthy and incomprehensible texts written in administrative language with precise, accessible forms and templates that include the purpose of data processing (Vojković, 2018). This allows citizens to understand precisely what data is collected, how long it is kept, and decide whether or not they want their data to be collected, impacting the development of the Smart City.

The influence of data protection regulations in Smart Cities is not unambiguously recognised; each author draws conclusions about the positive or negative impact based on their literature review and observations. For example, Vojković (2018) believes that regulation can positively affect development by enhancing security and protection, while other authors think that regulations limit the development of Smart Cities in various ways (Lučić, Boban and Mileta, 2018; Luckey et al., 2020; Franke and Gailhofer, 2021), thus negatively affecting their further development.

In a paper by Taeihagh and Ying Tan (2022), experts surveyed did not recognise data security and protection as crucial factors in developing smart cities. This might be explained by experts focusing on financial and technological development without considering protection or if they were not explicitly asked about it, as the posed questions were not disclosed in the paper. Progress can be assessed by conducting structured questionnaires with citizens or users of services provided by Smart Cities to determine if users consider their data protection essential for the further development of Smart Cities.

None of the authors of the analysed papers recognised the impact of international data protection standards and their connection with regulations, except for Vojković (2018), who mentions that the ISO/EIC 27000 standard can be helpful.

None of the reviewed papers suggesting the application of Blockchain technology for data protection El (Qian et al., 2018; Majdoubi, El Bakkali and Sadki, 2020; Makhdoom et al., 2020; Calzada, 2022) linked the technology with legal regulation or the necessity for it.

Discussion related to Table 1. For simplification, GDPR, CCPA, and PIPL will be collectively referred to as acts in this discussion. Table 1 shows that GDPR was the earliest to come into effect - in May 2018, while PIPL came into effect in November 2021. All three analysed acts do not necessarily geographically apply only to the areas where they were enacted; they also apply to organisations processing the personal data of a person, regardless of their geographical location, meaning they do not necessarily have to be in the geographical area where they were enacted. Furthermore, all three acts similarly define personal data as any data that can identify or help identify a specific person. Person rights are approximately similar in all three acts. The acts differ in the age at which children do not need parental consent for data processing. The lowest age limit according to CCPA is 13 years, while the highest limit is 16 years according to GDPR. The acts also differ in the time required to notify persons of a data breach. GDPR and PIPL state that this must be done without undue delay, although GDPR regulates a timeframe of 72 hours if there is a risk to persons' rights or freedoms, while CCPA allows a period of up to 30 days. According to GDPR and PIPL, data transfer outside the region is possible with certain restrictions, while no such possibility was found in CCPA. In conclusion, considering the impact on Smart City, it is evident that all three acts allow the use of data under certain conditions.

However, after reading and analysing the mentioned papers, no one argued the possibility of refusing service if the user does not want to use services from the Smart City concept. Section 2 of this paper states that the majority of Smart City services are, in fact, public services provided by public companies. On the one hand, since these companies provide a public service, they should not deny users access to services within the Smart City concept (for example, if users do not use mobile devices), but rather should offer an alternative (traditional) way to access these services. On the other hand, companies should communicate how and for what purpose they collect personal data from users and let them know how to process it and how long it is retained. Additionally, companies should inform users about how they can exercise their rights under the law (for example, the right to data deletion, etc.). For instance, in Italy, a company must clearly state whether it uses technology and for what purpose (Fabrègue and Bogoni, 2023). Finally, in developing Smart City services, companies should not adhere to the "the more data, the better" concept (Breuer and Pierson, 2020). Still, they should instead collect only the data necessary to provide a specific service.

Future research on this topic could explore the impact of Smart Cities on the information society as a whole and the impact of integrating best practices of Blockchain technology into the regulatory system on the development of Smart Cities.

REFERENCES

Act on the Implementation of the General Data Protection Regulation (Zakon o Provedbi Opće Uredbe o Zaštiti Podataka), NN 42/18, 2018. Republic of Croatia. Available at: https://narodne-novine.nn.hr/clanci/sluzbeni/2018_05_42_805.html (Accessed: 19 May 2024).

Breuer, J., Pierson, J., 2020. The Right to the City and Data Protection: complementary for Developing Citizen-centric Digital Cities?. *The 21st Annual Conference of the Association of Internet Researchers*. Virtual Event, pp. 1 – 3.

Calzada, I., 2022. Citizens' Data Privacy in China: The State of the Art of the Personal Information Protection Law (PIPL). *Smart Cities*, Vol. 5, No. 3, pp. 1129 – 1150.

California Consumer Privacy Act of 2018, 2024. California State Legislature. Available at: https://cppa.ca.gov/regulations/pdf/cppa_act.pdf (Accessed: 21 May 2024).

Cybersecurity act (Zakon o kibernetičkoj sigurnost), NN 14/24, 2024. Republic of Croatia. Available at: https://www.zakon.hr/z/3718/Zakon-o-kiberneti%C4%8Dkoj-sigurnosti (Accessed: 19 May 2024).

Daoudagh, S. et al., 2021. Data Protection by Design in the Context of Smart Cities: A Consent and Access Control Proposal. *Sensors*, Vol. 21, Art. No. 7154, pp. 1 – 21.

Directive (EU) 2016/1148 of the European Parliament and of the Council of 6 July 2016 concerning measures for a high common level of security of network and information systems across the Union, 2016. European Union. European Parliament and the Council, Available at: https://eur-lex.europa.eu/legal-content/EN/TXT/HTML/?uri=CELEX:32016L1148&from=EN (Accessed: 25 November 2022).

Directive (EU) 2022/2555 of the European Parliament and of the Council of 14 December 2022 on measures for a high common level of cybersecurity across the Union, amending Regulation (EU) No 910/2014 and Directive (EU) 2018/1972, and repealing Directive (EU) 2016/1148 (NIS 2 Directive). European Union. European Parliament and the Council, Available at: https://eur-lex.europa.eu/legal-content/EN/TXT/HTML/?uri=CELEX:32022L2555 (Accessed: 20 May 2024)

ECB Data Portal: Currency converter. European Central Bank (ECB), Available at: https://data.ecb.europa.eu/currency-converter (Accessed: 20 June 2024)

Ehwi, R. J., Holmes, H., Maslova, S., Burgess, G., 2022. The ethical underpinnings of Smart City governance: Decision-making in the Smart Cambridge programme, UK. *Urban Studies*, Vol. 59, No. 14, pp. 2968 – 2984.

El Majdoubi, D., El Bakkali, H., Sadki, S., 2020. Towards Smart Blockchain-Based System for Privacy and Security in a Smart City environment. *2020 5th International Conference on Cloud Computing and Artificial Intelligence: Technologies and Applications (CloudTech)*. Marrakech, Morocco, pp. 117 – 123.

Fabrègue, B. F. G., Bogoni, A., 2023. Privacy and Security Concerns in the Smart City. *Smart Cities*, Vol. 6, pp. 586 – 613.

Founoun, A., Hayar, A., 2018. Evaluation of the concept of the smart city through local regulation and the importance of local initiative. *2018 IEEE International Smart Cities Conference (ISC2)*. Kansas City, USA, pp. 1 – 6.

Founoun, A., Hayar, A., Haqiq A., 2019. The Textual Data Analysis Approach to Assist the Diagnosis of Smart Cities Initiatives. *5th IEEE International Smart Cities Conference (ISC2 2019)*. Casablanca, Morocco, pp. 150 – 153.

Franke, J., Gailhofer, P., 2021. Data Governance and Regulation for Sustainable Smart Cities. *Frontiers in Sustainable Cities*, Vol. 3, pp. 1 – 13.

Golubchikov, O., Thornbush, M. J., 2022. Smart Cities as Hybrid Spaces of Governance: Beyond the Hard/Soft Dichotomy in Cyber-Urbanization. *Sustainability*, Vol. 14, No. 16, pp. 1 – 12.

Herath, H. M. K. K. M. B., Mittal, M., 2022. Adoption of artificial intelligence in smart cities: A comprehensive review. *International Journal of Information Management Data Insights*, Vol. 2, No. 1, pp. 1 – 21.

Li, L., Taeihagh, A., Ying Tan, S., 2022. What factors drive policy transfer in smart city development? Insights from a Delphi study. *Sustainable Cities and Society*, Vol. 84, pp. 1 – 12.

Luckey D., Fritz H., Legatiuk A., Dragos K., 2020. Artificial Intelligence Techniques for Smart City Applications. *International Conference on Computing in Civil and Building Engineering*. Sao Paulo, Brazil, pp. 3 – 15.

Lučić, D., Boban, M., Mileta, D., 2018. An Impact of General Data Protection Regulation on a Smart City Concept, *MIPRO 2018*. Opatija, Croatia, pp. 390 – 394.

Makhdoom, I., Zhou, I., Abolhasan, M., Lipman, J., Ni, W., 2020. PrivySharing: A blockchain-based framework for privacy-preserving and secure data sharing in smart cities. *Computers & Security*, Vol. 88, pp. 1 – 34.

Moore, S., Lightcap, R. W., Butler, W. H., 2024. An Investigative Review: Smart Grid and Consumer Privacy Concerns. *International Journal of Applied Technology & Leadership*, Vol. 3, No. 1, pp. 1 – 19.

Personal Information Protection Law of the People's Republic of China, 2021. National People's Congress of the People's Republic of China. Available at: https://personalinformationprotectionlaw.com/ (Accessed: 21 May 2024).

Proposal for a Regulation of the European Parliament and of the Council Laying Down Harmonised Rules on Artificial Intelligence (Artificial Intelligence Act) and Amending Certain Union Legislative Acts, 2021. European Commission. Available at: https://eur-lex.europa.eu/legal-content/EN/TXT/HTML/?uri=CELEX:52021PC0206 (Accessed: 14 April 2024).

Proposal for a Regulation of the European Parliament and of the Council on European Data Governance (Data Governance Act), 2020. European Commission. Available at: https://eur-lex.europa.eu/legal-content/EN/TXT/HTML/?uri=CELEX:52020PC0767&from=EN (Accessed: 25 November 2022).

Regulation (EU) 2016/679 of the European Parliament and of The Council of 27 April 2016 on the protection of natural persons with regard to the processing of personal data and on the free movement of such data, and repealing Directive 95/46/EC (General Data Protection Regulation), 2016. European Commission. European Parliament and of the Council of European Union. Available at: https://eur-lex.europa.eu/legal-content/EN/TXT/HTML/?uri=CELEX:32016R0679 (Accessed: 21 May 2024).

Qian, Y., Liu, Z., Yang, J., Wang, Q., 2018. A Method of Exchanging Data in Smart City by Blockchain. *2018 IEEE 20th International Conference on High Performance Computing and Communications; IEEE 16th International Conference on Smart City; IEEE 4th Intl. Conference on Data Science and Systems*. Exeter, UK, pp. 1344 – 1349.

Srebalova, M., Peraček, T., 2022. Effective Public Administration as a Tool for Building Smart Cities: The Experience of the Slovak Republic. *Laws*, Vol. 11, No. 5, pp. 1 – 16.

Vandercruysse, L., Christofi, A., Buts, C., Dooms, M., Valcke, P., 2022. Data Protection in Smart Cities: Pre-Commercial Procurement as a Silver Bullet?. *European Procurement & Public Private Partnership Law Review*, Vol. 17, No. 2, pp. 81 – 93.

Vaniš, M., Holý, R., Vaniš, L., Ščerba, T., 2023. Possibilities of legal protection for autonomous mobility databases based on their characteristics. Smart Cities Symposium Prague 2023. Prague, Czech Republic, pp. 1 – 6.

Vojković, G., 2018. Will the GDPR slow down development of Smart Cities?. *MIPRO 2018*. 1295 – 1297.

INVESTIGATING THE IMPACT OF TREES AS NOISE BARRIER ON URBAN NOISE POLLUTION THROUGH 3D SCENE MODELLING

Seanglidet Yean[1], Zheng Ping Heng[1], Guan Wei Leow[1], Bu Sung Lee[1] and Peter Edwards[2]
[1]*College of Computing and Data Science, Nanyang Technological University, Singapore*
[2]*Manaaki Whenua – Landcare Research, Wellington, New Zealand*

ABSTRACT

Noise pollution is a cause for concern, especially since it has detrimental health effects. Due to Singapore's limited geographical size and rising population density, gas stations, roads, and even expressways are built close to residential neighborhoods. This situation is increasingly common in dense cities worldwide, making Singapore a relevant case study for global urban environments. Noise pollution in such areas is a significant problem, as a result of frequent traffic on these routes, especially during rush hours. The purpose of this project is to create a pipeline from a scene simulation to modelling noise propagation. This project aims to develop a scene modelling platform that will help bridge the gap between scene and noise modelling. It also aims to investigate different noise barriers and their effectiveness against vehicular noise pollution. Through experiments conducted in residential and park settings, we demonstrate the varying effects of trees on noise reduction, highlighting their potential to attenuate noise levels, especially in densely populated areas. Our findings reveal that while trees may not completely block noise like solid barriers, they offer additional benefits such as airflow and visual enhancement of the environment. Furthermore, we assess the impact of reforestation and strategic tree placement in reducing traffic noise, providing insights for urban planners and policymakers. Additionally, we present the case study modelling the scene after observed StreetView, showcasing the significant noise reduction benefits of planting trees in residential areas. Overall, our study contributes to the development of more effective strategies for mitigating noise pollution and creating healthier, more sustainable urban environments.
Demo Video: https://youtu.be/zi1R-1xDkwU

KEYWORDS

Noise Mapping, 3D Scene Modelling, Tree Modeling, Web-App, Urban Planning

1. INTRODUCTION

In densely populated urban environments like Singapore, traffic noise pollution in residential area has a significant concern affecting public health and well-being. With a population density of 8041 individuals per square kilometer (Department of Statistics, Ministry of Trade Industry, 2021) and rapid urbanization, noise pollution from roads, expressways, and residential areas has intensified, necessitating understanding of its sources, propagation, and mitigation strategies (Data.gov.sg, 2021). Chronic exposure to high noise levels can lead to various health issues such as hearing loss, sleep disturbances, and cardiovascular diseases (World Health Organization, Regional Office for Europe, 2021). Governments, including Singapore's National Environment Agency (NEA), have implemented measures like new noise standards for vehicles to mitigate noise pollution (Tan, 2021). However, addressing noise pollution comprehensively requires understanding its sources and propagation in complex urban landscapes characterized by high-rise buildings and dense vegetation.

While noise pollution studies have extensively examined sources like airplanes, less attention has been paid to noise propagation in urban areas using natural barriers such as trees. Conventional noise mitigation methods often employ sound barriers, effective yet obstructive to airflow and visual aesthetics. Conversely, trees offer a promising solution, attenuating noise while preserving airflow and enhancing visual appeal. Urban environments often lack diverse sound profiles present in nature (Truax, 2012), making tree integration crucial for creating positive, culturally significant soundscapes. The complexity of urban environments, however, has not been fully addressed in current studies, with a predominant focus on two-dimensional analyses and simplified vegetation models. For instance, while some studies have measured ground-level noise in relation to various road infrastructures, they have overlooked noise reception at higher elevations and have not incorporate vegetation as a variable in noise attenuation (Tan, Hasegawa, Lau, & Tang, 2022). Similarly, another study examining noise levels near a major road at height did not consider vegetation as a factor in their analysis (Chin, Thang, & Saju, 2019). Furthermore, existing scene modeling tools lack optimization for tree incorporation in noise simulations, highlighting a research gap.

To address this challenge, our study aims to develop a three-dimensional (3D) scene visualization platform that integrates realistic representations of urban landscapes, including trees, into noise modeling simulations. By incorporating trees as natural barriers to noise propagation, we seek to explore their effectiveness in mitigating noise pollution while minimizing adverse effects on airflow and visual aesthetics. This approach represents a novel contribution to the field of environmental acoustics and offers new insights into sustainable noise mitigation strategies for urban environments.

Existing studies on noise pollution have identified limitations in traditional 2D noise maps, particularly in accurately representing noise effects, especially at higher floors of buildings. These maps often overlook crucial factors such as noise barriers, buildings, trees, and terrain heights, resulting in inaccuracies in noise propagation modeling (Kumar, Ledoux, Commandeur, & Stoter, 2017). Recognizing these deficiencies, there has been a growing demand for more sophisticated modeling techniques, including the adoption of 2.5D and 3D noise maps, to better understand noise propagation dynamics, especially in densely populated urban areas like Singapore. To bridge this gap, our study integrates 3D scene visualization with noise modeling, presenting a significant advancement in environmental acoustics. However, while some studies, such as one conducted in Tehran (Ranjbar, Gharagozlou, & Nejad, 2012), have utilized 3D scene information to analyze traffic noise impacts and explore various noise barrier strategies, they primarily focus on noise mapping methodologies and solid barrier testing, overlooking other noise mitigation options. Similarly, another study in Singapore employed unstructured mesh representation and algorithm development for 3D noise mapping but highlighted the lack of standardized data collection, particularly in Singapore, hindering comparability among studies (Zhao et al., 2017). These scientific and technical barriers underscore the necessity for comprehensive scene modeling approaches to address the complexities of noise pollution accurately.

In this paper, we present our approach to scene modelling and noise propagation analysis, emphasizing the importance of integrating three-dimensional components in noise pollution research. We discuss the technical challenges associated with modeling noise propagation in urban environments and propose solutions to address these challenges. Furthermore, we demonstrate the potential applications of our platform for urban planning, environmental assessment, and public health research. Through our efforts, we aim to contribute to the development of more effective strategies for mitigating noise pollution and creating healthier, more sustainable cities.

2. METHODOLOGY

2.1 Noise Propagation Study Approach

Our approach in modelling the noise propagation in urban environment consists of three main steps: (1) modelling the environment of interest with 3D representation (polygons) pertaining tree and its species of interest, (2) carried out noise propagation estimation using noise modelling tools, traffic flow information and 3D representation (in step 1), and (3) conduct analysis with simulated environment of different settings, after modification and different tree species, collated with various data layers such as land use/ land cover, noise level and noise contour map. The experiments aims to find out how effective and feasible trees could block vehicular noise to buildings by answering the following research questions:

- How different noise barriers affect noise propagation?
- How different tree properties affect noise propagation?
- Would existing trees help to mitigate noise pollution?

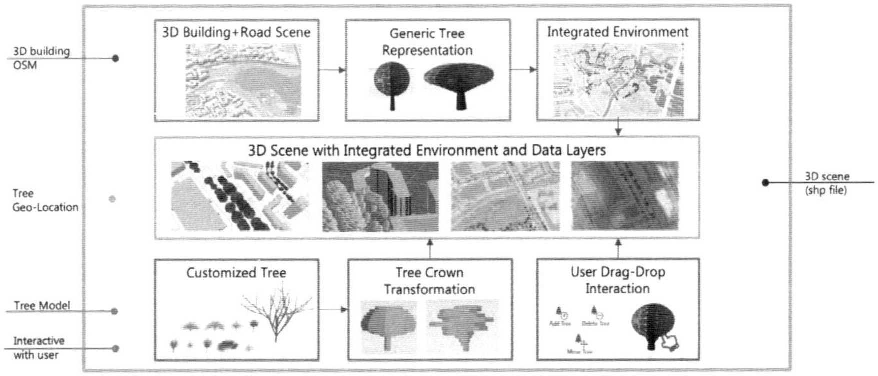

Figure 1. Scene modelling overview

2.2 3D Scene Modelling Platform Development

The importance of scene modeling for noise propagation simulations from traffic highlights the need for accurate representation in formats like GeoJSON or SHP files composed of polygons. Accurate three-dimensional modeling enables better assessment of noise levels, identification of potential barriers like buildings and vegetation, and analysis of noise impacts. However, current methodologies lack suitable representations for noise modeling studies, leading to compatibility issues with common software tools. For instance, the discrepancy between formats like CityGML and SHP files, as seen in large-scale virtual city models (Gobeawan et al., 2018), exemplifies the compatibility challenge, emphasizing the necessity for scene modeling platforms capable of generating representations in required formats. Bridging this gap requires a dedicated platform offering functionality to generate geometric polygons while ensuring compatibility with noise modeling software.

In this study, we develop a 3D scene modeling platform to efficiently represent integrated environments while addressing the challenge of simplifying complex objects like trees. Figure 1 illustrates the components and features of the web application, including direct scene exportation, user interactivity for tree manipulation, customized tree crowns, and the integration of new data such as noise levels and land-use cover for aggregated analysis.

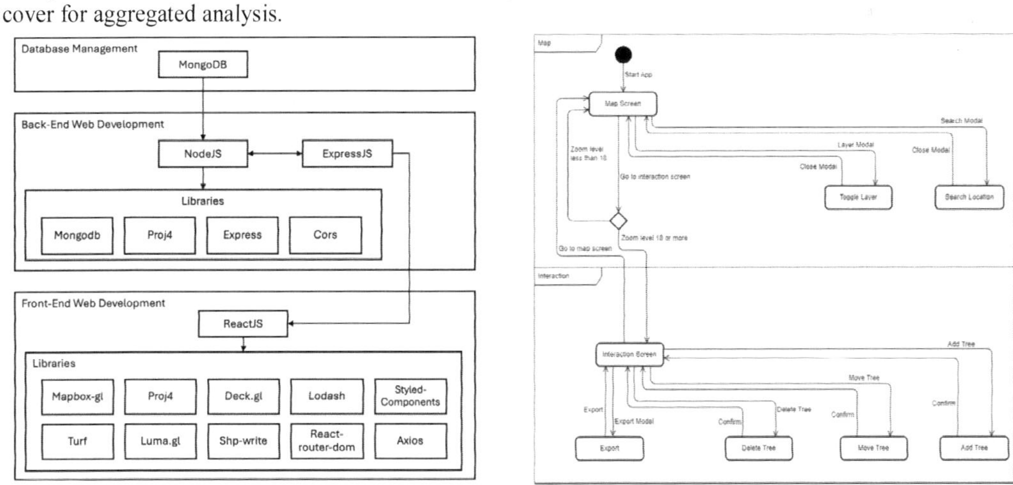

(a) System architecture

(b) Platform interaction dialog

Figure 2. 3D scene modelling web-app framework

To streamline data integration, we extract primary data required for noise calculation (detailed in Section 2.2.2) to enable single-source exportation from the web application. We employ JavaScript libraries to generate customized tree crowns (explained in Section 2.3), and integrate interactive services for on-the-fly scene editing along with geospatial functionalities like geocoding and layer configuration to enhance user experience. This interactive feature empowers users to demonstrate effects like reforestation and strategic tree placement. Additionally, we present a pipeline for scene simulation to bridge the gap between real and modeled environments for noise modeling. Further pattern analysis can be conducted within the platform by toggling between new associated data, including noise levels, land-use land-cover (LULC) data, noise contour maps, and the modeled tree scene (explained in Section 2.2.3 and discussed in Section 4).

2.2.1 Web-App Platform Architecture

The system architecture for 3D scene modeling, depicted in Figure 2a, comprises three primary components: frontend, backend, and database. The frontend, developed using ReactJS and illustrated in Figure 2b, engages users through an intuitive interface, leveraging Mapbox for mapping features. Express.js serves as the middleware for the backend, facilitating communication with MongoDB for data management. Docker containerization ensures easy deployment and scalability. The interaction and features of the tool include dynamic map page loading, interactive popups, streamlined data modification, tree manipulation, location search, data overlay toggling, and 3D scene export for noise modeling, enhancing user interaction and data utilization.

2.2.2 Base Map Data Retrieval and Integration

The base map of the application relies on three primary datasets: roads, buildings and trees. They provide comprehensive coverage of Singapore's urban landscape. While real-time data is not imperative, keeping the datasets updated at almost real-time is recommended. The QGIS platform, supported by the OSMDownloader plugin (Open Street Map, 2014), enables the extraction of OSM data for integration into the NoiseModelling application. The HDB Buildings dataset, generated using methods from the Urban Analytics Lab at the National University of Singapore (Biljecki, 2020), enhances the base map with detailed building information essential for noise analysis experiments. Trees data from TreesSG, extracted via iMaven (National Parks, 2022), augment the map with point coordinates and attributes crucial for accurate noise calculations, such as girth size, height, and species. The application's workflow centers on almost real-time loading of the base map and efficient data retrieval to manage large datasets. To optimize performance and address potential memory limitations, strategies like geoNear queries fetch only necessary data within the web browser frame. User interaction triggers data fetching with a debounce feature, preventing database overload. Additionally, useCallback functions cache operations to minimize redundant data retrieval, crucial for preventing errors. The useEffect hook ensures synchronized data retrieval, enhancing accuracy and reliability.

2.2.3 Auxiliary Data Layer

In addition to the base map, the web application integrates auxiliary data layers to enhance analysis. Noise levels recorded by the NoiseCapture community offer ground truth data beyond vehicle emissions (Picaut et al., 2019). Additionally, Land Use/Land Cover (LULC) data classified using machine learning techniques (Goh et al., 2022) provides insights into land coverage and green spaces distribution. By visualizing LULC data, users gain understanding of urban landscapes, aiding urban planning and environmental conservation. Furthermore, the application allows importing noise contour maps from specific experiments, enabling users to assess noise pollution spatially. Overall, these auxiliary layers improve the application's functionality, offering comprehensive insights into noise pollution and urban dynamics, with potential for future advancements in noise analysis and urban planning.

2.3 Tree Crown Generation and Transformation

The challenge of representing complex objects like trees in scene modeling platforms involves simplifying them while maintaining accuracy for noise calculation tools. This may require representing trees using multiple polygons to ensure recognition and realistic forms based on species characteristics, despite technological constraints. The proposed platform addresses these challenges, enhancing the efficiency and effectiveness of noise modeling studies for informed decision-making in urban planning and environmental management (National Parks, 2022). The dataset contains 715,396 trees stored in GeoJSON format, with attributes such as age, height, girth, and species, crucial for determining tree size and shape. Notably, focusing on top species like Samanea saman, Peltophorum pterocarpum, Hopea odorata, and Khaya senegalensis from the dataset facilitates a detailed analysis of tree representation and its impact on noise barrier effectiveness.

Tree Species	General	Samanea Saman	Peltophorum pterocarpum	Hopea odorata	Khaya senegalensis
Allometric Equation for Crown Shape		$Y_{width} = 10.98X^{0.59}$ $Y_{height} = 6.27X^{0.42}$	$Y_{width} = 10.36X^{0.48}$ $Y_{height} = 8.00X^{0.47}$	$Y_{width} = 8.76X^{0.80}$ $Y_{height} = 8.55X^{0.80}$	$Y_{width} = 9.39X^{0.65}$ $Y_{height} = 10.72X^{0.62}$
Web-App Representation					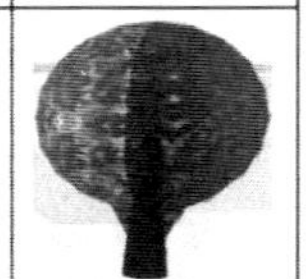
Exported Representation			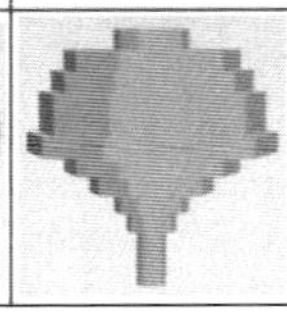		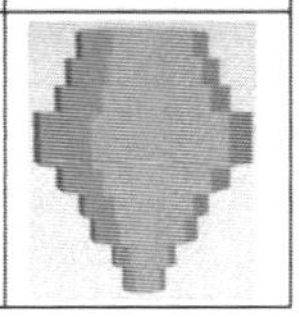

(a) Tree crown creation

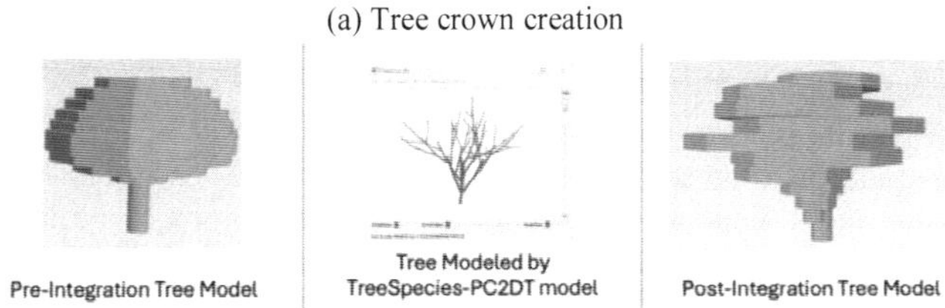

(b) Samanea Saman with TreeSpecies-PC2DT model

Figure 3. Crown creation methods

Given the height and girth provided, the crown size can be predicted with the allometric equations could be derived the form of $Y = aX^b$ (Lin, Teo, Yee, & Li, 2018), where a and b are constants and X and Y are tree attributes. Figure 3a depicts the transformation of tree representation from the original general looking tree to a representation based on the tree species characteristics. The application representation was modelled with the luma.gl library while the exported representation was modelled with multiple polygon GeoJSON objects. To improve the detail level of the representation, the application tree has assumed different form of sphere with an additional cone geometry as its tree crown, while the exported tree has increased layers of polygon with unique width at each level.

Furthermore, to model a more realistic tree representation, we extended the exploration to a more refined tree crown with the shape produced by TreeSpecies-PC2DT model (Gobeawan et al., 2024) since representation transformation in Figure 3a was modeled after description of the species, where the width of each layer was estimated by extrapolation in order to generate the described form of tree crown. Figure 3b illustrate one of many real life Samanea Saman tree modelled by TreeSpecies-PC2DT model. The model was constructed with multiple points that consisted of the x, y, z coordinates and radius. As the model was stored in Multiscale Tree Graph data structure (MTG) formatted file. To accurately capture the intricate details of the tree, the numerical data points were segmented into 10 layers. These layers were then represented by stacked cylinders, with the Euclidean distance and midpoint calculated between the two furthest points within each layer. The resulting cylinders were stacked at their respective midpoints. This extension enables us not only to represent any life-like trees reconstructed from point clouds but also to simulate various species profiling and growth modelling.

3. EXPERIMENT RESULTS AND ANALYSIS

3.1 Experiment Setup

In our experiments, the primary objective is to determine the most effective and practical method for mitigating vehicular noise around buildings with trees as the natural noise barrier. Hence, finding a suitable noise propagation modeling tool is equally crucial to scene modelling. One such tool is NoiseModelling (Noise Planet, 2019), an open-source application designed for mapping environmental noise in large urban areas. In addition to offering functionality and a user-friendly web interface, NoiseModelling is particularly adept at handling numerous spatial elements and is closely integrated with spatial databases like H2GIS or PostGIS. However, our investigation revealed a limitation: the CNOSSOS-EU calculation method within NoiseModelling does not support floating objects. Consequently, we turned our attention to the Softnoise predictor (Softnoise, 2023), a versatile tool for conducting noise experiments, especially regarding the influence of trees.

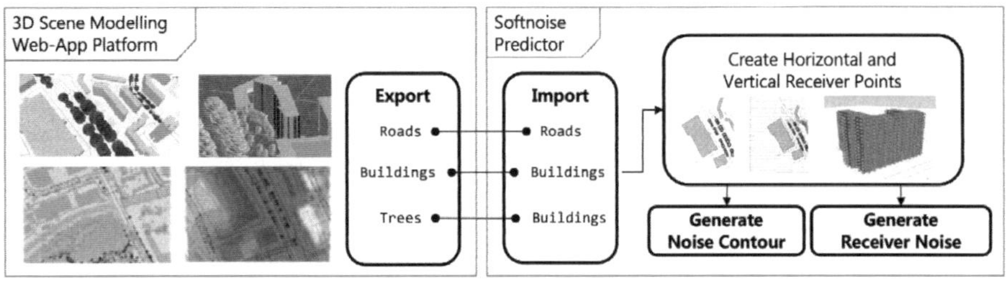

Figure 4. Crown creation methods

The next step involved preparing the data obtained from the web application for use in the Softnoise predicton. Figure 4 outlines the workflow for preparing the experiments, culminating in the generation of results suitable for analysis. Our project pursued two main approaches: creating noise contour maps to visualize noise propagation and assessing noise levels at specific receiver points. It's important to note that all noise propagation results discussed herein were generated using Softnoise. We deliver the findings in three sections that cover the experiments based on areas of interest, interactions, and differing in species.

3.2 Experiment at Area of Interest

For experiments based on area of interest, two locations of different settings:

- residential area: it features high noise emission due to high numbers of vehicle and speed, and high-rise buildings.
- park: lakeside that features a low noise emission road, large patch of flat land with a lake beside the park.

Figure 5a presents a comparison of noise contour maps generated along a highway adjacent to a residential area, with and without existing trees. Meanwhile, Figure 5b showcases a noise contour map from a skatepark environment, where trees near the lake were intentionally removed due to the absence of noise emissions from the water. Additionally, an experiment involving the construction of a 6-meter-high wall along the road was conducted to compare the effectiveness of trees versus traditional noise barriers commonly found on highways in Singapore. In our experiments, we observed notable noise reduction in both residential and park settings due to trees acting as noise barriers. In residential areas, trees reduced noise levels noticeably along the vertical grid, while in parks, a greater reduction was observed across the horizontal grid, attributed to the open space allowing noise to dissipate. However, trees may not fully block noise like solid barriers, such as walls, especially horizontally. Despite this, trees offer airflow and enhance visual aesthetics. Nonetheless, our findings emphasize the ongoing need for effective noise mitigation strategies in densely populated areas with busy highways.

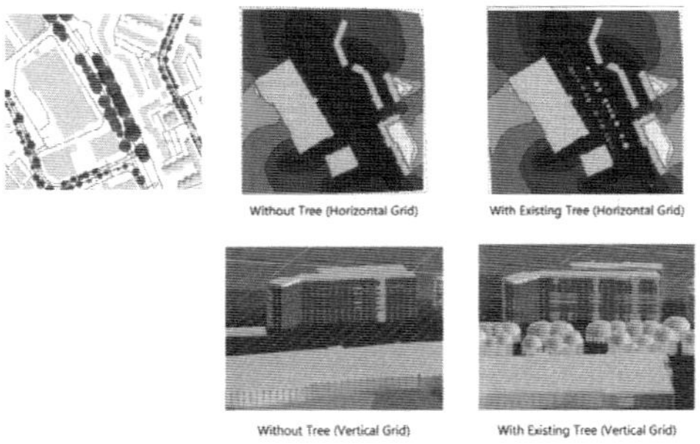

(a) The residential area

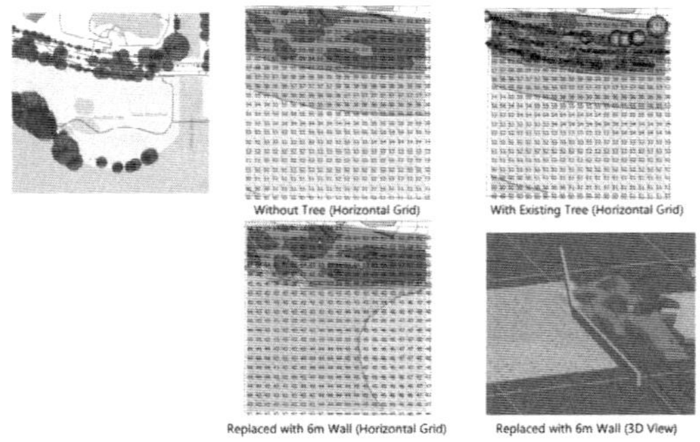

(b) The park

Figure 5. Noise propagation contour maps at the area of interest

Moving forward, our focus will be on the residential area experiments, aiming to explore the symbiotic relationship between trees and residential environments in mitigating noise pollution. By studying how trees can coexist with residential areas while effectively reducing noise levels, we hope to provide valuable insights for urban planners and policymakers in creating more livable and sustainable communities.

3.3 Experiment on Simulated Scenarios

This section presents experiments conducted to assess the efficacy of reforestation and strategic tree placement in mitigating traffic noise in residential areas adjacent to highways. The study focuses on a case study involving a residential area near a highway.

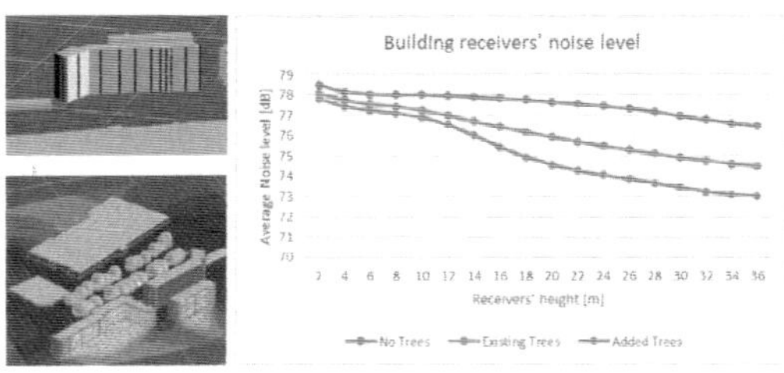

(a) Noise propagation in residential areas near highways

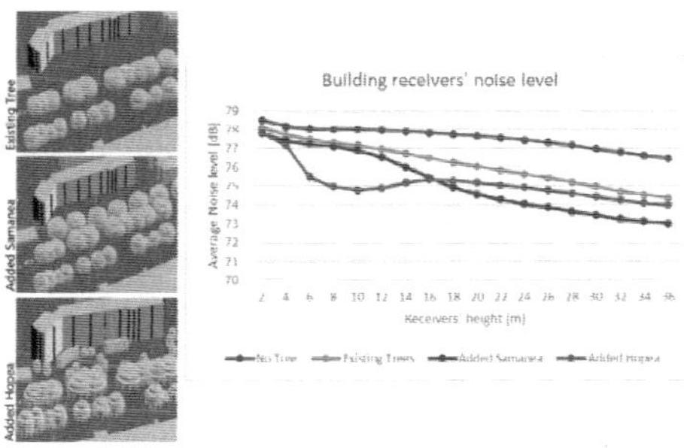

(b) Impact of reforestation with Samanea Saman trees

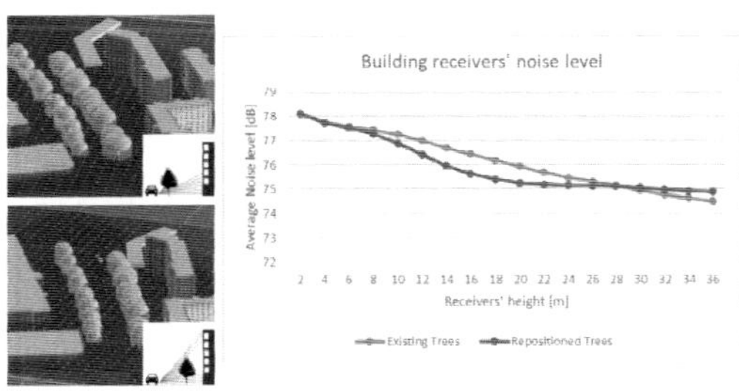

(c) Effect of tree relocation

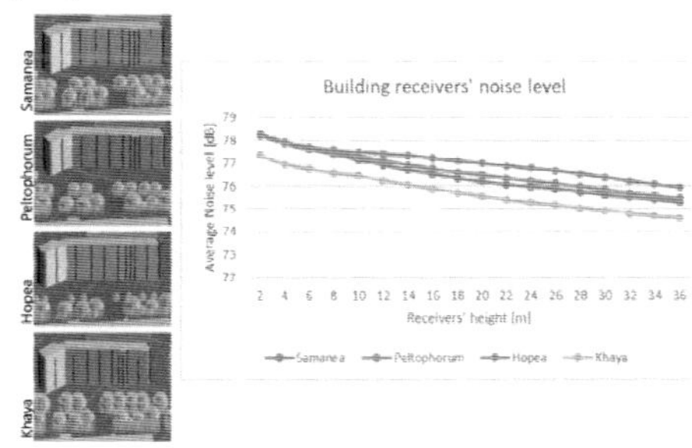

(d) Comparison of different tree species

Figure 6. Noise propagation in residential areas near highways with simulated reforestation strategies

The impact of trees on traffic noise propagation, particularly vertically, was evaluated through interaction features implemented in a web application. Figures 6a and 6b illustrate the effects of trees on noise levels, including scenarios with and without trees, as well as reforestation efforts involving the addition of 7 Samanea saman trees (between existing trees and building) and 7 Hopea (between existing trees). The results indicate a consistent reduction in noise levels with existing trees, with a reduction of approximately 2 dB from 18 meters onwards compared to no trees. The addition of Samanea saman trees further reduced noise by an additional 1 dB, resulting in a total reduction of 3 dB. However, to optimize noise reduction, additional green buffers and strategic placement of trees, such as incorporating smaller species like Hopea odorata, are recommended.

The relocation of trees, as depicted in Figure 6c, also demonstrated a reduction in noise levels when trees were moved closer to the building. This relocation strategy effectively scattered noise, leading to a reduction in noise levels projected onto the building. Specifically, the repositioning reduced noise by up to 1 dB from 8 meters to 26 meters above ground, with saturation in effect thereafter. Furthermore, Figure 6d illustrates the potential noise reduction benefits of replacing existing trees with other species such as Peltophorum pterocarpum, Hopea odorata, and Khaya senegalensis. While larger trees showed significant noise reduction, with an increase in size of tree crown correlating to higher noise reduction rates, it's noted that they require more land space. Hence, a combination of different tree species of varying sizes is proposed for practicality.

The proposed web application aims to facilitate informed decision-making by allowing users and city planners to experiment with different reforestation strategies and tree species in urban design, thereby contributing to the mitigation of noise pollution in residential areas near highways.

3.4 Scene Modelling after Observed Street View

In this experiment, our objective is to model the scene after observation using StreetView by incorporating tree elements into existing buildings and roads. While TreeSG data provides information on street trees, it does not include trees nearer to residences. Therefore, we utilize the proposed 3D scene modeling web app tool to add visually observed trees.

(a) Observed to scene modelling example

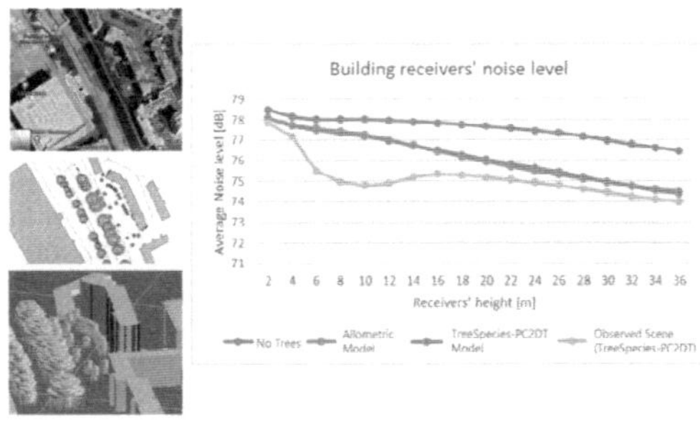

(b) Noise results

Figure 7. Noise propagation with observed street view

Figure 7b presents the simulated noise level of the scene created from observations on StreetView (an example shown in Figure 7a), compared to just trees documented in TreeSG data. We utilized the TreeSpecies-PC2DT model to represent the trees. By adding residential trees, we observed a significant reduction in noise, particularly between 4 meters to 14 meters in height, corresponding to lower floors closer to the traffic source. At higher floors, the noise continues to dissipate according to distance, resulting in sound dissipation.

This underscores the benefits of planting trees in residential areas, not only for visual aesthetics and heat comfort but also for mitigating the disadvantages of noise pollution in densely populated urban environments.

4. DISCUSSION

Addressing the research questions, the experiment compared trees and traditional barriers like walls for noise reduction in urban areas near highways. While walls were more effective horizontally, trees notably reduced noise levels, especially vertically. Additionally, the study examined different tree species' impact, finding larger trees significantly reduced noise but required more space. Strategic placement and relocation further reduced noise, highlighting the importance of species selection. Existing trees, particularly in residential areas near highways, significantly contributed to noise reduction, emphasizing their role beyond aesthetics and comfort.

The study not only provides valuable insights into the effectiveness of trees as natural noise barriers in urban environments but also sheds light on the broader impacts and challenges associated with urban greenspaces. By incorporating additional data sources (explained in Section II-B3 such as collected noise levels recorded by the community through platforms like NoiseCapture (Picaut et al., 2019), experimented noise levels layers, and Land Use/Land Cover (LULC) data, the study offers a comprehensive understanding of the interplay between trees, noise pollution, and urban landscapes. These data sources allow for a more nuanced analysis of the impact of 'layering' greenery in cities, revealing how greenspaces influence noise levels and urban movement patterns. Moreover, the study highlights the importance of considering diverse sources of noise and land cover classifications in assessing greenspaces' effectiveness in mitigating noise pollution and enhancing urban livability. Overall, the findings contribute to urban planning efforts aimed at creating more sustainable and pleasant urban environments by leveraging the benefits of greenspaces and trees.

Several project limitations were identified. One notable limitation relates to the representation of trees, where the current models may not accurately reflect real-world tree structures, potentially affecting their noise attenuation capabilities. Future iterations could address this by introducing more detailed polygonal representations of leaf-like structures and fine-tuning parameters such as the reflection factor in the Softnoise predictor for better alignment with actual measurements. Furthermore, the absence of a dedicated data management interface hindered efficient manipulation, emphasizing the necessity for a platform tailored for dynamic data management. Integrating concepts like Volunteered Geographic Information (VGI) could enhance data collection and validation, highlighting the benefits of trees in urban environments. Further studies may explore potential associations with social and environmental factors.

5. CONCLUSION

In this paper, we've presented our comprehensive approach to scene modeling and noise analysis, stressing the importance of integrating three-dimensional elements in noise pollution studies. We have introduced the 3D Scene Modelling Web-app Platform to tackle these challenges, highlighting its potential applications in urban planning, environmental assessment, and public health research. Our study aim to contribute to the understanding of noise propagation and the role of trees as natural noise barriers. Leveraging noise modeling tools like Softnoise predictor, we've visualized noise contour maps and assessed noise levels, revealing trees' effectiveness in mitigating vehicular noise around buildings. While trees may not completely block noise like solid barriers, they offer benefits such as airflow and visual enhancement. Our experiments in residential and park settings underscore trees' ability to reduce noise, particularly in densely populated areas. By exploring reforestation and strategic tree placement, we emphasize the importance of considering tree species and placement for optimal noise reduction. Given that many cities around the world face similar challenges with high population densities and limited space, the insights gained from this case study in Singapore are highly relevant. The development of a web application for scene modeling further highlights the practicality of integrating trees into urban landscapes for noise mitigation, providing valuable insights for urban planners and policymakers in creating more livable, sustainable communities.

ACKNOWLEDGEMENT

This research/project is supported by the Catalyst: Strategic Fund from Government Funding, administered by the Ministry of Business Innovation & Employment, New Zealand under contract C09X1923, as well as the National Research Foundation, Singapore under its Industry Alignment Fund – Pre-positioning (IAF-PP) Funding Initiative. Any opinions, findings and conclusions or recommendations expressed in this material are those of the author(s) and do not reflect the views of National Research Foundation, Singapore.

REFERENCES

Biljecki, F. (2020). Exploration of open data in Southeast Asia to generate 3D building models. *ISPRS Annals of the Photogrammetry, Remote Sensing and Spatial Information Sciences, 6*, 37–44.

Chin, C., Thang, Z. Y., & Saju, S. (2019). Study on impact of noise annoyance from highway traffic in Singapore City. *Proceedings of Meetings on Acoustics, 39*(1), 015001. https://doi.org/10.1121/2.0001116

Data.gov.sg. (2021). *Total land area of Singapore*. Retrieved February 10, 2022, from https://data.gov.sg/dataset/total-land-area-of-singapore

Department of Statistics, Ministry of Trade Industry. (2021). *Population trends 2021*. Retrieved February 18, 2022, from https://www.singstat.gov.sg/-/media/files/publications/population/population2021.pdf

Gobeawan, L., Lin, E., Tandon, A., Yee, A., Khoo, V., Teo, S., Yi, S., Lim, C., Wong, S., Wise, D., et al. (2018). Modeling trees for virtual Singapore from data acquisition to CityGML models. *The International Archives of the Photogrammetry, Remote Sensing and Spatial Information Sciences, 42*, 55–62.

Gobeawan, L., Liu, X., Lim, C., Raghavan, V., Chattoraj, J., Schindler, J., & Yang, F. (2024). Treespecies-PC2DT: Automated tree species modeling from point clouds to digital twins. In *Proceedings of the 19th International Joint Conference on Computer Vision, Imaging and Computer Graphics Theory and Applications - GRAPP* (pp. 81–91). INSTICC. SciTePress.

Goh, Y. S., Leong, J. W., Yean, S., Lee, B.-S., Ngo, K. M., & Edwards, P. (2022). A study on Singapore's vegetation cover and land use change using remote sensing. In *Proceedings of the 1st ACM SIGSPATIAL International Workshop on Spatial Big Data and AI for Industrial Applications* (pp. 1–12).

Kumar, K., Ledoux, H., Commandeur, T. J. F., & Stoter, J. E. (2017). Modelling urban noise in CityGML ADE: Case of the Netherlands. In *3D Geoinformation*, Delft University of Technology, the Netherlands. Retrieved August 8, 2022, from https://doi.org/10.5194/isprs-annals-IV-4-W5-73-2017

Lin, S. E., Teo, L. S., Yee, A. T. K., & Li, Q. H. (2018). Populating large scale virtual city models with 3D trees.

National Parks. (2022). *TreeSG*. Retrieved August 15, 2022, from https://www.nparks.gov.sg/treessg

Noise Planet. (2019). *Noisemodelling: An open source GIS based tool to produce environmental noise maps*. Retrieved August 15, 2022, from https://noiseplanet.org/noisemodelling.html

Open Street Map. (2014). *Open Street Map*. Retrieved from https://www.openstreetmap.org

Picaut, J., Fortin, N., Bocher, E., Petit, G., Aumond, P., & Guillaume, G. (2019). An open-science crowdsourcing approach for producing community noise maps using smartphones. *Building and Environment, 148*, 20–33.

Ranjbar, H. R., Gharagozlou, A. R., & Nejad, A. R. V. (2012). 3D analysis and investigation of traffic noise impact from Hemmat highway located in Tehran on buildings and surrounding areas.

Softnoise. (2023). *Predictor-Lima (Predictor Type 7810, Version 2023.)*. Retrieved May 10, 2023, from https://softnoise.com/products/predictor-lima/

Tan, C. (2021). NEA to lower vehicle noise levels, tighten motorbike emissions from 2023. *The Strait Times*. Retrieved February 18, 2022, from https://www.straitstimes.com/singapore/transport/nea-to-lower-vehicle-noise-levels-tighten-motorbike-emissions-from-2023

Tan, J. K. A., Hasegawa, Y., Lau, S.-K., & Tang, S.-K. (2022). The effects of visual landscape and traffic type on soundscape perception in high-rise residential estates of an urban city. *Applied Acoustics, 189*, 108580. https://doi.org/10.1016/j.apacoust.2021.108580

Truax, B. (2012). Music, soundscape and acoustic sustainability. *Moebius Journal, 1*(1), 1–16.

World Health Organization, Regional Office for Europe. (2021). *Burden of disease from environmental noise: Quantification of healthy life years lost in Europe*. Retrieved February 18, 2022, from https://apps.who.int/iris/handle/10665/326424

Zhao, W.-J., Liu, E.-X., Poh, H. J., Wang, B., Gao, S.-P., Png, C. E., Li, K. W., & Chong, S. H. (2017). 3D traffic noise mapping using unstructured surface mesh representation of buildings and roads. *Applied Acoustics, 127*, 297–304.

TOURISM SUPPORT SYSTEM TO UTILIZE VIRTUAL REALITY SPACE AND DYNAMIC INFORMATION IN REAL TIME

Yuro Koga and Kayoko Yamamoto
The University of Electro-Communications
1-5-1 Chofugaoka Chofu-shi Tokyo, Japan

ABSTRACT

It is important that both static and dynamic information is efficiently used to create a suitable tourism plan. However, collecting, accumulating and managing dynamic information can cost tourists time, money and energy. In the present study, an original tourism support system was designed and developed with the purpose of reducing the burden of tourists who are unfamiliar with the urban tourist destinations in particular. An original tourism planning support system and web-geographic information systems (Web-GIS) were integrated into a single system, and two original functions were implemented. The system was operated targeting Osaka City, Japan for a period of one month. The information concerning 529 sightseeing spots was collected from tourism-related web media and then saved to the database of the system beforehand. During the operation period, a cumulative total of 160 users accessed the system. Based on the access log analysis results, it was clear that the use of the system was in line with the purpose of the present study, as the number of visits to the page for the main functions were high.

KEYWORDS

Tourism Planning Support System, Web-Geographic Information Systems (Web-GIS), Virtual Reality (VR) Space, Dynamic Information, Social Media, Osaka City

1. INTRODUCTION

In the advanced information and communications network society of recent years, there is an abundance of a wide variety of information on the internet, and anyone can easily obtain the information they need. With the rise of various social media platforms, as anyone can become both a sender and a recipient, the information inundation on the internet has accelerated. This also applies to sightseeing information, as there has been an increase in the use of the internet as a method to easily and effectively obtain various information. Additionally, online reviews on tourism-related web media are seen as an important source of sightseeing information. However, tourists must appropriately sift through the information to find what they need. Tourists who are not familiar with the tourist destinations may find it hard to obtain necessary information and use this to create a satisfactory sightseeing plan that tours various sightseeing spots.

Regarding tourism, it is important that both static information such as the preferences of tourists and opening hours of sightseeing spots as well as dynamic information such as the congestion of tourist destinations and temporary events are efficiently used to create a suitable tourism plan. However, collecting, accumulating and managing dynamic information can cost tourists time, money and energy. Based on the above, a system that can provide support for the collection and systemization of various information and activity simulation before travelling as well as the efficient collection of necessary information while traveling is essential. Additionally, according to the survey on travel intentions after COVID-19 pandemic conducted by the Japan Travel Bureau Foundation (2022), there were many comments saying they wish to travel more frequently after COVID-19 pandemic, and that it is important to avoid congestion while traveling. Therefore, a system that visualizes and provides the information concerning the situations of sightseeing spots in real time is also necessary.

The present study aims to take the above academic and social background into consideration in order to integrate an original tourism planning support system with the web-geographic information systems (Web-GIS) to develop a system that provide tourism supports specifically for tourists who are not familiar with the urban tourist destinations. The system can be applied to urban tourist destinations especially in Japan. Moreover, two original functions are implemented in this system. One is the function to efficiently collect and update both static and dynamic information in real time from tourism-related web media as well as social media. The other is the function to visualize and display such information in a virtual reality (VR) space which was developed using the three dimension (3D) city model of the tourist destinations. Additionally, the area of Osaka City in Japan was selected as the operation target for the system among urban tourist destinations in Japan. Osaka City is an urban tourist destination with a concentration of many famous sightseeing spots, and it is necessary for tourists to efficiently create their sightseeing plans according to the purpose of travel.

2. RELATED WORK

As mentioned in section 1, the present study develops a unique system by integrating multiple systems such as a tourism planning support system and Web-GIS. Therefore, considering the system characteristics, the present study is related to four study fields, including (1) studies related to tourism planning support systems using Web-GIS (Ikizawa-Naitou et al., 2020; Koga et al., 2021). (2) studies related to tourism support systems using VR technology (Poux et al., 2020; Hou, 2021; Putra et al., 2021), (3) studies related to information sharing systems (Shibata et al., 2018; Kunieda et al., 2019), and (4) studies related to systems utilizing social media (Yamamoto et al., 2021; Yamada et al., 2022).

In addition to overcoming the issues of the previous studies as introduced above, the present study demonstrates the originality in developing a unique system that makes the following three elements possible. The first element of the originality is that the system can provide the continuous and flexible tourism supports by extracting social media information, having users update the database to produce dynamic information, and using the information as a reference. It can also contribute to resolve the issues related to sightseeing plan creation in Osaka City that are mentioned in section 1. The second element of originality is that the system can be developed and operated at low cost by effectively using the static and dynamic data and information collected from open data in addition to tourism-related web media and social media. The third originality is that the system enables users to have unique experiences of walking through an urban tourist destination in the VR space by adopting VR technology.

3. SYSTEM DESIGN

As shown in Figure 1, the system in the present study is made up of an original tourism planning support system and Web-GIS. Additionally, by efficiently collecting information from tourism-related web media and social media, the system can continuously provide information and estimate popular sightseeing spots. The collected information is saved to the database and can be viewed by users. Furthermore, according to the survey result of Sakiyama et al. (2022), a unique sense of scale, tours through the city in the VR space, efficient previews of tourist destinations, and recollections of travel experiences were a few possibilities for the VR travel content. Additionally, dizziness caused by VR as well as the insufficient replication of the real cityscape were raised as issues. As a solution to these issues, the present study will develop an online VR space without using head mounted display (HMD), and prepare an interface that can switch between the 2D and 3D digital maps. By doing so, users can have a pleasant experience when using the system. Additionally, the dynamic sightseeing information can be collected and saved to the database, and this information can be superimposed and displayed on the 2D and 3D digital maps.

Though it is assumed that this system will be mostly used on PCs or mobile information terminals including smartphones and tablet PCs, there is no difference between devices used, and both types of devices can use the same functions. PC users are assumed to be mainly indoors, and will use the system to collect sightseeing information to supplement their travel activities, and evaluate sightseeing spots that have already been registered in the system. On the other hand, mobile information terminal users are assumed to be both indoors and outdoors, and will use the system to collect sightseeing information to supplement their travel activities, submit new sightseeing spot information, and evaluate sightseeing spots that have already been registered in the system.

This system operates using a Web server, database server, and GIS server. Heroku, which was adopted for both the Web server and database server, is a PaaS provided by the Salesforce, Inc. ArcGIS Online provided by the Environmental Systems Research Institute, Inc. (ESRI) was adopted for the GIS server. Web applications developed in the system has PHP, JavaScript, and HTML as the main language, and Laravel of the PHP framework was adopted for the implementation.

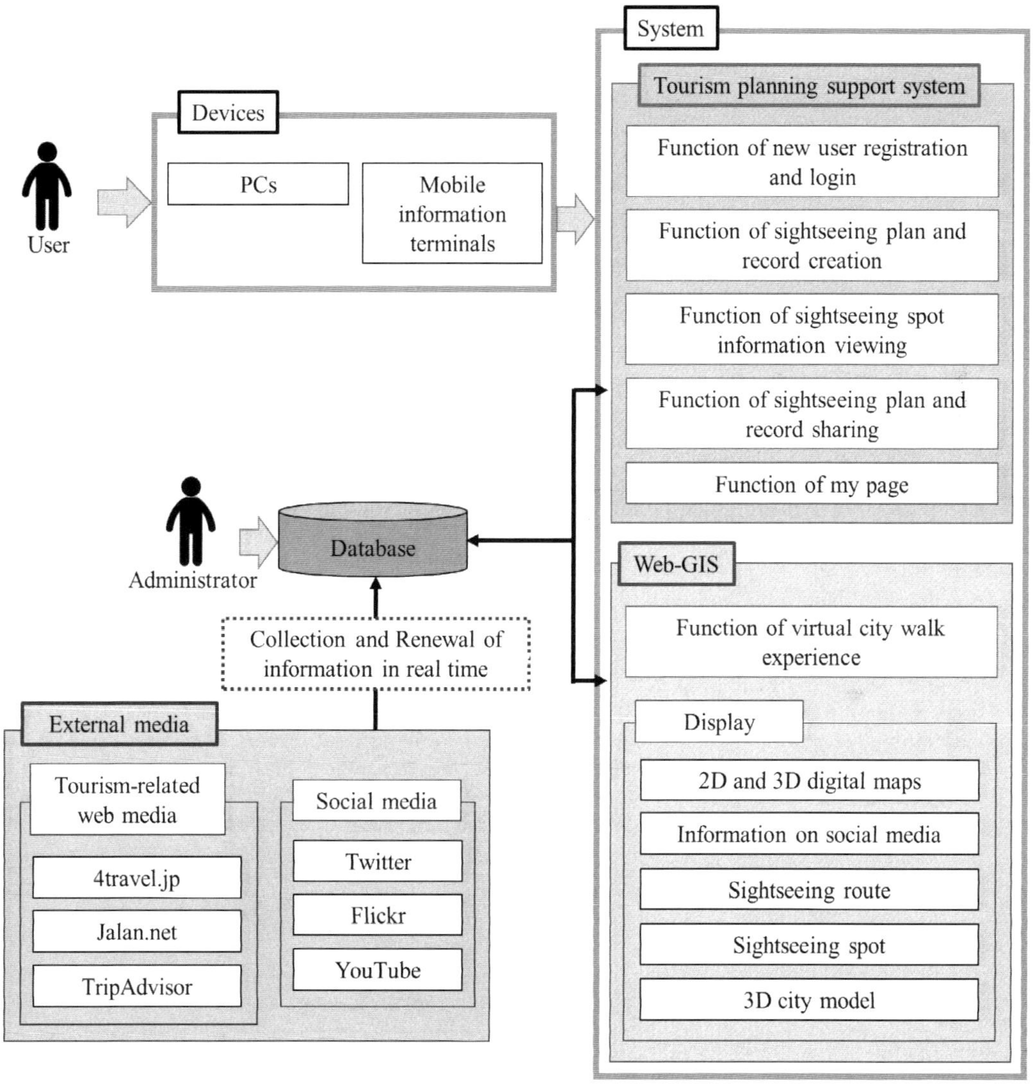

Figure 1. System design

4. SYSTEM DEVELOPMENT

4.1 System Frontend and Backend

The present study will implement six original functions of this system designed for users, including the function of new user registration and login, the function of sightseeing plan and record creation, the function of sightseeing spot information viewing, the function of sightseeing plan and record sharing, the function of virtual city walk experience, and the function of my page. Figure 2 shows the page for the details of sightseeing plan and record (3D digital map) related to the f unction of sightseeing plan and record sharing. Additionally, Figure 3 shows the page for the function of virtual city walk experience (3D city model). Additionally, the six processing for the function of new user registration and login, the function of sightseeing plan and record creation, the function of sightseeing plan and record sharing, the function of virtual city walk experience, the display of social media information, and the derivation of the shortest routes between sightseeing spots are handled by the system backend.

4.2 System Database

In order to make the functions of the system available for use immediately after the start of the operation, static data concerning sightseeing spots must be collected, processed in a layer adopting ArcGIS pro provided by the ESRI, and registered to the database beforehand. Static data concerning a total of 529 sightseeing spots in the operation target area (Osaka City) were collected from tourism-related web media such as 4travel.jp, Jalen.net and TripAdvisor. When users registering new sightseeing spots, the data are added to this layer. Dynamic data are collected in real time from social media such as Twitter, Flicker and YouTube, and also added to this layer. Users create sightseeing plans and records, these are processed in other layer adopting ArcGIS pro, and registered to the database. These layers are overlaid on the 2D and 3D digital maps of Web GIS. Additionally, users register their email addresses and passwords, these are registered to the database.

4.3 System Interface

There are two types of interfaces in this system. One is the screen for PC and mobile information terminals of users, and the other is the screen for PC of administrators. In this way, a responsive screen is adopted into the system, and two types of interfaces are used according to the screen size of the devices used. With the page for administrators, users, registered sightseeing spot information as well as submitted information can be managed. By adopting graphical user interface (GUI), users with malicious intent and inappropriate sightseeing spot information can be deleted without the influence of the information technology (IT) literacy of the administrators.

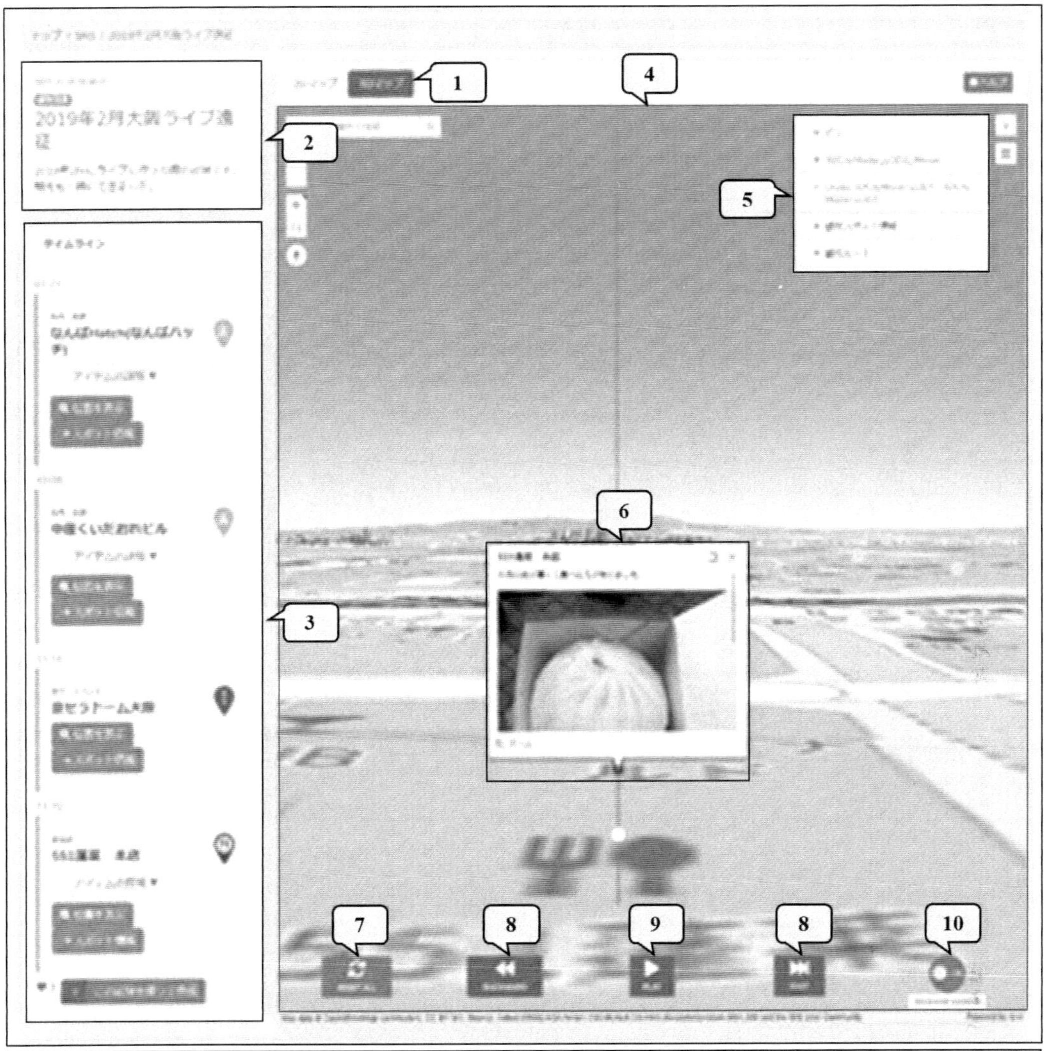

No.	Description
1	Switch between the 2D and 3D digital maps
2	Selected sightseeing plan
3	Timeline of sightseeing plan
4	Screen of the details of sightseeing plan and record on the 3D digital map
5	Items (Present location, 3D digital map, 3D city model, sightseeing spot information and sightseeing route)
6	Reverse of the direction of the course, and change of the perspective
7	Reverse of the direction of the course, and change of the perspective
8	Acceleration of the perspective movement
9	Stop
10	Insolation (Every time the button is clicked or tapped on, the time in the virtual space moves forward which changes the insolation.)

Figure 2. Page for the details of sightseeing plan and record (3D digital map)

No.	Description
1	Navigation bar (A: Questionnaire survey, B: Function of virtual city walk experience, C: Function of creation of sightseeing plan and record, D: Function of viewing sightseeing spot information, E: Function of sharing sightseeing plan and record, F: Log in and registration
2	Page for the function of sharing sightseeing plan and record
3	Selected sightseeing plan
4	Timeline of sightseeing plan
5	Switch between the 2D and 3D digital maps
6	Screen of virtual city walk experience
7	Reset of the perspective on the 3D map
8	Reverse of the direction of the course, and change of the perspective
9	Acceleration of the perspective movement
10	Stop
11	Insolation (Every time the button is clicked or tapped on, the time in the virtual space moves forward which changes the insolation.)
12	Change layer
13	Change of base maps (3D digital map and 3D city model)

Figure 3. Page for the function of virtual city walk experience (3D city model)

5. OPERATION AND EVALUATION

5.1 Operation

The operation of this system was conducted on people both within and outside the operation target area over a one-month period from December 21st, 2022 to January 24th, 2023. The system targeted Japanese and foreigner people who can understand Japanese in particular. The system use was advertised on the website, and Twitter and Facebook accounts of the authors' lab. The total number of users was 56, which included 36 men, 18 women and 2 others. 48% of users were in their 20s which was the highest percentage. 13% were in their 30s, 11% were under 20, 11% were in their 40s as well as over 60, and 7% were in their 50s respectively. Though the majority of users are a younger generation, it is used by a wide range of age groups. Additionally, during the operation period of the system, users registered 10 sightseeing plans and one travel record, and submitted seven sightseeing spot information. However, if the system were operated for a longer period of time, there would be a high potential for more multifaceted use.

5.2 Evaluation

An access log analysis was conducted using users' log data during the operation of this system. Google Analytics of Google was used in the present study. Specifically, the access log can be obtained by calling the Python Program, which contains the analysis code created with Google Analytics, from the HTML file read on each page within the website subject to the access analysis.

Table 1 shows the analysis result concerning the devices users used to access the system. A cumulative total of 160 users accessed the system during the operation period. In addition, regarding the devices used as a method to access the system, 89 used mobile information terminals while 71 used PCs. Furthermore, there were 155 sessions with engagement, the average ratio of engagement to users was 70%, and the average engagement time was 4 minutes 13 seconds. Based on these results, it was revealed that while mobile information terminals had a higher access number, the engagement rate was lower than that of PCs. Therefore, this showed that while many users easily used the system on their mobile information terminals, they use PCs to spend a long time to try out the various function. On the other hand, the responsive design that allows the same functions to be used on both mobile information terminals and PCs was effective.

Next, Table 2 shows the top 9 pages with the highest number of visits. According to Table 2, the function of virtual city walk experience, the list of sightseeing spots, the function of sightseeing plan and record creation, and the function of sightseeing plan and record sharing have many views. Therefore, there was a high number of visits to the pages of main functions aimed to provide tourism support, which is the purpose of the system. Consequently, it is clear that the use of the system was consistent with the purpose of the present study.

Table 1. Analysis results concerning the devices used by users to access the system

Accessed device	Number of users	Number of sessions with engagement	Engagement rate (%)	Average engagement time (second)
Mobile information terminal	89	75	63.0	101
PCs	71	80	78.0	436
Total or Average	Total 160	Total 155	Average 70.4	Average 253

Table 2. Number of visits for each page (top 9)

Ranking	Page name	Number of visits	Rate (%)
1	Top page	321	20.4
2	Function of virtual city walk experience	203	7.3
3	List of sightseeing spots	135	5.8
4	Function of sightseeing plan and record creation	97	5.0
5	Function of sightseeing plan and record sharing	74	4.6
6	Function of login	56	4.3
7	Sightseeing plan "From Tsutenkaku to Osaka Tenmangu"	45	3.0
8	Function of new user registration	42	2.8
9	Details of the sightseeing plan and record	17	2.7

6. CONCLUSTION

In the present study, after designing and developing the system (sections 3 and 4), it was operated and evaluated (section 5). Specifically, in the present study, an original tourism support system was designed and developed with the purpose of reducing the burden of tourists who are unfamiliar with the tourist destinations in particular. An original tourism planning support system and Web-GIS were integrated into a single system, and two original functions were implemented. One function enables the system to efficiently collect and update the static and dynamic information in real time from tourism-related web media and social media in real time (the function of sightseeing spot information viewing). The other function visualizes and displays such information in the VR space which was developed adopting a 3D city model of the operation target area (the function of virtual city walk experience).

The area of Osaka City was selected as the operation target area, and the information of a total of 529 sightseeing spots was collected from tourism-related web media, and then saved to the database of the system beforehand. The system was operated for a period of one month during December 21st, 2022 to January 24th, 2023. During the operation period, a cumulative total of 160 users accessed the system. Additionally, the system had 10 sightseeing plans and one travel record registered as well as seven sightseeing spots submitted by users. In order to evaluate the system, an access log analysis was conducted. According to the access log analysis results, it was clear that the use of the system was in line with the purpose of the present study, as the number of visits to the page for the main functions were high.

Regarding future study topics, the advancement of the system as well as the improvement of the utilization value by increasing the usage performance through the operation in other urban tourist destinations inside and outside Japan can be considered.

ACKNOWLEDGEMENT

In the operation of the tourism support system to utilize virtual reality space and dynamic information in real time, enormous cooperation was received from users of the system in Japan. We would like to take this opportunity to gratefully acknowledge them.

REFERENCES

Japan Travel Bureau Foundation. Trend of Japanese Tourists after the COVID-19 Pandemic. Available online: https://www.jtb.or.jp/wp-content/uploads/2022/05/covid-19-japanese-tourists-20_JTBF220513.pdf (accessed on 13 January 2023).

Ikizawa-Naitou, K., Yamamoto, K., (2020). A support system of sightseeing tour planning using public transportation in Japanese rural areas. Journal of Civil Engineering and Architecture, 14(6), pp.316-332.

Koga, Y., Yamamoto, K., (2021). A Tourism planning support system with Gamification. Journal of Geographic Information System, 13(4), pp.485-507.

Poux, F., Valembois,Q., Mattes, C., Kobbelt, L., Billen, R., (2020). Initial user-centered design of a virtual reality heritage system: applications for digital tourism. Remote Sensing, 12(16), 2583; https://doi.org/10.3390/rs12162583.

Hou, T., (2021). Virtual tourism simulation system based on VR technology. Proceedings of the 2nd International Conference on Computing and Data Science (CONF-CDS 2021), Stanford, CA, US, pp.28-29.

Putra, A., S., Dona, D., Akhirianto, P., M., Aisyah, N., (2021). Designing a virtual Jakarta tourism application on android OS and virtual reality tools. International Journal of Science, Technology & Management, DOI: 10.46729/ijstm.v2i5.322.

Shibata, T., Tominaga, T., Hijikata, Y., Sakata, N., Ito, K., (2018). Study of community support in evacuation shelters based on digital divide. Transactions of the Institute of Systems, Control and Information Engineers, 31(3), pp.101-111.

Kunieda, T., Izumi, R., Miyakawa, R., Kanaya, M., Yamada, S., Gotoda, B., Yoneya, Y., Yaegashi, R., (2019). Creation of guidebook to recored travel memories: Development of KadaPam and tourists' behaviors analysis using tourism guidebook in Shodoshima Island. IPSJ Digital Practice, 10(4), pp.829-849.

Yamamoto, K., Murayama, Y., (2021). Web portal to support remote island for sightseeing and disaster management. In Information Technology in Disaster Risk Reduction, Murayama, Y., Velev, D., Ziateva, P. eds., Springer, Berlin, Germany, pp.1-13.

Yamada, S., Imanishi, T., (2022). Transmission system for COVID-19 pandemic related news using tweet data. *Proceedings of the 84th National Convention of IPSJ*, Matsuyama, Ehime, Japan, pp.3-5.

Sakiyama, K., Sakuma, Y., (2022). A Study on the Actual Situation of CGVR in Tourism Contents and the Possibilities and Issues from the Viewpoint of Users' Evaluation. *Proceedings of the City Planning Institute of Japan, Kansai Branch*, Online, 20, pp.70-73.

BEHAVIORAL BASED DECISION-MAKING SCHEME FOR COOPERATIVE CONNECTED AUTOMATED MOBILITY

David Yagüe-Cuevas[1], Pablo Marín-Plaza[1], María Paz-Sesmero[2] and Araceli Sanchis[2]
[1]Intelligent Systems Lab, University Carlos III, Madrid, Spain.
[2]Control Learning and Systems Optimization Group, University Carlos III, Madrid, Spain

ABSTRACT

For some time now, the automotive industry has been undergoing major adjustments in recent years. Current technological solutions aimed to achieve an autonomous level 5 platform implement increasingly complex software and hardware systems. On the one hand, high-tech fields such as smart cities, the Internet of Things (IoT), and intelligent transportation systems (ITS) will benefit from connected and cooperative vehicles in the future. On the other hand, the more complex these systems are, the more knowledge, computing power and precision they require, and the more difficult they become to debug. There are still several challenges in the field of autonomous vehicles to address. From sensor perception to system localization, through arbitration and decision making. The goal of this paper is to address some of the problems regarding those latter fields by aggregating all the structured traffic knowledge extracted from the vehicle context. This knowledge will be used to deploy an effective decision-making system based on behaviors capable of choosing what to do when certain conditions are met. Actions such as changing its operation mode, selecting the most suitable trajectory or even adjusting the systems to the current situation for a better automated vehicle experience.

KEYWORDS

Connected Cooperative Automated Mobility (CCAM), Intelligent Transport Systems (ITS), Behavior Trees, Automated Decision Making

1. INTRODUCTION

Intelligent transportation Systems (ITS) have evolved since the mid-1960s producing technological solutions for the automation of ground platforms (Sussman, 2008), (Auer, et al., 2016). Nowadays, they have already turned into a huge technological field of research in which the industry is recently developing commercial solutions to exploit its potential of services (Miles, 2014). Therefore, they have become the basic pillar from which the industry intends to enhance modern mobility in future smart cities (Gohar & Nencioni, 2021), (Telang, et al., 2021). The aim of these advances is to develop better systems to improve aspects such as road congestion, fuel consumption (Szymanski, et al., 2021), (Yao, et al., 2021), resource usage (Nikitas, et al., 2021), and road safety (Tafidis, et al., 2022). However, the automation of any platform involves various fields of research and engineering whose objective is based on the production of these technological solutions to address the current problems of the sector. Consequently, state-of-the-art solutions have rapidly emerged around specific hardware implemented for real-world inference (Yeong, et al., 2021), (Oh & Yoon, 2019), and complex software systems to deliver increasingly automated capabilities for the research platforms used (Mallozzi, et al., 2019), (Vdovic, et al., 2019). Nonetheless, like any other system, they are prone to certain flaws that need to be addressed to find better and more efficient solutions. In addition to that, contextual information is also important, because depending on the situation, some systems may lose effectiveness. Thus, although this can occur in a wide variety of situations, new paradigms need to be created for automated systems to adapt to the current operating context, potentially increasing its efficiency.

Like any mechanical design, automated systems no matter the applications in the industry suffer from multiple hazards (Ahmed, et al., 2022). The ones that come from the internal status of modules operating the platform e.g., software exceptions, hardware anomalies, sensor failures, etc., and the ones that comes from the external environment where the vehicle is deployed, e.g., pedestrian, vehicles, wildlife, occlusions, etc.

Current automated vehicles are highly dependent on sensor reading, and those systems which process their outputs to produce real-world models. In addition, there is not a unique sensor setup solution to cope with every possible situation the vehicle may be involved in. Different sensors produce different measurements that varied depending on the environment and situation. In fact, each one is usually used to target specific perception areas because of its effectiveness when applied in those fields. Moreover, to the perception challenges, some other risks must be added considering the highly populated scenarios of current cities: mixed traffic vehicles, unperceived pedestrians, electric scooters or cyclists, environmental occlusions, etc. Because of these problems, performance of systems may vary depending on the situation, potentially making some of them useless in certain contexts. The proposal of this work is intended to solve these problems or at least mitigating its effects by adjusting the vehicle set up and operation mode dynamically thanks to the analysis of the context coming from a traffic scene representation.

However, for this solution to work, firstly, the platform evaluated needs to have a representation of how the world is, i.e. how the traffic scene is built and evolves over time (Yagüe-Cuevas, et al., 2023). Secondly, it must include an awareness system capable of advertising the internal state of the modules involved in the automation process (Nolte, et al., 2017), (Kanapram, et al., 2020a), (Kanapram, et al., 2020b). A system that detects software and hardware anomalies, and process sensor outputs to verify the correctness of their readings. Besides that, knowledge regarding the environment can be extracted directly from High-Definition maps prepared for autonomous road navigation, which need to be updated accordingly to reality (Seif & Hu, 2016), (Liu, et al., 2020), (Ebrahimi Soorchaei, et al., 2022), and environmental features detected while the vehicle is operating should be used for perception, planning and localization tasks. These environmental features can be organized into perceptual event data structures that typically come from deep learning models trained to detect, classify and track traffic objects in real time (Rao & Frtunikj, 2018), (Shengbo, et al., 2019) (Stroescu, et al., 2019). Hence, all this information aggregated with the traffic scene will construct in this work a valid virtual world full of knowledge from which to parameterize a decision-making system. Then the decision process integrated with high level planning capabilities will enable an automated vehicle to produce effective actions depending on the situation. And finally, once this system is enhanced with connectivity and advance point of behavior execution, collaboration and cooperation between intelligent traffic agents can be deployed easily at any time providing an arbitration between behaviors shared through a vehicular network.

With such purpose in mind, this work introduces the concept of Mission ABPs (Advanced Behavioral Points), i.e., important landmarks in which behavior, cooperation and collaboration are mixed to provide a complete solution for Cooperative Connected Automated Mobility. Therefore, the main aim of this work is to aggregate the formal description of any anomalous event, the current context of the traffic scene and a behavior-based system that takes them as inputs, to decide the mode of operation of the vehicle, what to do if there is any anomaly or emergency, and how to proceed on the roads with cooperation capabilities.

2. MOTIVATION

For some time now, the typical approach of current technological solutions for navigation and planning in intelligent transport systems are implemented based on reactive behaviors, which depend on the sensor readings that come from the ego perception models (Pisarov & Mester, 2021). However, most of them entail several conceptual flaws and problems that varied depending on the application (Sushma & Kumar, 2022). The industry is now realizing that to fully operate a hypothetical level 5 autonomous vehicle, ego information alone seems not enough. That is one of the main reasons why these systems are migrating now to better solution based on connectivity and cooperation (He, et al., 2019). Depending on the context, not all systems and sensors may function with the same accuracy and efficiency, which can lead to errors and dangerous situations. The connectivity and cooperation between traffic agents is meant to drastically change the safeness of the current roads, but what if these systems could be adjusted according to context inferred from the traffic scene. New solutions for self-driving decision-making lack of a standardized framework of knowledge from which to build the behaviors implemented and are often deployed in specific areas of the traffic network not giving a global solution to the problem (Hang, et al., 2021), (de Beaucorps, et al., 2017), (Orzechowski, et al., 2020). The behavioral patterns introduced in this work addresses the problem of behavior execution and connectivity by generating a framework to advertise this contextual information in generalized use case for real-time decision-making allowing sensors, modules, and operating mode to be adjusted depending on the vehicle needs. This will result in more effective systems that dynamically adjust their procedures and allocate resources efficiently when operating on certain conditions.

3. TRAFFIC ELEMENTS

For any automated system to decide the course of actions, it is necessary to formalize the current state of the world which in AV context is basically the Traffic Scene, i.e. the collection of static and dynamic elements with which the vehicle can interact while operating. From this representation the system will be able to generate behaviors producing valid and effective actions that adapt to the current situation. In (Yagüe-Cuevas, et al., 2023) a method to formalize and structure the traffic information is proposed. Furthermore, in (Yagüe-Cuevas, et al., 2024) this representation is extended to be used in an automated system able to decide what actions to perform considering current traffic information. In this work several changes were made to the last proposal to generalize the use case and enhance the representation for better behavior modeling.

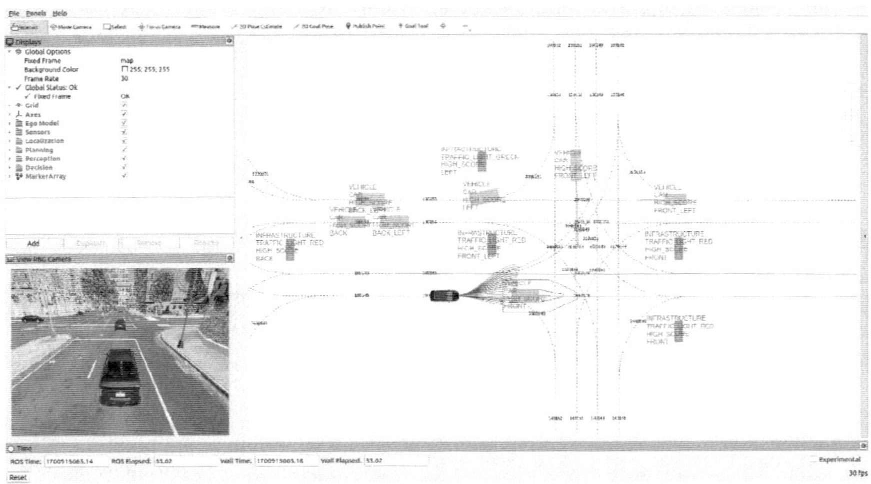

Figure 1. Traffic scene concept

Traffic scene static elements are those traffic items that do not change dynamically their position over time. Road elements such as traffic signs, traffic lights, roads, speed restrictions, intersections, etc. These are structured in data items which store information about the road and its relationship with other items. These are the *Road Events*, which also contain the current lane in which the vehicle is located. With *Road Events* the current road type and Frenet coordinates *D* and *S* are known, (Werling, et al., 2010), as well as the local Cartesian coordinates in which the vehicle is located. This information is used to follow a route plan by generating a set of possible trajectories to choose from considering the logical connections to other lanes.

Dynamic elements come from the processing of sensor readings, aggregated in detections inside the *Perception Events*. Each detection is generated based on the prediction of deep learning models, which are fused with other perception events from connected devices deployed on road. Fused perception events are processed to add semantic tags used by the decision-making module. Considering size, position and predicted trajectory, the system generates a set of polygons that are used by the planning system to generate a new set of semantic tags associated to each trajectory, which finally defines the action to perform. Since the current traffic situation may not allow to follow the estimated plan action, trajectory semantic tags give the system the opportunity to decide which one is the most appropriate depending on the situation, while also considering dynamic elements perceived. The *Perception Events* bounding box is used to create the primitive polygon that is inflated and combined with the current ego vehicle trajectory to check for possible collision.

Finally, awareness elements aggregate information about the internal state of the vehicle systems. Since in automated vehicles there exists several redundant approaches in case of failure, this representation allows the software ecosystem to have a notion of what is happening internally in every module deployed. This data structure is based on criticality. In an ideal situation, all systems should be working perfectly but there are some cases in which some sensor may not work optimally, or a system failure is detected in case of any software exception. The decision system needs to consider these situations and provide actions to recover from those errors if possible. The representation assigns validation flags to each of the layers involved in the automation process and advertises the subsystem with the specific failure, malfunction or underperformance.

4. BEHAVIOR DESIGN PRINCIPLES

The term behavior defines a set of actions performed in a particular way under certain conditions. These actions are generally executed in a specific order and with a clear objective. The main aim of this work is to provide these actions with the most accurate representation of the needed traffic conditions to make a decision that suits the current road situation. To accomplish this, the information explained in the previous section is formalized and then used in conjunction with a software framework able to deploy a custom set of actions when certain conditions are met. After generating the necessary traffic knowledge, this decision system will be able to select the most suitable action to perform next while the vehicle is operating. These actions are built based on human readable abstract processes that can take place in the automated vehicle. Actions such as *SelectTrajectory*, *ProcessObstacles*, *StopVehicle*, *SelectMission*, *SelectVehicleMode*... In addition to the main actions, backup actions are needed in case of emergency or failure, and selective actions that deploy custom behaviors based on mission information, which constitute a special type that can be changed, updated or remove by other connected systems capable of communicating with the ego vehicle.

4.1 Behavior Trees

A behavior tree (Colledanchise & Ögren, 2018) is a mathematical model to represent a behavior. Specifically in computer science, it is considered an execution model that performs tasks which can be decomposed into smaller ones until a goal is achieved. They are widely used in agent automation and NPC logic in video games and have been considered a modern approach to finite state machines. In this case, the tree nodes are going to be the vehicle actions, which are meant to change dynamically the internal state of the vehicle by executing the actions that decide what to do next considering the traffic scene. The preconditions and postconditions associated with these actions will be encoded from the state of the road by processing the perception, road and awareness events explained in previous sections. These can be represented in questions such as *where am I?*, *how many obstacles I am detecting and where are they?*, *what custom behavior if any do I have to execute for this landmark?*, *is there any additional task to be done considering the lane type I am in?*, *what is the most suitable trajectory to choose bearing in mind what I have ahead of me?*, etc.

In this work, the behavior tree notion is used to deploy two different yet collaborative execution models within the same software ecosystem. Firstly, a main tree that contains the principal directives of the automated system, the ones dealing with localization, navigation, planning, emergency and control. And secondly, a delegate tree in which custom behaviors are executed by selecting a specific sub-tree that adjusts the vehicle operation mode to what is ahead in the planning frame. For this, the advanced behavior point data structure must be mentioned because it stores important information for the last tree to properly work.

4.2 Advance Behavioral Point (ABP)

The ABP is a concept extracted from (Yagüe-Cuevas, et al., 2024) that defines a special point within a road plan, i.e., set of roads that must be traversed from current lane to goal lane. ABPs are used to generate a Mission, which contains a finite number of ABPs distributed across this plan. These points are treated as HD map landmarks by the main behavior tree, which selects them by computing the nearest one found in the current mission from the vehicle location. The idea behind the ABP is to determine what the vehicle will be doing when traversing important traffic locations, while also having the possibility to collaborate with intelligent components deployed in the infrastructure, able to change, remove or update them.

5. BEHAVIOR INTEGRATION

Once all traffic information is aggregated, a road plan is generated and a Mission is set up, the only thing left to do is to implement the rules the decision system needs to follow while the vehicle is operating. These decision rules are composed of a set of actions that take place within a modular software ecosystem based on ROS2 (Foundation, 2021). In (Yagüe-Cuevas, et al., 2023) this ecosystem can be seen. Each ROS2 module deployed uses information provided by the traffic scene to produce a wide variety of outputs that assist the decision system in generating a feasible set of actions. The implementation is based in two different execution modes using several behavior patterns, one for the main behaviors the platform should execute globally, such as follow a trajectory or check system status; and another different set of behaviors to deploy specific tasks depending on the mission information that the decision system is managing dynamically.

The main tree, shown in Figure 1, is composed of several global tasks like *SelectTrajectory* or *SelectABP*. These tasks do not need a context since they are executed always to control and check the vehicle status. However, when the main tree detects an ABP is near, it generates and checks periodically a local behavior tree that executes concurrently the tasks provided in that specific ABP. ABPs come in the form of subtrees (blue nodes) that in real time are adjusted and selected depending on what the ecosystem stores in its behavior database, and the mission information available. This behavior database can be modified to improve the behaviors defined for each ABP. Furthermore, each ABP can be updated dynamically through the *behavior* field which can store custom behaviors for cooperation scenarios. For the time being, the automated system always selects the cooperation behavior from that field if any is given. Nonetheless, this entails some security problems that need to be addressed in the future. Finally, backup systems as well as alternative actions must be implemented in case of a system failure. The bottom of the tree tries to address those situations. By selecting the emergency level, the tree can perform a sequential check of what actions to perform depending on the planning, localization and sensor layers status. Thanks to the modularity of actions in a behavior tree these checks can be updated, revised and improved at any time by behavior designers.

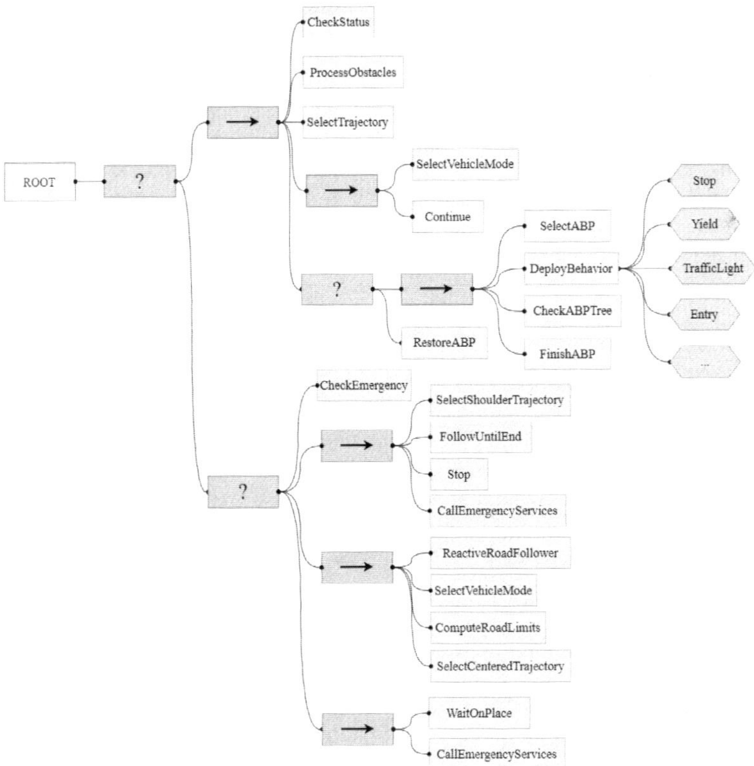

Figure 2. Behavior tree for generalized decision-making

6. RESULTS

The proposal of this work is continuously developed and integrated inside a software ecosystem based on ROS2 found in (Yagüe-Cuevas, et al., 2024), to which this new behavioral scheme was added for decision making capabilities. The decision scheme proposed is implemented in the *Mission Manager* module, which is responsible for processing the information given by traffic scene events and acting according to the situation. This ecosystem was mostly designed and tested in Carla simulator (Dosovitskiy, et al., 2017).

In Figure 3 some tests carried out in Carla to evaluate the behavior trees performance can be seen[1]. In Figure 3a a road plan is given to the mission system with no obstacles in the path. In 3b a new goal is requested, but with a stop signal in the plan, and later, a dummy obstacle that acts as a threat. In Figure 3c the awareness system reports a failure in localization and sensor layer, and in 3d only the sensor layer failed, a situation in which the vehicle can still use its planning system to select a shoulder trajectory to follow and stop once safe.

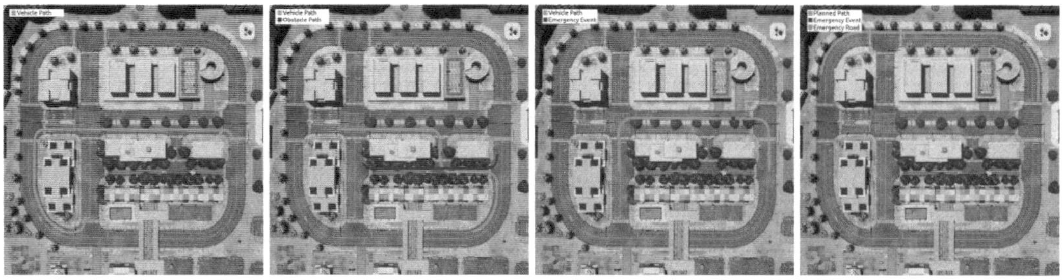

(a) Custom ABP (b) Stop ABP (c) Critical Failure (d) Sensor Failure

Figure 3. Carla world experimentation use cases

In Figure 4 the road plan followed in each use case can be seen[1]. First, in 4a the ABP is designed to approach with a controlled speed, stop the vehicle, wait some time, and when finished, continue with the road plan until the *PlanEndABP* signaling the plan end is reached. Second, in 4b the *StopABP* tree waits until the obstacle agent passes the intersection by approaching the point and stopping until the perception system detects no obstacles in path. Once a free course flag is received, the behavior is finished, and the vehicle continues until the *PlanEndABP* is reached. In 4c the awareness system reports a failure in localization and sensor layer forcing a critical stop. On the contrary, in 4d only the sensor layer reports a failure, a situation in which the vehicle can use its planning system to select a shoulder trajectory and perform an emergency stop.

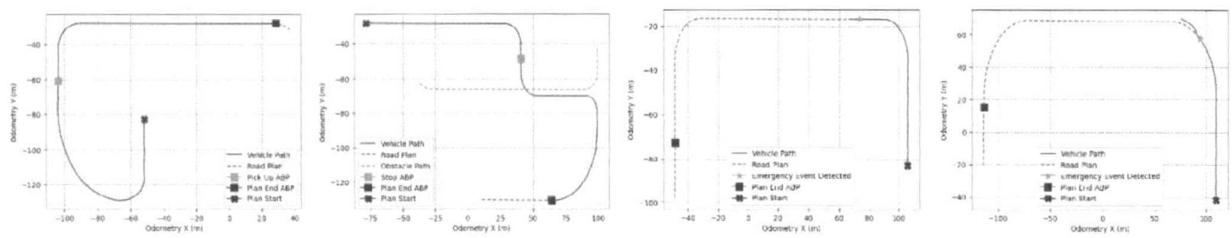

(a) Custom ABP Plan (b) Stop ABP Plan (c) Critical Failure Plan (d) Sensor Failure Plan

Figure 4. Result road plans followed with odometry and ABPs

In Figure 5 the internal status of all control variables can be seen[1]. In Figure 5a the vehicle achieved the desired speed approaching the ABP, reaching it and stopping. Once on the ABP, the system waits for some time, and when finished, resumes the plan until its completion. In 5b the same pattern is repeated but instead of waiting a certain amount of time, it waits until the obstacle passes the intersection, and a free course flag is received. Then, it continues with the plan until its end. On the other hand, in Figure 5c a critical failure is detected by the awareness system in sensor and planning layers. An emergency mode is set, and the behavior

[1] Enlarged Figures can be seen at (Yague-Cuevas, et al., 2024)

tree stops abruptly the plan and waits for help. In 5d, the awareness system is only detecting a failure in the sensor layer. Thus, the *Mission Manager* can use the planning capabilities to generate an emergency shoulder road, which is followed until the vehicle is aligned, at which it stops softly and waits for help.

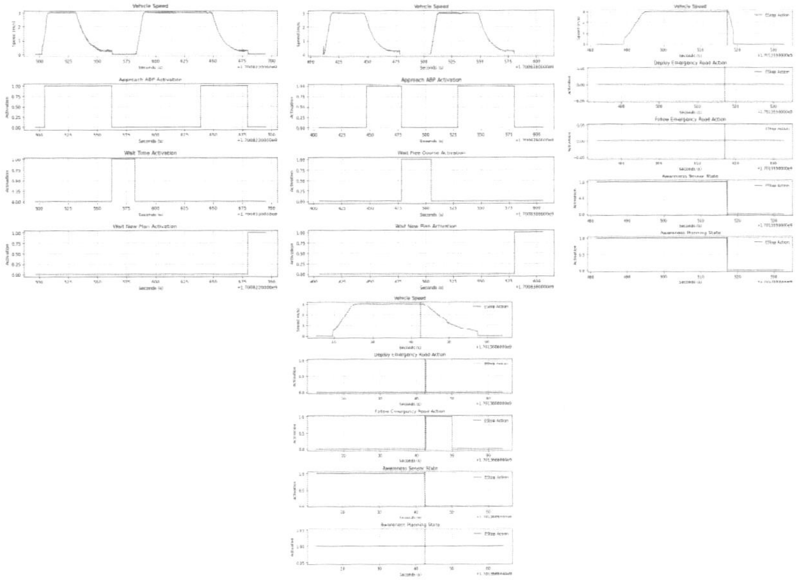

(a) Custom ABP (b) Stop ABP (c) Critical Failure (d) Sensor Failure

Figure 5. Vehicle speed and behavior activation during testing

7. CONCLUSION

Behavior trees are widely used execution paradigms in the video game industry, as well as in robotics solutions and high-level control tasks. Their modular design with the ability to easily revise, update and improve actions enhances the production cycle to develop autonomous vehicle decision systems easy to maintain and audit. Although this paradigm is still in its infancy in terms of technological solutions in the field of autonomous vehicles, it is now showing huge potential for the implementation of decision-making systems in which the responsibilities are clear and well defined. The proposal for this work is a step towards this kind of solution. It addresses the possibility of deploying customized behaviors for the governance of an automated vehicle as well as the possibility of collaboration and behavior sharing. In this work, a double execution paradigm based on behavior trees was implemented effectively to deploy custom behaviors based on the traffic scene context as well as govern a vehicle while operating on the road addressing unexpected emergencies and failures.

In the future, this design is expected to evolve providing more robust mechanisms to address these emergencies and/or failures, and allowing deeper control during overtaking, lane changing and specific driving maneuvers. Specifically, by providing new actions for the control of the selected trajectory, and techniques for overtake detection and assistance, the behavior tree is expected to produce more complex decisions based on the context. Thus, if the system decides to overtake or perform a lane change, the actions will be validated and tracked, if performed, until completion. Finally, more testing regarding cooperation behavior and arbitration needs to be done, as secure mechanism and cooperative behavior sharing is the key to solving complex traffic situations in the future. The ABP aims to provide this collaboration that is suitable for improvements once shared with intelligent traffic infrastructure like intersections and roundabouts.

ACKNOWLEDGEMENT

Grant PID2021-124335OB-C21, PID2022-140554OB-C32 and PDC2022-133684-C31 funded by MCIN/AEI/10.13039/5011000 11033.

REFERENCES

Ahmed, H. U., Huang, Y., Lu, P. & Bridgelall, R., (2022). Technology Developments and Impacts of Connected and Autonomous Vehicles: An Overview. *Smart Cities,* Volume 5, p. 382–404.

Auer, A., Feese, S., Lockwood, S. & Hamilton, B. A., (2016). *History of intelligent transportation systems.,* s.l.: s.n.

Colledanchise, M. & Ögren, P., 2018. *Behavior trees in robotics and AI: An introduction.* s.l.:CRC Press.

de Beaucorps, P. et al., (2017). *Decision-making for automated vehicles at intersections adapting human-like behavior.* pp. 212-217.

Dosovitskiy, A. et al., (2017). *CARLA: An Open Urban Driving Simulator.* s.l., s.n., p. 1–16.

Ebrahimi Soorchaei, B. et al., (2022). High-Definition Map Representation Techniques for Automated Vehicles. *Electronics,* Volume 11.

Foundation, O. S. R., (2021). *ROS: Robot Operating System.* s.l.:s.n.

Gohar, A. & Nencioni, G., (2021). The Role of 5G Technologies in a Smart City: The Case for Intelligent Transportation System. *Sustainability,* May, Volume 13, p. 5188.

Hang, P. et al., (2021). Decision Making of Connected Automated Vehicles at an Unsignalized Roundabout Considering Personalized Driving Behaviours. *IEEE Transactions on Vehicular Technology,* Volume 70, pp. 4051-4064.

He, J. et al., (2019). *Cooperative Connected Autonomous Vehicles (CAV): Research, Applications and Challenges.* pp. 1-6.

Kanapram, D. et al., (2020). *Self-awareness in Intelligent Vehicles: Experience Based Abnormality Detection.* Cham, Springer International Publishing, p. 216–228.

Kanapram, D. T. et al., (2020). Self-awareness in intelligent vehicles: Feature based dynamic Bayesian models for abnormality detection. *Robotics and Autonomous Systems,* Volume 134, p. 103652.

Liu, R., Wang, J. & Zhang, B., (2020). High Definition Map for Automated Driving: Overview and Analysis. *The Journal of Navigation,* Volume 73, p. 324–341.

Mallozzi, P. et al., (2019). Autonomous Vehicles: State of the Art, Future Trends, and Challenges. In: *Automotive Systems and Software Engineering: State of the Art and Future Trends.* Cham: Springer International Publishing, p. 347–367.

Miles, J. C., (2014). Intelligent transport systems: overview and structure (History, Applications, and Architectures). *Encyclopedia of Automotive Engineering,* p. 1–16.

Nikitas, A., Thomopoulos, N. & Milakis, D., (2021). The environmental and resource dimensions of automated transport: a nexus for enabling vehicle automation to support sustainable urban mobility. *Annual Review of Environment and Resources,* Volume 46, p. 167–192.

Nolte, M. et al., (2017). *Towards a skill- and ability-based development process for self-aware automated road vehicles.* pp. 1-6.

Oh, C.-s. & Yoon, J.-m., (2019). *Hardware Acceleration Technology for Deep-Learning in Autonomous Vehicles.* pp. 1-3.

Orzechowski, P. F., Burger, C. & Lauer, M., (2020). *Decision-Making for Automated Vehicles Using a Hierarchical Behavior-Based Arbitration Scheme.* s.l., s.n., pp. 767-774.

Pisarov, J. & Mester, G., (2021). The future of autonomous vehicles. *FME Transactions,* Volume 49, p. 29–35.

Rao, Q. & Frtunikj, J., (2018). *Deep Learning for Self-Driving Cars: Chances and Challenges.* New York, NY, USA, Association for Computing Machinery, p. 35–38.

Seif, H. G. & Hu, X., (2016). Autonomous Driving in the iCity—HD Maps as a Key Challenge of the Automotive Industry. *Engineering,* Volume 2, pp. 159-162.

Shengbo, L. I. et al., (2019). Key technique of deep neural network and its applications in autonomous driving. *Journal of Automotive Safety and Energy,* Volume 10, p. 119.

Stroescu, A., Cherniakov, M. & Gashinova, M., (2019). *Classification of High Resolution Automotive Radar Imagery for Autonomous Driving Based on Deep Neural Networks.* s.l., s.n., pp. 1-10.

Sushma, R. & Kumar, J. S., (2022). Autonomous vehicle: Challenges and implementation. *Journal of Electrical Engineering and Automation,* Volume 4, p. 100–108.

Sussman, J. S., (2008). *Perspectives on intelligent transportation systems (ITS).* s.l.:Springer Science & Business Media.

Szymanski, P. et al., (2021). The future of road transport in Europe. Environmental implications of automated, connected and low-carbon mobility. *Combustion Engines,* Volume 60.

Tafidis, P., Farah, H., Brijs, T. & Pirdavani, A., (2022). Safety implications of higher levels of automated vehicles: a scoping review. *Transport reviews,* Volume 42, p. 245–267.

Telang, S., Chel, A., Nemade, A. & Kaushik, G., (2021). Intelligent transport system for a smart city. *Security and privacy applications for smart city development,* p. 171–187.

Vdovic, H., Babic, J. & Podobnik, V., (2019). Automotive Software in Connected and Autonomous Electric Vehicles: A Review. *IEEE Access,* Volume 7, pp. 166365-166379.

Werling, M., Ziegler, J., Kammel, S. & Thrun, S., (2010). *Optimal Trajectory Generation for Dynamic Street Scenarios in a Frenet Frame.* s.l., s.n., pp. 987-993.

Yagüe-Cuevas, D., Marín-Plaza, P., Paz-Sesmero, M. & Sanchis, A., (2023). Towards a Robust Traffic Scene Representation in Cooperative Connected Automated Mobility. In: *Proceedings of the 9th International Conference on Vehicle Technology and Intelligent Transport Systems.* s.l.:INSTICC, p. 265–272.

Yagüe-Cuevas, D., Marín-Plaza, P., Paz-Sesmero, M. & Sanchis, A., (2024). Modeling the Traffic Scene in Intelligent Transport Systems for Cooperative Connected Automated Mobility. In: *Communications in Computer and Information Science.* s.l.:Springer International Publishing, p. to appear.

Yague-Cuevas, D., Marin-Plaza, P., Sesmero-Paz, M. & Sanchis, A., (2024). *Complementary Material.* [Online] Available at: https://www.uc3m.es/ss/Satellite/DeptInformatica/es/TextoDosColumnas/1371350946990/ [Accessed 05 04 2024].

Yao, Z. et al., (2021). Fuel consumption and transportation emissions evaluation of mixed traffic flow with connected automated vehicles and human-driven vehicles on expressway. *Energy,* Volume 230, p. 120766.

Yeong, D. J., Velasco-Hernandez, G., Barry, J. & Walsh, J., (2021). Sensor and Sensor Fusion Technology in Autonomous Vehicles: A Review. *Sensors,* Volume 21.

A DIGITAL HEALTH PLATFORM FOR SHARED EHR SERVICES: REAL WORLD CASE STUDIES

Ricardo Gonçalves, André Toffanello and Ricardo Puttini
Core Consulting
SIG qd. 1, lt. 385, sala 404 - 71610-410 - Brasília, DF, Brazil

ABSTRACT

This paper demonstrates a practical digital health platform (DHP) used to provide digital services, including a shared electronic health record service (S-EHR), to heterogeneous health ecosystems. The main features of the DHP architecture are: (1) patient-centric design; (2) service-oriented; (3) polyglot interoperability; (4) normalization of electronic health records (clinical data repository); (5) use of connectors and adapters (integration layer); and (6) GDPR-aware privacy and data protection. The DHP architecture is described as a set of decoupled and distributed components that interact together to provide digital services to patients, practitioners, and managers in health care ecosystems. Real world case studies are presented to establish proper validation for the DHP design: (1) network of 41 public hospitals in Brazil, established in 34 cities along 23 (out of 27) States in Brazil; and (2) national public health care network in Jamaica.

KEYWORDS

Digital Health Platform, Shared Electronic Health Record, Patient-Centric Health Care, Interoperability, Service-Oriented Architecture

1. INTRODUCTION

Digital Health Platforms (DHP) are becoming increasingly useful to connect several disparate clinical information systems that are used in heterogeneous health care ecosystems and establishing digital shared services. According to the World Health Organization (WHO) DHP Handbook (WHO/ITU, 2020), the concept of the DHP emerged from a recognition that most digital health progress has arrived in the form of silo applications and information systems, resulting in fragmentation of the health information and services, clinical and administrative likewise.

In this paper we present a DHP that aims at establishing a shared electronic health record (S-EHR) service in large (national-wide) public health ecosystems. We demonstrate how a set of carefully selected design goals enables successful deployment and scaling of the DHP and its digital services.

The remining of the paper is organized as follows. Section 2 presents the DHP and its digital services, discussing the design goals and resulting architecture. Section 3 presents the results from 02 practical (real world) use cases. Section 4 analyses the DHP implementation in the reported use cases, while Section 5 presents our conclusions for the study.

2. DIGITAL HEALTH PLATFORM AND SERVICES

Electronic health records (EHR) tend to be produced and managed at the health organization level. As a matter of fact, each health organization deals with electronic medical information with proprietary processes, applications, formats, semantics, and policies. This creates "information silos" that restricts the EHR access inside the organization that manages it, creating gaps and fragmentation in the health care process.

In a nutshell, our Digital Health Platform (DHP) main feature reduces this fragmentation – and its effects. It consists of a set of digital services that allows the electronic health information to be accumulated and shared in an opportunistic, secure, and consented manner, across organization borders.

The DHP provides health interoperability services. These services are designed to accomplish a set of design goals described in section 2.1. The complete architecture of the platform comprises a set of decoupled components that work together through mediations within an orchestration engine, as described in section 2.2.

2.1 DHP Design Goals

2.1.1 Patient-Centric

Modern health care business models tend to place the patient at the center of its processes and workflows. In fact, several studies show that engaging and empowering the patient in monitoring and maintenance of her/his health and well-being leads to improved health outcomes (Schofield et al., 2019). This is particularly important with patients that presents chronic health conditions, as these patients are in permanent care and are frequent users of health care services (Brands et al., 2022).

Our DHP enables patient engagement through carefully designed patient-facing eHealth tools, i.e. the patient portal and patient Mobile App that implements personal and shared electronic health record (PHR/sEHR) services. These applications also provide several digital services for the patients, including: (1) online access to digital EHRs, prescriptions (meds, lab tests, and others), lab results and others; (2) self-referred health information (such as, allergies, meds, health conditions and others); (3) online EHR sharing – which allows a patient to securely share all her/his EHRs that are interoperated in the DHP among physicians, health organizations and related persons easily; (4) appointments; and (5) push-notifications, among others.

2.1.2 Service-Oriented

Our DHP has a fully service-oriented architecture. Therefore, all the services and components that are implemented in the platform are exclusively provided, in the backend, by a set of standardized APIs, forming an API collection (service catalog) that provides agnostic EHR and workflow services. Most of our APIs deliver services that implement the HL7 FHIR RESTful API (https://hl7.org/fhir/http.html). Selected APIs implements IHE infrastructure profiles (PIXv3, PDQv3, XDS.b, MHD, APPC, among others). Some custom services and a reduced set of microservices are also provided. For a complete list of the public APIs supported at the DHP, please visit https://dev.fluxmed.com.br/devportal.

Leveraging a standard-based API layer is relevant to delivering intrinsic interoperability and increased federation capabilities, together with other strategic goals and benefits from service-oriented approach (Erl, 2016), meaning that solutions based on such architecture not only are capable of exchanging data from birth, but also that they can be coordinated to operate together as one in a way that users and systems can reach multiple DHPs even if they are interacting with a single one.

2.1.3 Polyglot Interoperability

In our DHP, the medical information produced in EHR applications is transformed to a consensual and standard-based format and then loaded into the platform. Therefore, the platform holds a set of records sourced from different systems but represented in the same format. Different health information standards are used in this process. The clinical information representation follows the OpenEHR (https://openehr.org/) reference model. This is complemented with information resources (clinical and non-clinical) representations using the HL7 Fast Healthcare Interoperability Resources (FHIR) (https://hl7.org/FHIR).

Supporting both OpenEHR and HL7 FHIR is one of the salient features of the presented DHP. In short, the platform supports detailed clinical information models in OpenEHR format, while HL7 FHIR is used to maintain clinical and administrative data (FHIR resources), summarizing patient clinical information and supporting several digital services and workflows. The API layer of the DHP provides the 'horizontal' foundation for the 'vertical' applications.

The platform service-oriented architecture leverages a modern service infrastructure (middleware), comprised of: (1) service bus (integration and orchestration engine); (2) API management and gateways; and (3) identity management. This is implemented using opensource tools from WSO2 (https://wso2.com).

2.1.4 Normalized Electronic Health Records

Our DHP enables a system-wide approach to application and architecture design based on clinical data repositories (CDR). This means that any record that is interoperable by the platform is persisted in normalized format in one of our CDRs. DHP has CDRs supporting HL7 FHIR (HAPI JPA servers, https://hapifhir.io/hapi-fhir/docs/server_jpa) and OpenEHR (ERHBase servers, https://www.ehrbase.org/). The DHP's common infostructure serves as the foundation for a cohesive system.

2.1.5 Connectors and Adapters

Currently, these form the great majority of systems that are integrated through the DHP. In fact, most medical record systems (provider's facing EHR applications) do not have interoperability capabilities, and just a few have limited service interfaces (APIs), usually in proprietary data/message formats. This is a major challenge for the platform. To overcome such limitations, the DHP enables fast integration using connectors and adapters.

Connectors and adapters provide a flexible integration layer for the DHP. These components are based on specialized integration software that abstracts the communication with every medical/clinical information system that cannot interoperate natively. Connectors are software components that act as autonomous software agents that connect directly to data sources (persistence layer) from information systems and communicate remotely with the DHP APIs through a regular internetwork (usually, the Internet).

Connectors are designed to: (1) extract data from an external data source (database, file system and others); (2) transform data into a DHP standardized data format (FHIR or MHD+OpenEHR); (3) send data (patient data and clinical data) to DHP APIs (MPI and MHD); (4) control the data flow between the data source and the digital health platform; and (5) enable data integration audit (audit track).

Adapters, on the other hand, are implemented directly in the integration engine (service bus). The most common use case for adapters is the integration of systems that provides public integration/interoperability APIs.

Connectors and adapters have several reusable components that perform operations (3) to (5) listed above, along with a wide set of supporting controls and mechanisms for handling both legacy and real-time data, remote management and monitoring. However, the functions related to (1) data extraction (data abstraction layer), and (2) data transformation (data transformation layer) should be specifically designed for each use case.

2.1.6 Privacy and Personal Data Protection

Sharing health data in the context of a person-centric digital health ecosystem and for the purpose of public interest should be encouraged with the patient's consent, when undertaken in a manner that is built on trust, protects patient privacy, secures digital systems, and protects against malign or inappropriate use (WHO, 2020).

The DHP provides a specialized consent service (API), that implements patient consent management along with attribute-based access control. S-EHR access (API call) is mediated by the consent service. Established from the APPC foundation for Advanced Patient Privacy Consents, it enables a highly expressible solution for defining, asserting and auditing both individual policies (in which users manage if/how their data should be handled by the DHP with regards to information collection and sharing) and consents (in which users manage how other users may request and use shared data). Therefore, the patients' privacy preferences and consent directives are applied in real-time. Consent service also provides a detailed audit trail for every S-EHR access.

Consent functionalities are available in the patient-facing applications enabling the patient to: (1) opt-out; (2) share (consent access) records with other users and organizations; (3) visualize who has accessed their clinical information and with which purpose (routine care, emergency, or secondary usage); and (4) control information access at the record level, for instance, selecting to maintain the access to select records/information private only to users (practitioners) that are directly related to the care provisioning.

2.2 Architecture

The DHP (digital health platform) is permanently in evolution and several components are built and incorporated to the platform continuously. This section describes the core components of the platform. Figure 1 illustrates the DHP architecture.

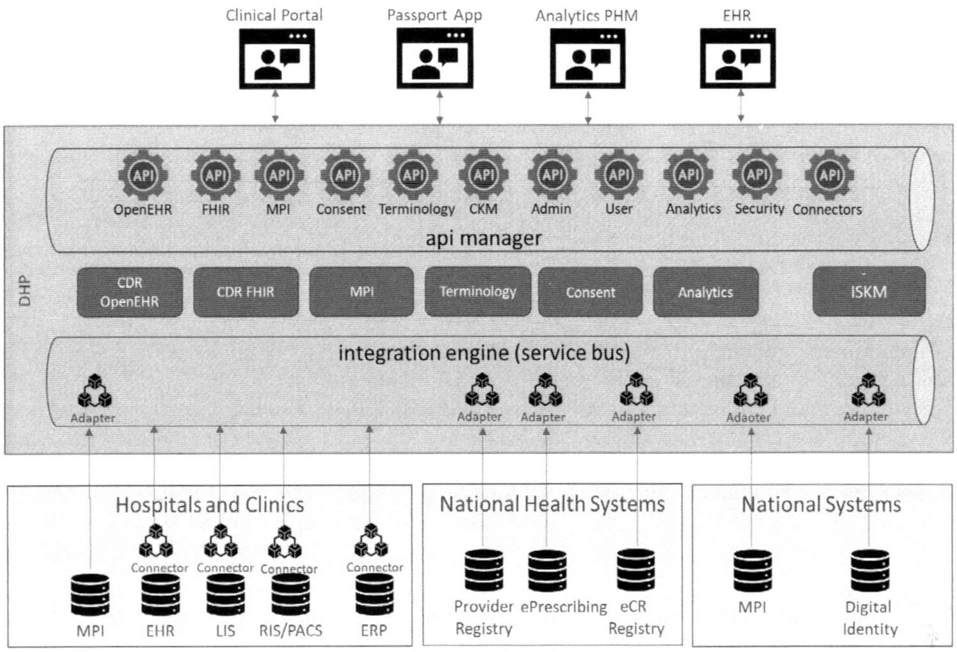

Figure 1. DHP architecture

2.2.1 Platform Components and APIs (back-end)

MPI: *Master Patient Index* (MPI) implements a patient/person registry. This component integrates patient information (demographics and identifiers) from different sources, provides functionality for data linkage and deduplication and is exposed through both an administrative application (admin portal) and APIs. MPI operates using FHIR R4 representation (Patient resource).

CDR: DHP persists and retrieve information from Clinical Data Repositories (CDR), based on a federated open standards-based architecture. Two CDRs are used: OpenEHR CDR and FHIR CDR.

Terminology Service: DHP has a standards-based terminology server capable of serving terminologies and value-sets (FHIR CodeSystem, ValueSet, ConceptMap and NamingSystem resources). These are used to serve terminology resources coming from standard terminologies and classifications (e.g., SNOMED-CT, LOINC, ICD-10, ICPC2, ICPM) and local value sets.

Consent: DHP provides a consent service capable of managing Advanced Patient Privacy Consents (APPC) and authorization.

Analytics: DHP implements a data lake infrastructure. Different analytics frameworks can be used to process and aggregate data in the data lake. According to legal regulation and local requirements, analytical data can be ingested not only from the S-EHR but also external data sets, which can be processed either in full detail or through anonymized events and indicators complying to the standardized OMOP Common Data Model (https://www.ohdsi.org/data-standardization/) format.

Service Infrastructure: DHP is established on a service infrastructure middleware comprised of: (1) integration engine/service bus; (2) API Management (APIM, includes API gateway); and (3) Identity Service and Key Management (ISKM).

2.2.2 Platform Applications (front-end)

Passport App (patient-facing): Passport App is intended to be used by patients. Users will be able to view their S-EHRs, to control privacy and consent preferences, and engage to the health care process by accessing different digital services. In some cases, the mobile app functionality is delivered through third-party applications that provides other services to users (super apps).

Clinical Portal (provider facing): Clinical Portal is an S-EHR Viewer web-based application. This aims at allowing for clinicians to log in securely, search for patients and view their S-EHR – whenever authorized by the consent policy and directives. There are several digital services available to providers, as well.

Analytics & Population Health Management (PHM): S-EHR data are unvaluable source for data analytics and, more specifically, population health management (PHM). Population health management (PHM) has increasingly been mentioned as a concept to realize improvements in population health and quality of care while reducing cost growth (the so-called Triple Aim) (Steenkamer et al., 2017). Besides PHM, S-EHR secondary use should support different decision support mechanisms related to monitoring, evaluation, programming, and controlling the health care process, both at clinical and administrative domains.

DHP provides data analytics and reporting capabilities through connecting data analytics frameworks to the DHP analytics APIs and data lake. DHP has several data sources that are accumulated in different repositories, including the Clinical Data Repository (CDR) and Master Patient Index (MPI). The data coming into these repositories are also available in a specialized data infrastructure called DHP data lake. Besides, there are data sets that can be delivered directly into the data lake.

DHP data lake is the primary source for data analytics. In practice, the data lake presents several data collections, which encompasses clinical, administrative, demographic, and statistical data collected from the different data sources. The data lake is established as a vendor neutral data infrastructure, allowing for different data analytics frameworks to couple into it. In this work, we demonstrate our Analytics & PHM using Apache Superset™ (https://superset.apache.org), an open-source data exploration and visualization platform.

3. RESULTS

The DHP described in this paper was developed in Brazil and has been deployed to several distinct heath care settings, since 2019, in Brazil and abroad. The results obtained by applying our DHP are reported in this study for the first time. We have selected two relevant use cases to report the results in this paper.

3.1 Use Case 1: Brazilian Federal University Hospitals Network (Ebserh)

The Brazilian Federal University Hospitals (UH) Network is controlled by Ebserh (https://www.gov.br/ebserh), a public company funded and managed by the Ministry of Education, in Brazil. The network of federal UH comprises 51 hospital in total, among which, 41 hospitals are managed by Ebserh. These are spread over 23 (out of 27) States in Brasil, as shown in Figure *2*.

Ebserh is responsible for a large share of hospitalization in Brazilian public health and provides referenced health care services to Brazilian Unique Healthcare System (SUS), since its creation in 2011. In 2023, Ebserh has performed 203,540 surgeries, 1,638 organ/cells/tissues transplantations, while providing medical residency for 3,169 MDs and multiprofessional residency for other 2,018 practitioners.

Deployment of the DHP in Ebserh started in March 2022, aiming at providing interoperability for 41 hospitals and establishing digital services to patients. The first results were obtained six months later, when the main EHR of each hospital was fully connected to the DHP, using a customized connector. One instance of the connector was deployed at each UH, which was able to collect episode-of-care summaries (in-patient/discharge summaries and out-patient/encounter summaries) and send these records into the DHP using HL7 FHIR R4 resources. By December 2022 (nine months project), the DHP was already integrated to several other systems, using both connectors and adapters, providing interoperability for discharge summaries, encounter summaries, lab test results, image study reports (with PACS viewer integration), appointments and teleconsultation. The program is called Digital University Hospital (DUH) – a free translation to *HU Digital*, the original name in Portuguese (see https://hudigital.ebserh.gov.br). The public APIs are available at APIs Dev Portal: https://hudigital.ebserh.gov.br/devportal.

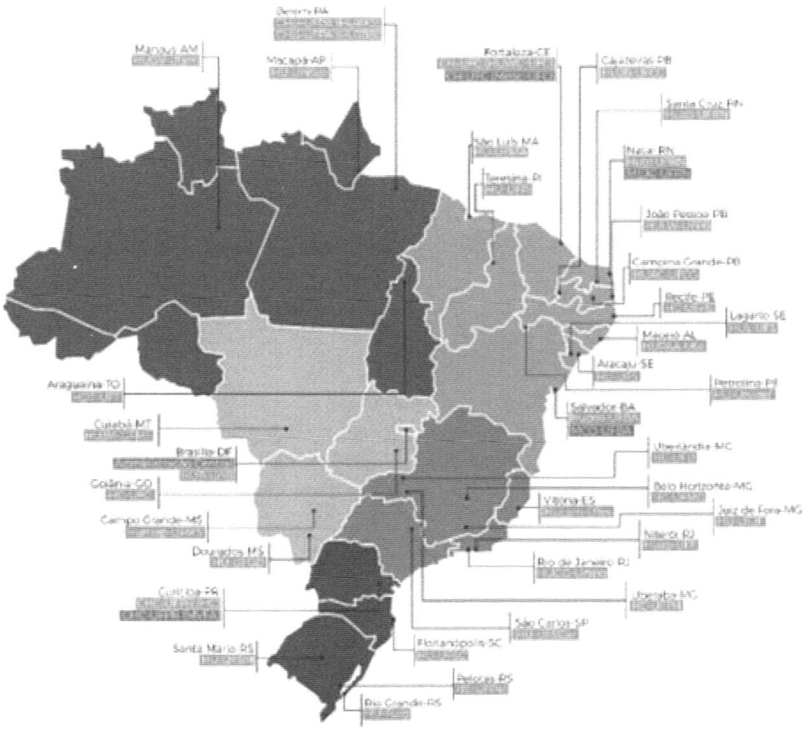

Figure 2. Ebserh UH network (41 hospitals)
Source: https://www.gov.br/ebserh/pt-br/hospitais-universitarios/sobre-os-hospitais-universitarios-federais (accessed 2024-04-23T21:32:03-03:00)

A major achievement in the DUH program was the federation of S-EHR services from Ebserh and the city of Recife (1.5M people). This started to operate in June 2023, providing complete access to primary and secondary care records from Recife to Ebserh patients and practioners. Communications between the Ebserh DHP and Recife's platform are performed solemnly by means of HL7 FHIR messages. Consent and privacy preferences were also federated, allowing for patients to grant and revoke consent, opt-out, and monitor S-EHR accesses from both platforms, interchangeably.

Currently, the results of adopting the DHP platform are summarized in Table 1.

Table 1. DHP at Ebserh UH network (41 hospitals in 34 cities) as of December, 2023

Source system	Type	S-EHR use case	Connection type	Records in DHP
RNDS	National S-EHR	PDQv3 providers registry	Adapter	Online transactions
GovBR	National SSO (citizens)	Digital identity (patients)	Adapter	Online transactions
Microsoft	ERP SSO (practitioners)	Digital identity (practitioners)	Adapter	Online transactions
AGHUX	EHR (multiple instances)	Health Summary (IPS-like)	Adapter &	5,000,000+
		Discharge summary	connectors	1,200,000+
		Encounter summary	(multiple	12,000,000+
		Lab/image results	instances)	19,500,000+
		e-Prescription		1,000,000+
		e-Appointment		500,000+
Vivace	RIS/PACS	Lab/image results	Adapter	Online transactions
Recife	DHP	All	Federation	Online transactions

Digital services for patients are provided by the patient portal and the HU Digital mobile app (available from usual mobile app stores). These services are: (1) privacy and consent; (2) S-EHR access and sharing online); (3) lab results; (4) e-Prescribing; (5) teleconsultation; (6) e-Appointment (teleconsultation only).

Figure 3 illustrates the Clinical Portal (providers-facing, Portuguese only) showing the Patient Health Summary (partial).

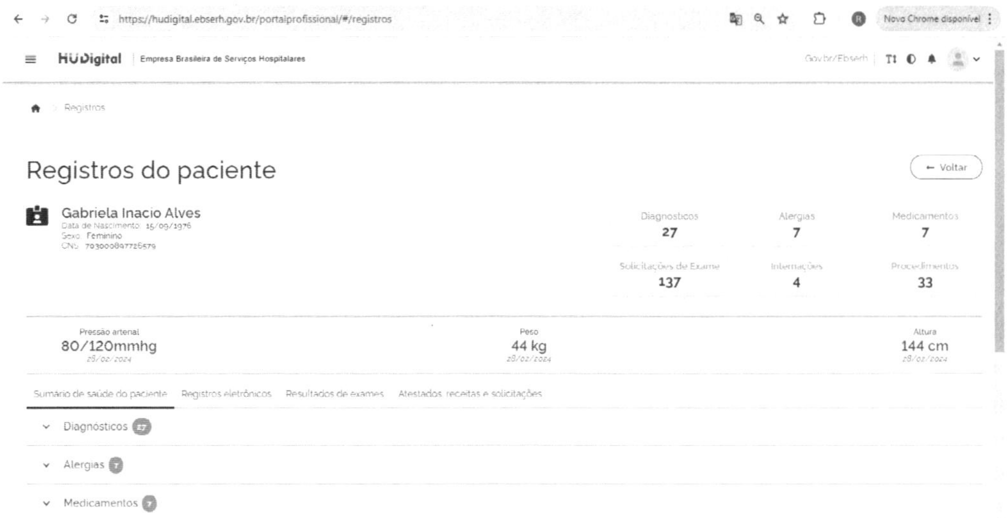

Figure 3. HU Digital Clinical Portal (test patient/synthetic data)

3.2 Use case 2: Jamaica Ministry of Health and Wellness

The public healthcare network in Jamaica is structured over a set of 25 hospitals and 319 health centers, which provide specialized care and primary care, respectively. The National Public Health Laboratory is the National Reference Laboratory, and together with some laboratory services located in several hospitals provide diagnostic procedures (lab tests). These are the primary source of health records in the Jamaica's public health system. The location of public hospitals is shown in Figure 4.

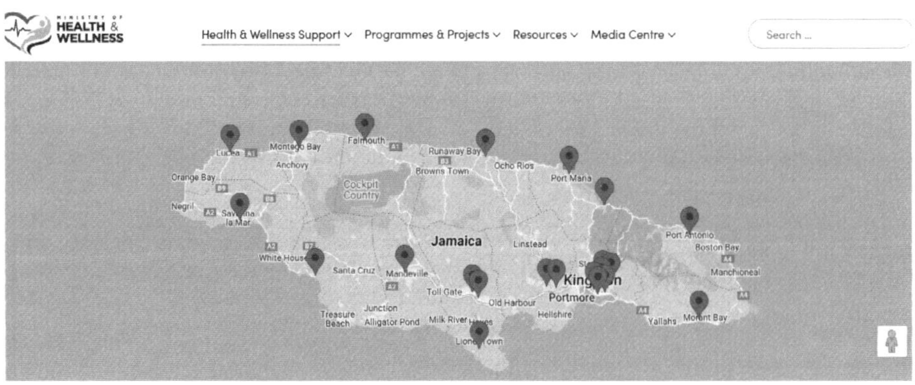

Figure 4. Jamaica public hospitals (25 hospitals)
Source: https://www.moh.gov.jm/location-of-hospitals-and-health-centres/#
(access 2024-04-21T18:07:53-03:00)

Jamaica Ministry of Health and Wellness (MOHW) launched the Shared Health Records Open Platform – SHROP (this is the local name for the DHP) – initiative in early 2022, as part of the digital transformation strategy for the health sector in Jamaica. SHROP aims at providing a technical infrastructure capable of normalizing, aggregating, and sharing health information that is produced by different actors and applications in the healthcare ecosystem. The SHROP goals are related to reducing the current fragmentation of information in the sector, and producing enhanced decision support, both for providers (clinical/practice) and managers. The strategy also encompasses the creation of digital health services, which will enable the engagement of Jamaica citizens in the health care process. The SHROP roll out as a pre-production environment has completed by August 2023.

The DHP is available through different applications (Clinical Portal, Patient Portal, Analytics Portal, and Patient Passport App) as well as through a set of standardized APIs, also publicly available:

- Clinical Portal and Patient Portal: https://myhealth.moh.gov.jm/portal
- APIs Dev Portal: https://myhealth.moh.gov.jm/devportal
- Analytics Portal: https://myhealth-analytics.moh.gov.jm

Although the platform deployment was fast, the first data set became available several months later, relating to MPI (patients registries from different hospitals) and COVID-19 Lab test results. These data sets were integrated into the DHP (SHROP) by using adapters. The adapters and connectors development time (time to integrate new data sets into DHP) was 4 to 6 weeks, for each developed component.

Currently, the results of adopting the DHP platform are summarized in Table 2.

Table 2. DHP at Jamaica MOHW as of December, 2023

Source system	Type	S-EHR use case	Connection type	Records in DHP
PAS/GNU Health	ERP (hospital)	MPI	Adapter	300,000+
OpenLDR	LIS	MPI, lab results (COVID-19)	Adapter	515,000+
CommCare	Immunization	Immunization	Connector	2,000,000+
HMSR	Hospital indicators	Hospital indicators	ETL	N/A

Digital services for patients are provided by the patient portal and the My Health mobile app (available from usual mobile app stores). These services are: (1) privacy and consent; (2) S-EHR access and sharing online); (3) lab results.

Figure 5 illustrates the Clinical Portal (providers-facing) showing the Patient Health Summary (partial).

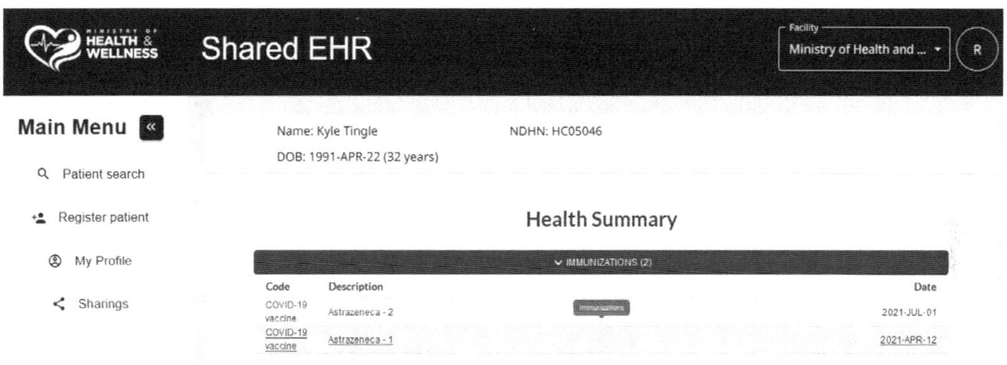

Figure 5. HU Digital Clinical Portal (test patient/synthetic data)

4. ANALYSIS

The DHP was deployed and customized for two different practical (real world) use cases, featuring large public health ecosystems. The implementation is usually fast, and first results are usually available in 3-6 months, depending on how soon personal identified data becomes available for the development team. Fast implementation is usually a result of (1) leveraging DHP pre-built components, and (2) use of best-practices for project management and execution.

This use of common components has proven to streamline the interoperability process and investments are more cost effective. Instead of investing and re-investing in the development and deployment of application components that can be provided more efficiently by a DHP, the digital health budgets can focus on innovation and ongoing sustainability.

There are common and specific challenges for each reported use case. The demonstrated scenarios shared the following challenges:

- Source systems are not interoperable. We have had to develop and deploy connectors or adapters for each connected system shown in Table 1 and Table 2.
- Data access and data governance in general are complex, as data is usually managed at different jurisdictions, each one having their own data protection rules and not always available or computable.
- Data assessment was performed for each data set (usually one data set from each system and each S-EHR use case). Data quality (deduplication, completeness, indexing, syntaxis and semantics) varies considerably from system to system.
- Poor data semantics: management of code systems and value sets in source systems is often poor, without proper standardization and governance of clinical terminologies.

At EBSERH (section 3.1), there were additional challenges related to exposure of S-EHR to patients, including concerns about the exposure of internal, subjective or non-validated data flowing out of the original systems, as these records contain extracts of narratives written by physicians, especially for anamnesis and evolution of patients. Therefore, EBSERH has decided to request physicians to elaborate and digitally sign event summaries (discharge summary and encounter summary) that are made available for patients as a way of delivering curated data. Thus, the S-EHR that is seen by patients is somewhat different from the S-EHR that is available for practitioners and providers.

At Jamaica MOHW (section 3.2), patient registries often do not contain government issued identifiers (e.g., TRN the local taxpayer identifier). Therefore, patient identification is based on poor demographics data, which does not allow for accurate and unique identification. Several duplications were documented. Demographic information is limited to some basic fields (last name, first name, date of birth and sex). Although there is a MOWH-defined data dictionary for demographics in the health sector, this is not used or mandated in the field. We have established our MPI for this use case as a national MPI in heath sector, as there are current regulations that prevented the DHP to have access to government citizen registries, such as TRN data base/APIs. We have introduced the Digital Health Number (My Health number), which is now used as a national identifier for patients in the public health ecosystem. These numbers are centrally generated and managed by the DHP.

Another aspect that is worth mentioning is the S-EHR use case development. Although there are common use cases defined internationally for health data interoperability, building the S-EHR requires customizations to local and regional conditions. In Brazil, there is a national initiative that is developing HL7 FHIR national profiles (br-core), also called RNDS (https://rnds-guia.saude.gov.br). However, the development is severely delayed (2-3 years) and we had to define many of the FHIR profiles that are in use at EBSERH currently. This is not ideal, as when other interoperability networks become available, adaptation will be required. In Jamaica MOHW, the initial definition of S-EHR use cases and respective infostructure were delivered by an independent group of OpenEHR experts. The defined models, however, needed significant simplification during implementation, as the data sources have poor data quality. Thus, mapping between the source data sets and the target S-EHR data models was generally not feasible.

Despite the differences in dimension, infrastructure and language, due to the adherence to interoperability standards for both clinical data modeling and messaging, along with the reusability and robustness provided by following a Service-oriented Architecture (SOA), the DHP was quickly implemented in both projects with minimal changes building up from the same foundation, which enabled the development of localized customizations and use cases in a sustainable fashion. Core components, such as the CDRs, MPI and consent were entirely unchanged, while new features were implemented by adding layers and modules to the platform.

Regardless of the challenges, both DHP were successfully delivered on time (only minor delays in internal phases of the project were observed). Currently, the deployments are scaling, in both cases.

The features of the DHP highlighted in Section 2 can be evaluated in both practical use cases as follows:

- Patient-centric: A patient portal and mobile app enabled by the DHP are part of the implementations in each reported use case. Digital services for patients are becoming available (and scaling), enabling patient engagement.
- Service-oriented: The DHP applications were fully customized (white label design) for each reported use case. Several capabilities are being used by different systems, starting with MPI (but not limited to it). In EBSERH, a custom S-EHR viewer was developed and deployed as an embedded feature of AGHUX (the hospitals' EHR). This was only possible because the S-EHR service is 100% available through an standardized FHIR API.

- Polyglot interoperability: The DHP were execute using both HL7 FHIR (EBSERH) and OpenEHR+FHIR (Jamaica MOHW). Proper functioning of workflows and service orchestration is demonstrated for both "languages".
- Normalized EHR: In both cases, the CDR components (FHIR and OpenEHR CDRs) were used to persist, index and retrieve S-EHR information. The health data in DHP only exists in a normalized format, regardless of the proprietary data structure (and semantics) from source data sets and systems.
- Connectors and adapters were key to allow fast implementation of interoperability and connectivity with source systems and data sets.
- Privacy and Personal Data Protection: In both cases the application of data protection acts were analyzed against the DHP capabilities, especially in regards to the consent service and its parametrization. In both cases, the DHP were successfully configured to comply with local regulations.

5. CONCLUSIONS

In this paper we have demonstrated the deployment of S-EHR services in heterogeneous healthcare ecosystems by usage of a DHP. The salient features of the DHP architecture are: (1) patient-centric design; (2) service-oriented; (3) polyglot interoperability; (4) normalization of electronic health records (clinical data repository); (5) use of connectors and adapters (integration layer); and (6) GDPR-aware privacy and data protection. These features were analyzed against the challenges to implement interoperability services.

The reported DHP is currently being used as a main interoperability infrastructure in EBSERH (national-wide hospital network) and Jamaica (national public healthcare system). Although the initial focus during the implementation was targeted to providing normalized data and insights from the entire ecosystem to patients and practitioners, further digital services for patients and analytics for PHM are scaling out in both use cases.

ACKNOWLEDGMENTS

The authors would like to acknowledge RNP (https://rnp.br) (grant #ADC/10747/2022) and EBSERH (https://ebserh.gov.br) for the long-lasting partnership in developing the HU Digital platform.
The authors would like to acknowledge Jamaica MOHW and the Inter-American Development Bank (IDB) for supporting of SHROP development (loan # JA-L1049)

REFERENCES

Brands MR, Gouw SC, Beestrum M, Cronin RM, Fijnvandraat K, Badawy SM, 2022. Patient-Centered Digital Health Records and Their Effects on Health Outcomes: Systematic Review. *J Med Internet Res. 2022 Dec 22*; 24(12):e43086.

Erl, T, 2016. Service-Oriented Architecture: Analysis and Design for Services and Microservices. *The Pearson Service Technology Series from Thomas Erl*. Pearson, Dez. 2016.

Schofield P, Shaw T, Pascoe M, 2019. Toward Comprehensive Patient-Centric Care by Integrating Digital Health Technology With Direct Clinical Contact in Australia. *J Med Internet Res 2019*; 21(6):e12382.

Steenkamer BM, Drewes HW, Heijink R, Baan CA, Struijs JN, 2017. Defining Population Health Management: A Scoping Review of the Literature. *Population Health Management Vol. 20, No. 1.* Research Article published Online: 1 February 2017.

World Health Organization and International Telecommunication Union – WHO/ITU,2020. Digital health platform handbook: building a digital information infrastructure (infostructure) for health. 2020.

TOWARDS PRECISE DETECTION OF PERSONAL INFORMATION LEAKS IN MOBILE HEALTH APPS

Alireza Ardalani, Joseph Antonucci and Iulian Neamtiu
New Jersey Institute of Technology, USA

ABSTRACT

Mobile apps are used in a variety of health settings, from apps that help providers, to apps designed for patients, to health and fitness apps designed for the general public. These apps ask the user for, and then collect and "leak" a wealth of Personal Information (PI). We analyze the PI that apps collect via their user interface, whether the app or third-party code is processing this information, and finally where the data is sent or stored. Prior work on leak detection in Android has focused on detecting leaks of (hardware) device-identifying information, or policy violations; however, no work has looked at processing and leaking of PI in the context of health apps. The first challenge we tackle is extracting the semantic information contained in app UIs to discern the extent, and nature, of personal information. The second challenge we tackle is disambiguating between first-party, legitimate leaks (e.g., the app storing data in its database) and third-party, problematic leaks, e.g., processing this information by, or sending it to, advertisers and analytics. We conducted a study on 1,243 Android apps: 623 medical apps and 621 Health&Fitness apps. We categorize PI into 16 types, grouped in 3 main categories: identity, medical, anthropometric. We found that the typical app has one first-party leak and five third-party leaks, though 221 apps had 20 or more leaks. Next, we show that third-party leaks (e.g., advertisers, analytics) are 5x more frequent than first-party leaks. Then, we show that 71% of leaks are to local storage (i.e., the phone, where data could be accessed by unauthorized apps) whereas 29% of leaks are to the network (e.g., Cloud). Finally, medical apps have 20% more PI leaks than Health&Fitness apps, due to collecting additional medical PI.

KEYWORDS

Personal Information Leaks, Medical Apps, Health&Fitness Apps, Android, Information Flow Analysis

1. INTRODUCTION

Mobile apps collect billions of users' data each day: Android alone has in excess of 3 billion monthly active users (Samat et al., 2022). Apps designed for medical, health, or fitness purposes, are particularly important, because the data they collect contains personally identifiable information, and medical/health data. However, one major downside of apps collecting (and users providing) this data lies in the possibility of its misuse. Data is routinely sent to advertisers; stored in the Cloud where it can be accessed by malicious actors/ PI leaks are consequential: leaked PI can be sold for profit, (mis)used to identify individuals, etc. (Van Alstin et al., 2024).

Little effort has been put into understanding and exposing leaks of data that users supply via the app's UI, especially in the context of health, fitness, or medical apps. Prior work on PI leaks have used dynamic analysis (McClurg et al., 2013) but only analyzed up to 100 apps. Some efforts used differential cryptanalysis (Continella et al., 2017) or network traffic analysis (Jia et al., 2019) (Ren et al., 2016) but only examined data that left the device, without tracking sensitive data that could be exposed in device storage. One line of prior work aimed to find leaks of hardware device identifiers, such as MAC addresses, serial numbers, device location, etc. (Arzt et al., 2014). However, we aim to find leaks of personal information, or personal identifiers such as name, weight, height, date of birth, or postal (Zip) code. Another line of work on detecting sensitive input in UI (Huang et al., 2015) has included identity and weight/height information as we do, but did not look for medical information, and did not track where the information is flowing, or who is responsible for the leak (first-party vs third-party). Work that has distinguished between first-party and third-party leaks (Rahaman et al., 2021) has only focused on hardware identifiers, and not analyzed the destination of leaks, e.g., network, logs, files, local DB storage, as we do.

2. APPROACH

Our approach consists of two analyses. First, we run a GUI analysis that extracts and categorizes the PI collected. Second, we run a taint (information flow) analysis to determine where the PI flows (leaks).

2.1 GUI Analysis

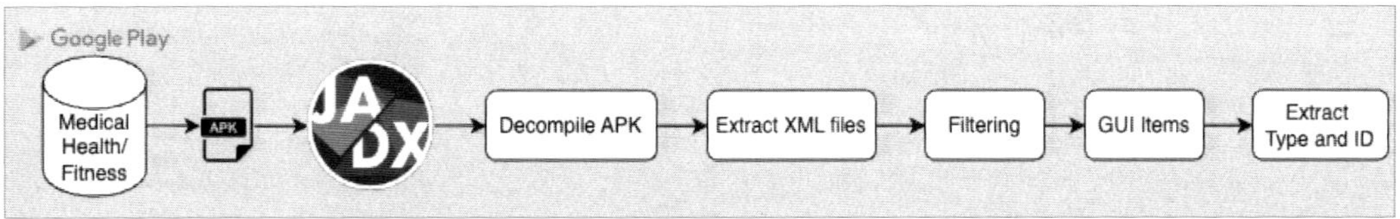

Figure 1. Overview of the GUI analysis process

Overview. Figure 1 shows our analysis process. Android apps are distributed as ".apk" files, so we name them APKs for short. An APK bundles app code with app resources (e.g., strings, icons, images). The dataset for this study consists of 1,243 APK files, downloaded from the Google Play Store's Medical, as well as Health&Fitness, categories. The APK files are then decompiled using JADX (JADX, 2024), yielding the source code and associated resources, including XML files.

Extracting GUI elements. Next, the extracted XML files, which contain the GUI layout definitions and other critical information about the application's GUI components, are processed. In Android, all GUI elements descend from class *View*. We filter out non-*View* elements, ensuring that only relevant GUI components, such as *EditText*, *Spinner*, *CheckBox* and *RadioButton*, are retained.

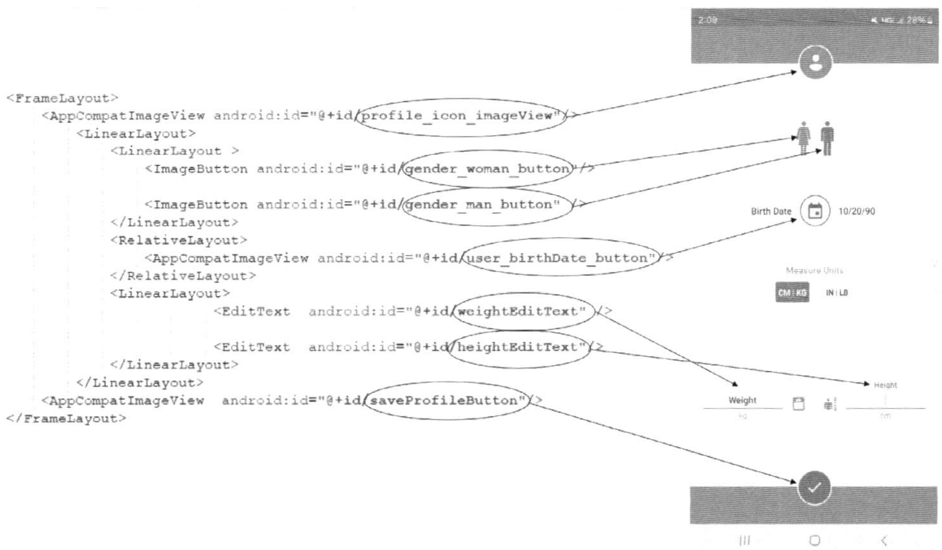

Figure 2. UI-XML mapping for app *Weight Loss Tracker*

In Figure 2 we show the mapping between XML code and UI for app *Weight Loss Tracker*. This app contains several *Views*: *EditText*, *Button*, etc. Note how in this case the *android:id* associated with UI elements facilitates analysis, because the *id* name, e.g., *id/weightEditText* or *id/user_birthdate_button*, directly encodes the PI semantics. However, there were two main additional challenges associated with UI extraction. First, there are UI elements where the *android:id* is non-suggestive, which required us to analyze other XML attributes, e.g., text, or hints associated with that *View*. Second, especially for medical information, we had to look for adjacent terms, as explained under "PI-based categorization" below.

Personal Information (PI) definition. Regulatory frameworks such as HIPAA in the US define "protected health information" to include health conditions, care provided, and information that can be used to identify the individual (e.g., name, phone number, social security number, birthdate, etc. (HIPAA, 2024)). The focus of this paper is on information that is collected from the user, via the GUI. Specifically, based on initial analysis of the most frequent information present in GUIs, we focus on 16 PI grouped into:

- *Identity*: email, name (first, last), address, zip code, credit card number, social security number.
- *Anthropometric/biopsychosocial*: age (birthdate), height, weight, gender.
- *Medical*: medical history, medication, blood-related, mental health, smoking or alcohol use.

PI-based categorization. Mapping textual information attached to UI elements onto a PI semantics was nontrivial. We used techniques from Information Retrieval along with an iterative refinement process to increase precision (gradually reduce False Positives and False Negatives). Whereas Identity PI are relatively straightforward to find, other PIs require searching for adjacent terms; in Table 1 we show some examples.

Table 1. Information retrieval: adjacent terms for PI extraction

PI	Adjacent terms
Medical history	Surgery, Allergy, ...
Medication	Prescription, Dosage, Dose, Drug, ...
Blood	Glucose, Cholesterol, Oxygen, Pressure, ...
Mental health	Stress, Panic, Anxiety, Depress, ...

2.2 Leak Analysis

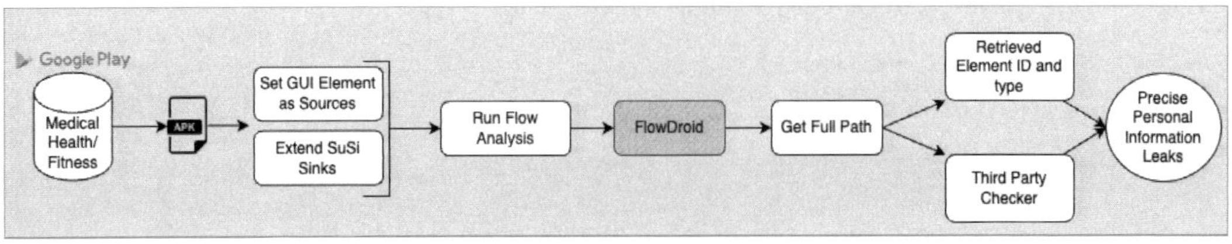

Figure 3. Overview of the leak (information flow) tracking process

Flow analysis. Figure 3 shows the pipeline we use, leveraging flow analysis, a standard program analysis that tracks flow of data from a *source* (origin) to a *sink* (destination). We define a *leak* as a path from a source to a sink; when multiple paths exist from one source to one sink, we only count the shortest path. Prior work on security has typically set trusted device identifiers as sources, and untrusted destinations (network, SMS) as sinks. However, in our approach, we set UI components as *sources*, while as *sinks*, we set any method that can potentially leak user information, e.g., methods that send data to the network, or write to local storage. To create the list of sink methods, we started with the SuSi list (widely-used in security research (SuSi, 2024)) as our baseline and expanded it by manually adding more method signatures, tripling the count of the baseline. As flow analyzer, we used FlowDroid (Arzt et al., 2014), which requires specifying Java methods as sinks and sources. However, in Android, UI elements are specified as XML objects, not as Java methods. Therefore, setting UI elements as sources required addressing this challenge, as follows: we used the JADX tool to map object identifiers in XML to their Java creation code in the Android-specific *R.java* file, and from there we set the *findViewById* method as a source (*findViewById* connects a UI element to its corresponding code section).

First-party vs. third-party flows. Another crucial aspect of precise leak attribution is distinguishing between first-party and third-party code. First-party code is identified by the package name. For example, in app *com.gotokeep.yoga.intl.apk*, code whose package is *com.gotokeep.** or *com.gotokeep.yoga.** is considered first-party, whereas code in packages *io.branch.** or *com.facebook.** is considered third-party. Of course, code in package *com.facebook.** would be considered first-party in the Facebook app itself.

3. RESULTS

App dataset. We ran our approach on 1,243 Android apps: 623 from Google Play's Medical category and 621 from Google Play's Health&Fitness category. We used two criteria for including apps in our analysis: (1) a minimum popularity threshold (≧ 500 installs) and (2) the apps had to be in English.

3.1 What is the Prevalence of PI Collection?

Figure 4 shows the prevalence of PI collection. We found that *Email* is by far the most collected information, with 44% of Health&Fitness apps ('Fitness' for short), and 39% of Medical apps collecting it, respectively. Next, we found that *Age, Name (first, last), Phone number, Address, Gender, Height*, are collected by about 15%–20% of the apps. As expected, the collection of specialized medical information (e.g., *current medications, medical history, use of smoking/alcohol*) is more prevalent in medical apps. However, we were surprised to see that fitness apps have a comparatively higher prevalence of collecting identity information such as *email, weight, age, gender, height*, compared to medical apps.

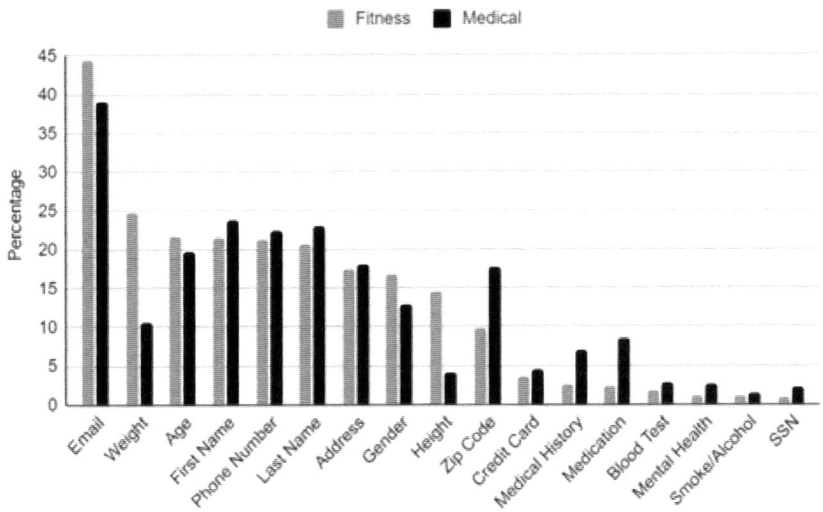

Figure 4. Prevalence of PI collected

3.2 First-Party vs. Third-Party Leaks

We begin with two motivating examples that illustrate the difference between first- and third-party leaks.

Motivating example: first-party leak. The mental health app *Panic Shield* (*com.panic.shield*) is designed to protect users from panic attacks. The list of user phobias, labeled fear8name, is among the PI it collects; the app collects this information via an *EditText* and stores this information in the local *SharedPreferences* database. The result of flow tracking, that starts at the source and ends at the sink, indicates the methods and VM registers involved in the flow:

```
                                                               SOURCE
$r1=virtualinvoker0.<com.p...CreateH'chy:android.view.View findViewById(int)> ⇒
$r23 = r0.<com.panic.shield.exposure.CreateHierarchy: java.lang.String l> ⇒
Interfaceinvoke$r25.<android.cont.SharedPs$Editor:android.cont.SharedP$Editor
putString(java.lang.String,java.lang.String)>("fear8name", $r23)        SINK
```

We consider this a *first-party leak* since all the code involved belongs to the app (or Android itself).

Motivating example: third-party leak. The *Keep Yoga* fitness app (*com.gotokeep.yoga.intl*) collects the user's gender (encoded as a boolean) and uses the *io.branch* third-party library, which will store the gender in the local *SharedPreferences* database. The result of flow tracking is:

```
                                                                      SOURCE
r11=virtualinvk$r6.<android.app.Activity:android.v.View findViewById(int)>($i0) ⇒
$r1 = <io.branch.ref.PrefHelper: io.branch.ref.PrefHelper prefHelper_> ⇒
putBoolean(java.lang.String,boolean)>($r1, $z0)                         SINK
```

Note how in the flow from collecting the user's gender (source) to storing it in the database (sink), there is a method from the *io.branch* third-party library. This library is primarily used for deep linking and user engagement (Table 3). The presence of third-party methods, such as those from *io.branch*, makes this *third-party flow*. Note how third-party flows increase the risk of data leakage: the data, once handled by third-party code, can be stored locally but also potentially shared with third-parties (Reardon et al., 2019). Therefore, this second leak scenario, involving the sharing of data with third-parties, is potentially more dangerous due to the additional risk of exposing personal information to external entities.

Table 2. Leak statistics

	First-party	Third-party
Median	1	5
Average	13.46	39.42
Max	320	2,856

Table 2 shows statistics: the typical app has 1 first-party leak and 5 third-party leaks (the average values are affected by apps that have a large number of leaks, up to 320 first-party, and 2,856 third-party, as can be seen in the last row). We believe that these results are concerning, especially from the standpoint of third-party leaks. *When the ratio of leaks induced by third-party libraries to an app's own code is 5:1, essentially the developer has long lost control over who collects user data, and how this data is processed or sent.*

Overall leaks. Figure 5 shows histograms for the overall number of leaks, as well as first-party vs third-party leaks. The histogram on the left shows that 266 apps have less than 20 leaks. However, please note that, to be included in the figure, an app had to have at least one leak. The histogram shows that dozens upon dozens of apps have a substantial number of leaks: 62 apps have 20–40 leaks, 63 apps have 40–60 leaks, and *52 apps have more than 100 leaks*. These numbers indicate that a high number of leaks is not an isolated incident; this provides an impetus to find and analyze apps with a high number of leaks.

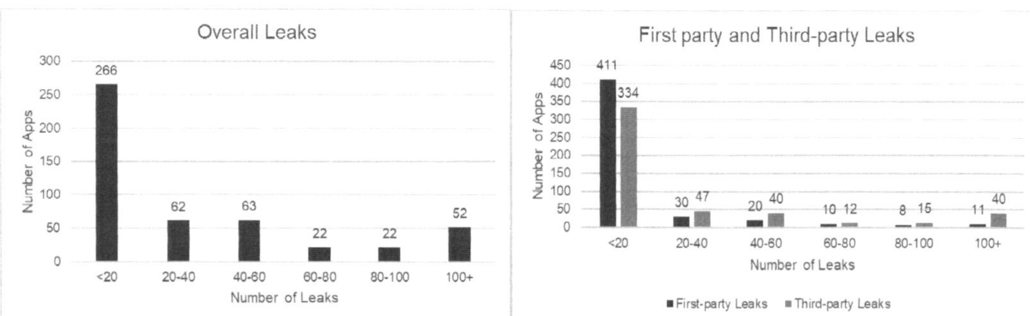

Figure 5. Leak distributions: overall (left) and first- vs -third party leaks (right)

Most prevalent third-party libraries. In Table 3 we show the nature and prevalence of top third-party libraries. For example *com.facebook* (which implements analytics and social media integration services) appears in 87 apps; *io.fabric*, which provides analytics, crash reporting, and user engagement services, appears in 38 apps. Note that an app may contain more than one of these third-party libraries, and there is a lot of fragmentation. While, for space reasons, we only present the top-9 libraries, there are dozens of other third-party libraries that we found in apps.

Table 3. Most frequent third-party libraries and their nature

	#Apps	Ads	Analytics	Payment Processing	Crash reporting	User engagement	Deep linking	Authenti-cation	Social Media Integration
com.facebook	87		•						•
com.squareup	58		•	•					
io.fabric	44		•		•	•			
io.branch	38					•	•		
net.hockeyapp	34				•	•			
com.crashlytics	28		•		•				
com.appboy	26		•			•			
com.mopub	24	•							
com.applovin	20	•	•						

3.3 Where does the PI Leak?

We now discuss each leak destination, quantified in Table 4. *Net* indicates that the information is leaked over the network; these leaks are the most concerning, because the moment the information has left the phone, the user has de facto lost control over where the information propagates. Please note that characterizing the network *domain* where information leaks to over the network – e.g., the app's own servers, or analytics/advertisers – is challenging and requires a dynamic analysis (Wei et al., 2012). *Log* indicates that PI information is leaked to system logs; this is problematic because system logs are visible to any app. *Local DB/Bundles/Shared Pref* indicates that the information is leaked to local persistent storage (*Bundle* is the Android serialization mechanism, and *SharedPreferences* store user preferences). *File/IO* indicates that the PI is saved in local files. Note that leaks to local storage (DB, files) have been proven to be used by malicious apps and malicious libraries to exfiltrate data (Reardon et al., 2019). The table reveals that, depending on the destination, third-party leaks are 2.8x–13.4x more numerous than first-party leaks. For space reasons we omit showing the distribution for each destination, however we found that apps that have first-party network leaks typically have few (<20) whereas 40 apps had $\geq$ 20 network leaks (they are all third-party leaks).

Table 4. Leak destinations

	#Leaks	% of overall leaks	First-party	Third-party
Net	7534	28.90	592	6942
Local DB/Bundle/SharedPreferences	7998	30.68	1490	6508
Log	8637	30.12	2272	6365
File, I/O	1903	7.30	132	1771

Table 5. Fine-grained PI leak information (#of apps exhibiting *one or more leaks* of that PI)

	Fitness					**Medical**				
	Net	DB	Log	FileI/O	*Total*	Net	DB	Log	FileI/O	*Total*
Email	23	67	47	7	*144*	56	153	124	13	*346*
First name	9	6	30	5	*50*	5	12	7	2	*26*
Last name	12	7	31	5	*55*	14	11	4	2	*31*
Phone	14	15	9	0	*38*	20	8	36	3	*67*
Address	55	94	28	0	*177*	163	94	49	1	*307*
Zip	0	0	0	0	*0*	3	0	0	0	*3*
Gender	7	18	2	0	*27*	16	17	7	0	*40*
SSN	0	0	0	0	*0*	0	0	0	0	*0*
CCard	0	0	0	0	*0*	0	0	2	0	*2*
Age	13	10	12	0	*35*	6	36	6	0	*48*
Weight	93	137	71	28	*329*	68	47	22	0	*137*
Height	39	45	31	17	*132*	18	13	9	0	*40*
Medical hist.	0	0	0	0	*0*	62	6	14	2	*84*
Medication	2	35	0	0	*37*	10	29	11	0	*50*
Blood	0	4	0	0	*4*	0	7	0	0	*7*
Mental health	0	0	0	0	*0*	4	37	4	0	*45*
Smoke/Alcohol	0	7	0	0	*7*	0	0	0	0	*0*
Total	*267*	*445*	*261*	*62*	*1035*	*447*	*470*	*295*	*30*	*1233*

3.4 Fine-Grained PI Leak Analysis

Table 5 shows fine-grained results: the number of apps in each category that exhibit at least one leak for that PI, as well as totals for each PI and each destination. Note that an app can have multiple leaks of the same PI to the same destination, e.g., the *Email* is saved into two different files, or sent onto the network via two different connections. *In this subsection only, when referring to "leaks" we count apps, not individual leaks; in other words, any app that has more than one PI→destination leak is only counted once.* Overall, Medical apps have about 20% more leaks than Fitness apps (1233 vs. 1035) in part due to medical apps collecting specific medical PI; nevertheless, some Fitness apps still collect and leak medical PI such as medication, blood, or smoking/alcohol use. *Email*, *Address*, *Weight*, and *Height* are the most frequently leaked PI. Interestingly, while 44% of Fitness apps collect the *Email*, only about 14% leak it (see Figure 1); whereas 39% of Medical apps collect the *Email*, and 34% leak it. Another interesting point is that Fitness apps have a much higher tendency to collect and leak *Weight* and *Height*. Finally, Medical apps have 67% more network leaks than Fitness apps do (447 vs. 267), making them higher risk (and inviting more scrutiny) than Fitness apps.

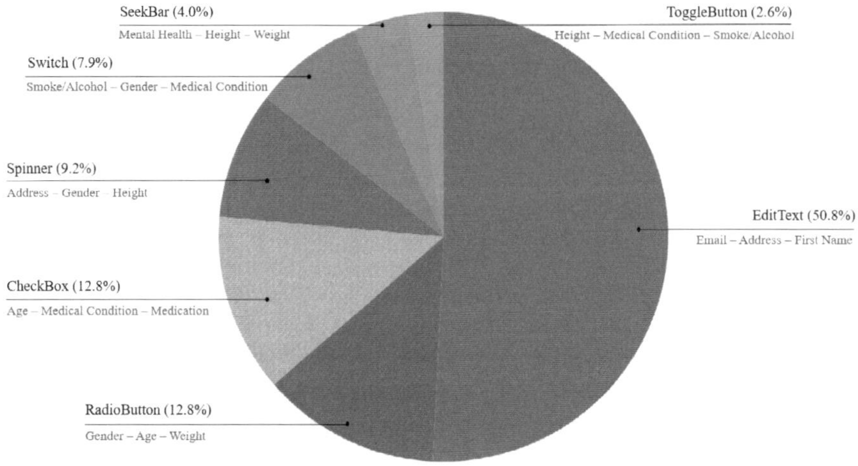

Figure 6. UI element distributions and the PI they collect

3.5 Characterizing the UI Elements that Collect PI

In total, our apps contain 47,749 Android *View* objects (i.e., GUI elements) that collect PI. Figure 6 shows the *View* types, along with the collected PI semantics. For each type, we list the top-3 most frequent collected PI in that *View*. The most prevalent UI object is *EditText* (50.8%) which allows arbitrary text input. Naturally, *EditText* is used to collect emails, addresses, first names, and so on. However, the flexibility of *EditText* can be a downside, when the collected information has a certain type (e.g., numeric) or range, e.g., 0–100; in such cases, developers need to add input validation, whereas other controls, e.g., *Spinner* or *SeekBar*, can directly enforce discipline on values or ranges. *RadioButton* and *CheckBox* are tied for the second most frequent PI collectors, typically used to select the gender, an age range, or a list of medications/medical conditions.

3.6 Discussion

We believe that (1) such apps should move toward collecting and leaking less, and (2) users, developers, app markets, and regulators can all play a role. First, users should question developers' over-collection, e.g., an app that simply computes the BMI should just ask for weight and height, and not ask for the user's medical history. Second, as developers typically use third-party code for monetization (ads), they should balance revenue with the leaks that third-party code induces; similarly, users can put pressure on developers to reduce ads and leaks. Third, app markets can be more transparent about the data apps collect, e.g., offer a detailed description of the data collected, and where this data is stored/sent. Fourth, regulators should be much more aggressive in enacting measures to make apps transparent about collection, and protective of user data.

4. CONCLUSIONS AND FUTURE WORK

Our study has revealed that medical, health, and fitness apps collect and leak a plethora of personal information. Future work directions include: a dynamic analysis to categorize the destination of PI data; reporting PI data collection and exfiltration for a given individual app to the end-users, e.g., as a browser extension or on-the-phone; informing users of the regulatory environments the app complies with (or conversely, violates), e.g., GDPR in the EU or CCPA in California.

ACKNOWLEDGMENT

We thank the anonymous reviewers for their feedback. This material is based upon work supported by the National Science Foundation under Grant No. CCF-2106710.

REFERENCES

Arzt, S. et al. (2014) "FlowDroid: Precise Context, Flow, Field, Object-Sensitive and Lifecycle-Aware Taint Analysis for Android Apps". PLDI '14: Proceedings of the 35th ACM SIGPLAN Conference on Programming Language Design and Implementation.

Continella, A., et al. (2017) "Obfuscation-Resilient Privacy Leak Detection for Mobile Apps Through Differential Analysis" Network and Distributed System Security (NDSS) Symposium 2017.

HIPAA (2024): US Department of Health & Human Services: Guidance Regarding Methods for Deidentification of Protected Health Information in Accordance with the Health Insurance Portability and Accountability Act (HIPAA) Privacy Rule https://www.hhs.gov/hipaa/for-professionals/privacy/special-topics/deidentification/index.html#protected.

Huang, J. et al. (2015) SUPOR: precise and scalable sensitive user input detection for android apps. In Proceedings of the 24th USENIX Conference on Security Symposium (SEC'15). USENIX Association, USA.

JADX (2024) Dex to Java decompiler https://github.com/skylot/jadx.

Jia, Q., et al. (2019). "Who Leaks My Privacy: Towards Automatic and Association Detection with GDPR Compliance." In: Biagioni, E., Zheng, Y., Cheng, S. (eds) Wireless Algorithms, Systems, and Applications. WASA 2019. Lecture Notes in Computer Science, vol 11604. Springer, Cham.

McClurg, J., et al. (2013) "Android Privacy Leak Detection via Dynamic Taint Analysis" Northwestern University https://jrmcclurg.com/papers/internet_security_final_report.pdf.

Rahaman, S., et al. (2021) "Algebraic-datatype taint tracking, with applications to understanding Android identifier leaks." 29th European Software Engineering Conference and Symposium on the Foundations of Software Engineering.

Reardon, J., et al. (2019) "50 Ways to Leak Your Data: An Exploration of Apps' Circumvention of the Android Permissions System" USENIX Security Symposium 2019.

Ren, J., et al. (2016) "ReCon: Revealing and Controlling PII Leaks in Mobile Network Traffic" MobiSys'16: The 14th Annual International Conference on Mobile Systems, Applications, and Services.

Samat, S (2022) Living in a multi-device world with Android. https://blog.google/products/android/io22-multideviceworld/ May 11, 2022.

SuSi (2024). https://github.com/secure-software-engineering/SuSi.

Van Alstin, C (2024) Massive data trove from Change Healthcare hack now for sale on dark web; April 17, 2024 https://healthexec.com/topics/health-it/cybersecurity/massive-data-trove-change-healthcare-hack-now-sale-dark-web.

Wei, X., et al. (2012) ProfileDroid: Multi-layer Profiling of Android Applications. The 18th Annual International Conference on Mobile Computing and Networking (MobiCom 2012), August 2012.

INTEROPERABILITY PLATFORMS: A COMPONENT BASED APPROACH TO DERIVE FUNCTIONAL CLASSIFICATION

Martin Staemmler
Health Informatics
Department of Electrical Engineering and Computer Science
University of Applied Sciences, Stralsund, Germany

ABSTRACT

Interoperability platforms (IOPs) have been and are continuously designed, deployed and used for a variety of scopes, from simple data integration, reducing heterogeneity between data sources, data management systems without and with archival function up to allowing for process support and supporting data analysis. Furthermore, the scope depends on the intended use in acute care or research or both of them as well as on the stakeholders involved from health care professionals, external cooperating organizations and up to the involvement of patients.

The objective of the paper is to derive a functional classification reflecting not only the IOP development over the last years but also anticipating upcoming designs. As such, four functional levels which build on top of each other have been identified: (i) management of unstructured data, (ii) management of structured data, (iii) process support, and (iv) data analysis. For each level, the respective system architecture is sketched, including functional components and references to relevant standards, profiles, and means for semantic annotation. Furthermore, the functional classification is applied to a set of projects and initiatives and allows assigning them to functional levels, thus enabling functional discrimination.

KEYWORDS

Interoperability, System Architecture, Unstructured and Structured Data, Images, Semantic Annotation

1. INTRODUCTION

An interoperability platform (IOP) can be defined as an institutional platform facilitating the integration and interaction of diverse and often incompatible IT-systems by providing standardized protocols, interfaces, and tools and by acting as a façade for IT-systems with limited, proprietary interfacing capabilities. These features would render an IOP into a so-called "Communication Server" or more formally "Enterprise Application Integration (EAI)", however there is a key difference. Whereas an EAI only allows for temporarily keeping data (e.g. in a table for managing a status) an IOP includes data storage capabilities which may also provide long-term archival functions. In addition, an IOP usually supports interactions for cross-institutional data sharing or health information exchange or with centralized platforms. These may reuse some functionality of an IOP but usually are dominated by a strong focus on data privacy and security issues as well as managing patient consent and access control.

Despite the initial definition of an IOP, its role and scope does not appear well-defined. As such, the objective of this paper is to establish a clear positioning of an IOP in the context of IT-systems typically present in a hospital environment and derive a functional classification for IOP planning, scope, design, deployment and usage. IOPs are on the rise for various reasons (Berger, R., 2020), in particular (i) getting in charge of patient data through IOP based consolidation (Torab-Miandoab A., 2023), (ii) as a kind of tool for prompt reaction to new requirements in reporting, cooperation or complying with new legislation (Dixon B.E., 2020), (iii) reduction of vendor lock-in (Malm-Nicolaisem, K., 2023), (iv) risk aversion for discontinued IT-system, (v) cost reduction with regard to interface licenses, (vi) versatile support for health information exchange and business processes and (vii) as a basis for federated networks and data access (Prokosch U. et al., 2018) and following the FAIR principles (Wilkenson MD., 2016).

2. MATERIAL AND METHODS

Figure 1 shows the typical building blocks of a hospital information system architecture, some of which might not be present everywhere depending on the type of hospital (basic and standard care up to university hospitals, which besides providing care, conduct studies and perform research).

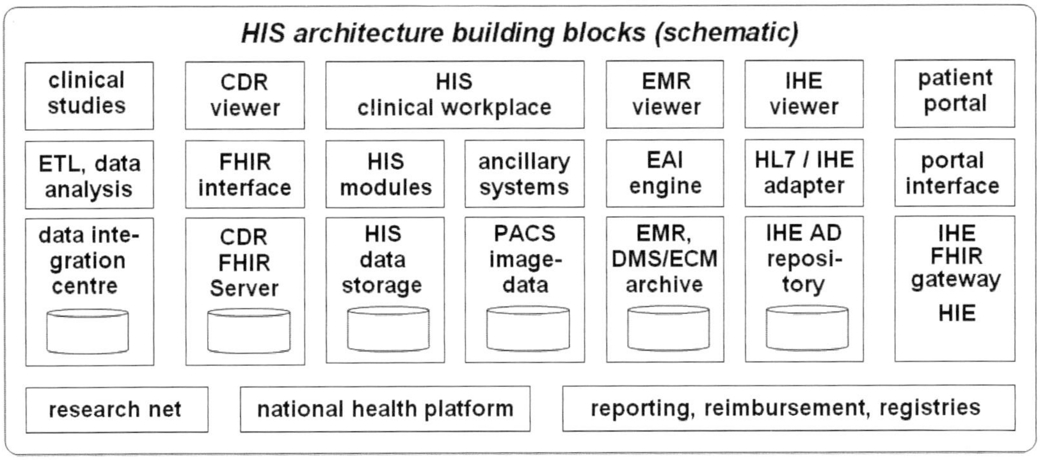

Figure 1. Schematic HIS architecture with typical building blocks

From an IT perspective, Figure 1 depicts a 3-tier architecture comprising the presentation / user interface layer at the top, the logical processing layer in the middle and the physical data management layer at the bottom. Furthermore, an interface has been added below to reflect the communication necessities for research, national health platforms and means for population health reporting, reimbursement and quality control using registries (e.g. for trauma, endoprosthesis, and dialysis).

The central columns refer to the hospital information system (HIS) with its modules reflecting its functionality and attached ancillary systems (laboratory, intensive care, radiology, …). For example, data storage is shown for the HIS and for the Picture Archiving and Communication System (PACS) for managing image data. Providing user access to the Electronic Medical Record (EMR) is reflected by the EMR viewer. The required integration of information into the EMR uses an EAI engine. The persistence of the information in the EMR is accomplished through the utilization of a Document Managing System (DMS), an Enterprise Content Management (ECM), and / or an archive. The columns embracing these central ones refer to managing highly structured data using a Clinical Data Repository (CDR) and Fast Healthcare Interoperability Resources (FHIR) as well as handling unstructured documents using profiles of the Integrating the Healthcare Enterprise (IHE) in a locally trusted environment, named as IHE Affinity Domain (AD). Ultimately, the left outer column pertains to research utilizing the Extract, Transform and Load (ETL) process and means for managing and sharing research data via the research net. It also allows for local and federated data analysis. The column on the right addresses the topic of population health by involving the patient through a portal and Health Information Exchange (HIE) with other organizations for informed decisions in patient treatment.

The schematic architecture is well suited to represent the role and scope of an IOP. Managing unstructured data, namely documents (e.g. pdf) and images (e.g. gif, tif, jpg) is easily illustrated by the box (dashed line) in Figure 2. The box contains not only the storage but also the interfacing and the viewers of such data. This box can be extended to include medical images available from the PACS as DICOM objects. Combining at the user interface level, the document viewer transforms into a so called multi-format viewer, allowing juxtaposition of DICOM images with the radiological report.

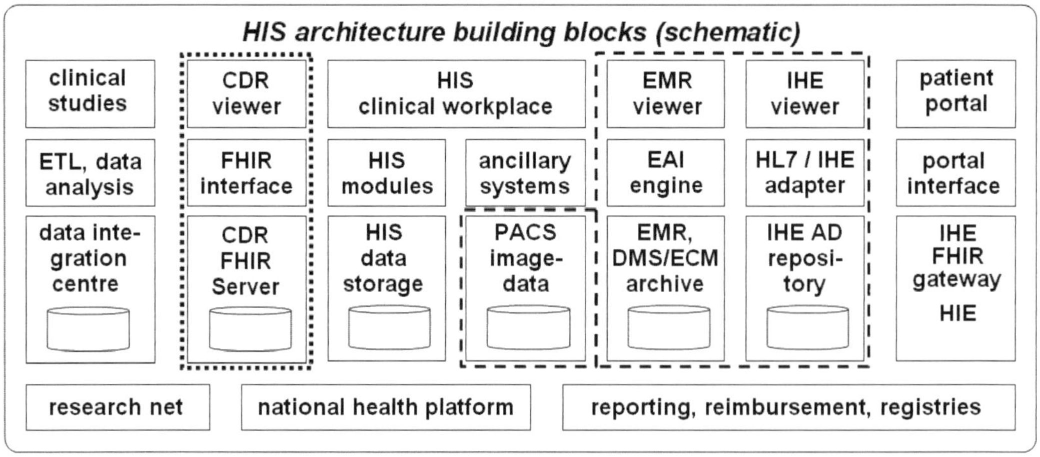

Figure 2. IOP scope covering unstructured (dashed line) and structured (dotted line) data

Figure 3 shows the interfacing between the EMR, DMS/ECM and PACS with the depicted IHE infrastructure. For documents, it relies on the IHE XDS actor DocumentSource with the transaction to provide and register documents. For DICOM images either typical DICOM functions like Web Access to DICOM Objects (WADO) using procedure calls, RESTful communication or the appropriate transactions of an IHE XDS actor ImagingDocumentSource can be used.

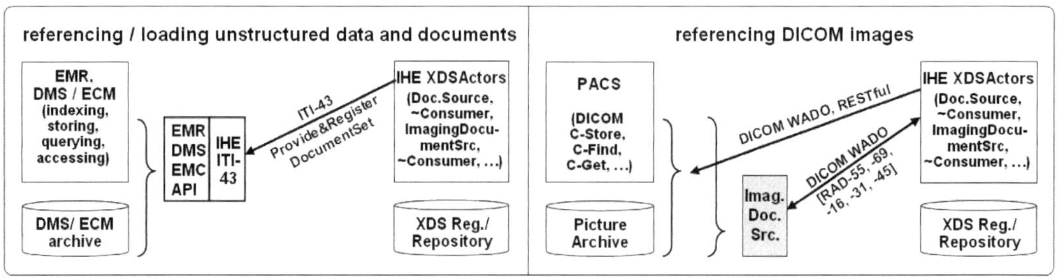

Figure 3. Referencing unstructured documents and DICOM images in an IHE infrastructure

The box (dotted line, left side) in Figure 2 refers to the management of structured information using the CDR over the three identified layers, including a dedicated viewer for e.g. representation of a timeline of measurements or observations.

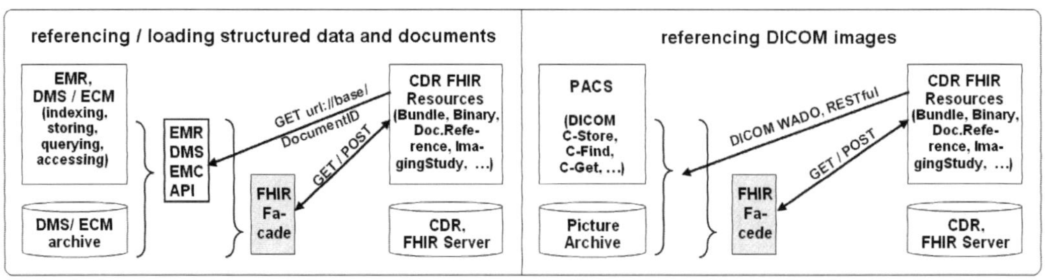

Figure 4. Referencing structured data and documents and DICOM images in a FHIR based CDR

As FHIR provides resources for referencing unstructured documents (DocumentReference) and for images (ImagingStudy) the interfacing shown in Figure 4 from the EMR, DMS/ECM can utilize either an Application Programmers Interface (API) or a FHIR Façade. DICOM objects in the PACS are referenced with standard DICOM functions or a FHIR Façade.

Although the schematics above provide an explanation of the IOP building blocks and their interfacing, it is imperative to address the desired IOP functionality, role and scope separately.

2.1 IOP Administrative Functions

Managing and referencing unstructured documents, structured data and DICOM objects has been demonstrated using IHE and a FHIR based CDR. Usually, each solution would have overlapping functions with regard to access control, protocol and audit and an adapter for standards based and proprietary interfaces (Figure 5, left side). To avoid double administration effort, an IOP needs to have these functions commonly implemented (Figure 5, right side). For implementation, it may be beneficial to take a "best of breed" approach by taking advantage of the individual approaches existing for an IHE infrastructure or a CDR as well as from the FHIR enabled IHE profiles, namely from the Mobile Health Document Sharing (MHDS) ecosystem.

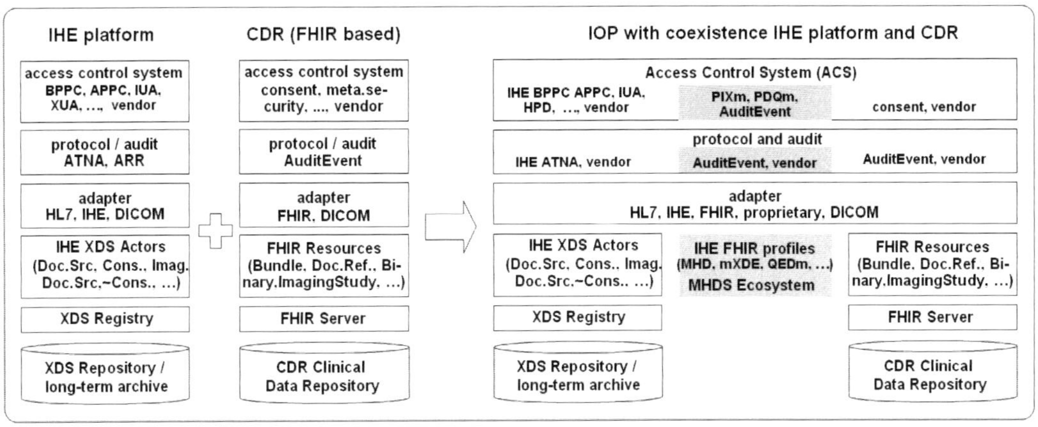

Figure 5. IOP with coexistence IHE platform and FHIR based CDR

2.2 IOP Process Support

An IOP provides a variety of functions for health data exchange including validation, transformation, terminology binding, mapping, and support of various representations of information as well a communication protocols. Including process support in an IOP allows

- using an interactive, graphical editor to design the process flow. This includes typical programming constructs (if … else, for … each, do … while) and taking advantage of the variety of functions
- exposing the defined business process towards attached IT-systems using the typical order entry metaphor and allowing invocation based on HL7 V2 order entry messages, FHIR ServiceRequest or IHE XDW profile.
- instantiating a business process for a specific patient from an attached IT-system using an order and to receive notification on successful execution or failure

Such process support would add a layer on top of the adapter layer as shown in Figure 5 (right side).

2.3 IOP Data Analysis Support

With the IOP being able to provide an overarching view of all data available from different legacy and ancillary systems for one patient or a selected group of patients an IOP may be used to support data analysis in various contexts. For example, (i) finding a comparable patient to the one currently treated, (ii) selecting a cohort of patients matching the inclusion criteria of a study, (iii) conducting Patient Related Outcome Measures (PROMs) for patients of department, with a specific condition or procedures or (iv) to support a hypothesis for research purposes.

In relation to Figure 1 or 2 this would extend the IOP into the left column when addressing research or utilizing the data analysis of that column to support the contexts (i) to (iii) from above.

3. RESULTS

Based on the considerations made in the previous section Table 1 shows the derived IOP functional classification with four levels.

Table 1. IOP Functional classification using four levels

no	IOP level	description	
1	management of unstructured data	user interface	multi format viewer for documents and images
		data objects	documents and DICOM images
		semantics	metadata for documents, DICOM attributes for images predefined values sets
		functions	query, retrieve, store data objects
		standards	IHE XDS profiles, DICOM
		IT systems	DMS/ECM, IHE repository for data storage, IHE XDS registry for metadata
2	management of structured data	user interface	- structured data viewer providing preconfigured views - multi format viewer for documents and images
		data objects	- structured information and documents, - references for unstructured data (documents / images)
		semantics	granular annotated content of structured data, using nomenclatures, classifications and terminology support
		functions	query, retrieve, store information
		standards	FHIR, FHIR enabled IHE profiles
		IT systems	CDR, FHIR façade as wrapper
3	process support	user interface	- interactive, graphical process designer - process invocation, order entry, monitoring process flow
		data objects	any of the data objects supported by the IOP
		semantics	metadata and annotated content to manage process flow
		functions	typical programming constructs
		standards	- process representation (BPMN, UML, ...) - HL7 V2 ORx, FHIR ServiceRequest, IHE XDW
		IT systems	HIS for order entry, IOP functional components
4	data analysis	user interface	- interactive dashboard and display designer - visualization
		data objects	any of the data objects supported by the IOP
		semantics	use of metadata and annotated content
		functions	local / federated query, data analysis (statistical - AI based)
		standards	data and metadata standards for documents, images, resources, biomedical / genetic information
		IT systems	ETL support, data integration centre

It is worth noting, that the levels build on each other, partly because the respective lower level functionality serves as prerequisite for the higher level functionality. For example, allowing for referencing unstructured data objects in level 2 requires their management in level 1. Similarly, data analysis at level 4 is contingent upon the availability of structured and accessible information from level 2.

4. DISCUSSION

Although the positioning of the levels 1, 2 and 4 in Table 1 is obvious, the positioning of process support at level 3 might be questionable, but it is fully justified due to its needs regarding available data objects and meta data from level 1 and information from level 2. The potential exchange with level 4 does not make sense based on the functionality ranking, as process support is more prevalent than data analysis in order to facilitate and establish innovative clinical business processes.

There have been other approaches for assessing IOP and in particular health data platforms (HDP) as some kind of overarching IOP, but also serving a different purpose, mainly health information exchange between multiple independent organizations in a regional, national or multinational setting.

The rise of health care platforms [Berger R., 2020] identifies "pure data platforms" for data management, encompassing patient involvement, platforms with integration at a "horizontal level" for a specific medical condition or at a "vertical level" to reflect a patient journey. In addition, "meta platforms" are used to connect between different health service providers to obtain large amounts of data for AI based analysis or national eHealth systems. This approach covers the identified levels 1, 2 and 4, but does not sufficiently take into account level 3. Specifically, a level 2 IOP may effectively facilitate support "horizontal integration" for a medical indication, which was a crucial requirement during the COVID-19 pandemic [Sass J., 2020].

The assessment of patient-centric health record sharing platforms [Azarm M., 2023] takes a governance oriented viewpoint to control and steer the design of platforms. It discriminates between governmental or commercial governance for providing the health information exchange framework, covering regulatory, data privacy and ontology based semantic annotation. Furthermore, a distinction is made based on whether health care organizations employ a common design or act on their own, nevertheless adhering to the established, overarching framework. This assessment resides somewhat superior to the one presented in this paper and may be utilized to enhance the potential individuality in IOP design thereby potentially reducing design effort and costs.

Table 2 shows the functional classification applied to different projects and initiatives targeted to establish an IOP or a HDP.

Table 2. Functional classification applied to projects and initiatives

projects and initiatives	type IOP, HDP	number of hospitals	extent	status	functional classification
Rhineland-Palatinate	IOP, HDP	5	federal	implementating	level 1, 2, 3
Health Harbor Hamburg H³	IOP, HDP	ca. 20	federal	planning	level 1, 2
Hospital Groups [Matzerath I.,Bosk J.,2024]	IOP, HDP	up to 80	national	operational	level 1, 2
Mein-Krankenhaus.Bayern	federated IOP	ca. 100	federal	implementating	level 1, 2, 3
Medical Informatics Initiative Germany (MII) [Gulden et al., 2021]	data integration center (DIC) research IOP	ca. 35	national	implementating and operational	level 1, 2, 4
German National Health Telematics Infrastructure (TI) [Rau et al., 2024]	HDP	ca. 1500 + 100000 practices	national	implemented, HDP operational by 01/2025	level 1, 2

In some instances an IOP is utilized locally to enhance the integration with the regional, federal or national HDP. In the project Mein-Krankenhaus.Bayern, over 100 participating hospitals implement a federated IOP, which interacts directly with the institutional IT-systems. Because of this, it relies using their decentralized repositories instead of an overarching central repository, which would result in central patient data retention and be in conflict with European and national Data Protection Regulations. The Medical Informatics Initiative (MII) includes all German university hospitals and focuses on the research perspective by establishing so-called Data Integration Centers (DICs) which hold fully structured data based on multiple, defined basic data sets represented by profiled FHIR resources. DICs allow distributed data analysis, so they are the only ones that meet level 4. The German national Health Telematics Infrastructure (TI) implements a nationwide, centralized HDP under the governance of the patient, thereby avoiding data retention issues. The connection to that HDP is primarily based on IHE. However, it does not include the flexibility of adapters typically available in an IOP. Also, both the MII and the TI do not support DICOM encoded medical images. These examples not only demonstrate the different approaches, but also the application of the effective functional classification consisting of four levels developed in this paper.

5. CONCLUSION

The paper describes the scope, role and function of an IOP and provides details for interfacing the HIS and ancillary systems with the IOP. The functionality classification consists of four levels, which build on each other. Well-founded decisions on the scope and role of an IOP within an organization and also to plan for future extensions are served by these levels. Decisions are based on the IOP functionality, but each organization is advised to identify their particular needs and map them onto the proposed functionalities. The list of needs might encompass one or more of the subsequent objectives and it is not conclusive: (i) consolidation of a heterogeneous IT-landscape, (ii) integrated view on patient data in the sense "single point of truth", (iii) local backup in case of a temporarily non-functional HIS, (iv) archival function, (v) versatile tool for designing and implementing innovative business processes meeting dedicated needs, (vi) reduction of vendor lock-in due to having the IOP based on international standards and profiles, (vii) facilitating mapping and assuring terminology binding, (viii) partial takeover of functions from existing IT-systems e.g. to mitigate risks of discontinuation, to lower license fees, (ix) IOP as platform for interfacing to external health data platforms and reporting demands, (x) allowing for extensive data analysis locally and in federated setting and (xi) supporting clinical studies and research.

ACKNOWLEDGEMENT

The author is grateful to several hospital trusts and many individual hospitals provided insight into their IT system architecture, which allowed for planning, designing, preparing calls for tender, and supporting the subsequent implementation of IOPs.

REFERENCES

Azarm M. et al., 2023, Towards a universal patient-centric health record sharing platform, *In Health Policy and Technology*, 12 100819

Berger R., 2020, *Future of health 2 | The rise of healthcare platforms*, available from https://e-health-com.de/fileadmin /user_upload/dateien/Downloads/Roland_Berger_Future_of_health_2_-_The_rise_of_healthcare _platforms.pdf (last access 21.5.2024)

Dixon, B. et al., 2020, Interoperability and Health Information Exchange for Public Health, *in Public Health Informatics and Information Systems*, pp 307-324, Springer, New York

Gulden C. et al., 2021, Prototypical Clinical Trial Registry Based on Fast Healthcare Interoperability Resources (FHIR): Design and Implementation Study, *in JMIR Med Inform 2021,* vol. 9 | iss. 1 | e20470 | pp 1-13

Matzerath I., Bosk J., 2024, *Use of a stretegic Health Data Plattform within the AMEOS Group*, Health Data Management, Springer 2024, pp 405-416, doi 10.1007/978-3-658-43236-2_21

Malm-Nicolaisen K. et al.,2023, Exploring the Emergence of Open Platforms in Healthcare: Design Considerations and Experiences from an Initial Case in Norwegian Primary Care, *Proceedings of the 56th Hawaii International Conference on System Sciences, pp 2799 – 2808*

Prokosch U., 2018, MIRACUM: Medical Informatics in Research and Care in University Medicine: MIRACUM: Medical Informatics in Research and Care in University Medicine, *Methods Inf Med 2018*; 57(Open 1): e82– e91

Torab-Miandoab A. et al., 2023, Interoperability of heterogeneous health information systems: a systematic literature review. *BMC Med Inform Decis Mak. 2023 Jan 24*; 23(1):18

Rau E. et al., Implementation of the electronic health record in the German healthcare system: an assessment of the current status and future development perspectives considering the potentials of health data utilisation by representatives of different stakeholder groups, *Front Health Serv. 2024 May 10*, doi 10.3389/frhs.2024.1370759

Sass J. et al., 2020, The German Corona Consensus Dataset (GECCO): a standardized dataset for COVID-19 research in university medicine and beyond, *BMC Med Inform Decis Mak 20,* 341 (2020)

Wilkinson MD. et al., 2016, The FAIR Guiding Principles for scientific data management and stewardship, *Scientific Data 3:160018*

VERIFICATION OF BODY SURFACE AREA-CALCULATING ANDROID APPS

Ian Almutawa and Iulian Neamtiu
New Jersey Institute of Technology, USA

ABSTRACT

Smartphone applications ("apps") are seeing rapid adoption in healthcare because they offer numerous benefits, e.g., tailoring patients' treatments to their specific body measurements. Despite the importance of accurate calculation in this domain, currently such apps do not undergo formal correctness verification, opening the door to incorrect treatment or dosages, and ultimately negative treatment outcomes. This paper takes a step towards correctness verification, focusing on Body Surface Area (BSA) calculation; BSA is used in scenarios where accuracy is crucial, e.g., to determine drug dosages, chemotherapy dosages, or fluid replacement for burn patients. First, we show that the two most common formulas for BSA, Mosteller and Du Bois, can produce divergent BSA values for the same individual, differing by as much as 14%. Next, we verify BSA-calculating apps from Google Play; we identified 23 such apps, and performed a systematic testing of their user interaction and calculation. Our study has revealed several kinds of errors: incorrect calculations, crashes, or failure to specify a reference formula.

KEYWORDS

Body Surface Area, Medical Apps, Android, Digital Health, Mobile Computing

1. INTRODUCTION

Mobile apps have the potential to advance healthcare, e.g., via precision/personalized medicine (Yau, 2019). The appeal of using smartphone medical apps stems from portability (a provider has access to vast reference and computational resources that fit into their pockets), as well as UI simplicity: encapsulating complex formulas down to a simple set of parameters. However, this convenience has had negative consequences. The lack of scrutiny and formal verification in medical apps can lead to severe errors that ultimately affect outcomes, from incorrect insulin dosage calculations (Huckvale et al., 2015) to variation in opioid conversion (Haffey et al., 2013), or incorrect emergency score calculations (Rahaman et al., 2023). In general, apps tend to overstate their functionality, claiming to diagnose or address conditions for which capabilities way beyond current smartphones would be needed (Samuel et al., 2022).

What is BSA and why is it important? The Body Surface Area (BSA), an estimate of a patient's body surface area, is a critical parameter used by providers and pharmacists to make treatment decisions or compute dosage in chemotherapy, transplantology, and toxicology (Redlarski et al., 2016). The BSA is also used in emergency medicine for burn treatment, using the *Rule of Nines* method, which assigns a percentage to different body areas to estimate the burn surface area; for example, the anterior or posterior surface of an arm is 4.5% of the BSA, whereas the groin represents 1% of the BSA (Moore, 2022); hence accuracy within 1% is crucial. Redlarski et al. have stressed the importance of BSA computation accuracy, as BSA-based dosing is used to address conditions that can be fatal if untreated or mistreated. The BSA formulas use the patient's *weight* and *height* as input, and produce a scalar value, typically between 0.5 and 3 square meters, as output. While BSA formulas only provide a body surface estimate (not an exact measurement) BSA-based dosing is considered safer than fixed- or weight-based dosage (Faisal et al., 2016). Therefore, improper computation or use of BSA formulas can be life-threatening. When apps are used to compute the BSA, accurate implementation is critically important. Therefore, the focus of this paper is to verify the reliability of BSA calculation in Android apps. In *Section 3* we discuss specific formulas and their prevalence in our examined Android apps.

We scraped Google Play to first identify a set of BSA-calculating apps, then looked for the specific BSA formulas each app implements. Surprisingly, we found four apps that do not reference the formula they use, which can lead to confusion and misinterpretation. Next, we conducted a series of numerical analyses to determine input parameters that would maximize the likelihood of revealing errors. We used Android automation to input those parameter values to Android apps automatically and therefore test the correctness of BSA formula implementations. We found errors in another four apps, where the errors in calculation ranged from 1% to an astonishing 8.6% from what a normal BSA value should be.

Our work differs from prior efforts in this area in several ways. Before the rise of smartphones, medical staff employed BSA calculating devices; Vu's study found inconsistencies between devices that implemented the same BSA formula (Vu, 2002). Due to the increasing adoption of smartphone-based healthcare, there is a stringent need to verify apps that implement BSA calculations. Verification of medical scoring calculator apps (Rahaman et al., 2023) has a different scope: discrete score calculations used in emergency settings, not continuous calculations as is needed in BSA. Testing of insulin dosing calculator apps (Huckvale et al., 2015), although related to dosing, focuses on a different formula, does not use a rigorous space exploration approach as we do, and is less general. Testing of opioid conversion apps (Haffey et al., 2013) was specific to one type of treatment that is general and not personalized for patient attributes.

Responsible Disclosure. We have disclosed our findings and error details to the app developers. As of Apr. 26, 2024, one developer has responded to our findings.

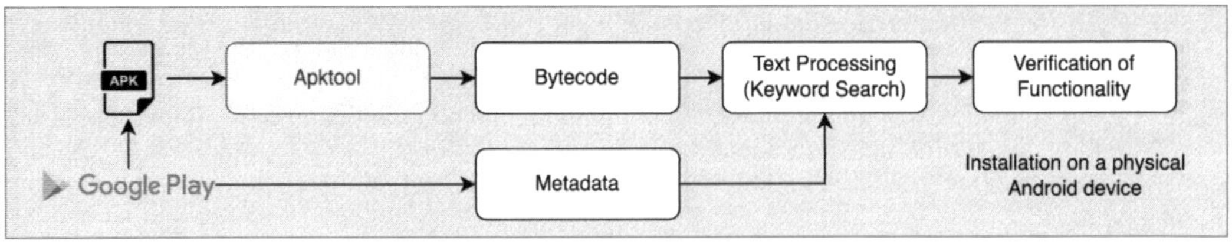

Figure 1. Overview of the selection process

2. APPROACH

Our study had two thrusts: (1) finding Google Play apps that offer BSA-calculating functionality and extract the formula(s) they implement; (2) verifying the correctness of the BSA calculation on each selected app.

The app discovery process is illustrated in *Figure 1*. To compile our list of candidate apps, we searched Google Play's *Medical*, and *Health & Fitness* app categories using keyword terms such as '`dose`', '`dosage`', '`BSA`', etc. We also used the *Similar Apps* feature in the Google Play app. Next, we examined apps' metadata: app name, package IDs, and description. To further extend the number of apps we conducted an in-depth keyword search on apps' bytecode, yielding 4,536 apps. This initial set of apps was overly broad, as many apps were fitness trackers, BMI calculators, etc. Using Apktool (Tumbleson, 2021), we decompiled the apps and found 1,120 keyword matches, though a large number of the matches were false positives. Finally, we narrowed down the list to 77 candidate apps that supported various dosage calculations. The 77 "finalist" apps were installed on a physical Android phone, where we thoroughly investigated the dosage calculation functionality: ultimately, 23 apps implemented BSA calculators.

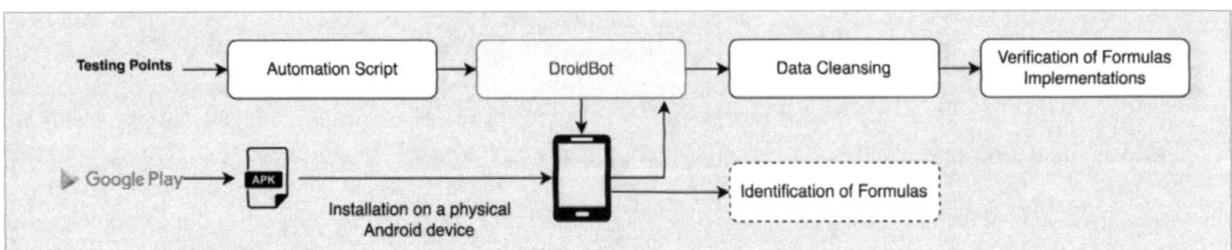

Figure 2. Overview of the verification process

Next, we describe the verification step, illustrated in *Figure 2*. This step involved identification of the formulas, and feeding a set of testing inputs to a physical Android device, then extracting the calculation result.

The formula identification was done manually by humans, navigating the app's UI and locating the reference to the formula. For those apps that did not reference any formula, we reverse-engineered the formula by comparing their calculation results to ground truth. All the apps were tested using automation scripts in full or in part, except for two apps that were verified manually (due to their implementation using Android's WebView, which rendered them unfit for automation). For automation we leveraged the Droidbot Android automation tool (DroidBot, no date) along with shell scripts we wrote. JQ (Jq, no data) was also used to manipulate DroidBot's JSON scripts to perform bulk input testing (i.e., all 25 inputs sequentially) in a single app run, hence performing calculations for 25 pairs of our testing points. Due to technical challenges, some apps had to be restarted in between tests, so to test 25 points, we used a single pair per app run and restarted the process 25 times. This impacted the efficiency of performing some tests. The variety and lack of inconsistency between the user interfaces added the need for result values to be cleaned and unified. Other complications included the use of advertising, which makes UI interaction nondeterministic, as well as inconsistency in the units of measurement which had to be converted to SI (International System of Units).

3. EVALUATION

The 23 Android apps that formed the scope of our evaluation are shown in *Table 1* and discussed shortly. Apps fall roughly into three categories:

1. standalone apps, designed solely for BSA calculation
2. apps that offer a variety of medical calculators besides BSA
3. apps that included a list of medications and computed a corresponding BSA-based dosage, while along the way calculating the BSA value.

The computation of BSA is a century-old problem. Du Bois's BSA formula (Du Bois, 1916) was one of the first BSA formulas proposed, and is still popular today, followed in popularity by Mosteller's formula (Mosteller, 1987); the reasons for their popularity are their simplicity and their computation requires only a basic calculator (Vu, 2002). We focused on verifying the implementation of Du Bois and Mosteller as we discovered their popularity still prevails today among Android apps. We also examined the Haycock formula, more suitable for children (Haycock et al., 1978), whereas Mosteller and Du Bois are most suitable for adults. Although the selection of the appropriate formula for a treatment is in itself a process that requires careful consideration, Redlarski et al. have examined 25 formulas and found that there is no formula general enough for all treatments and for all individuals (Redlarski et al., 2016).

The Mosteller formula was implemented in 16 apps, the Du Bois formula was implemented in 15 apps, with 10 apps supporting both. The Haycock formula was only implemented in 3 apps, which is alarming since only one app directly indicated an age segment (adults and children) and referenced the right formula. Another app, for example, had three age segments but did not refer to any formulas. Many of the apps required extra input (besides the BSA-required height and weight), e.g., gender, age, or other parameters that did not affect the calculation in any way, in what we see as a deceitful attempt to provide the user reassurance that the BSA value is carefully tailored. While 4 apps did not reference which BSA formula is used in their implementations (see the 'NO' entries in Table 1), we were able to determine the formulas after performing the calculation against our testing points.

We now list the formulas, which use *W*eight in kg and *H*eight in cm. The Du Bois formula is:

$$BSA = 0.007184 \times W^{0.425} \times H^{0.725}$$

The Mosteller formula is:

$$BSA = \sqrt{\frac{W \times H}{60}} = 0.016667 \times W^{0.5} \times H^{0.5}$$

Besides these formulas, we found that apps implement additional formulas: Boyd (4), Gehan and George (4), Haycock (3), Fujimoto (2), Takahira (2), Schlich - men (1), Schlich - women (1), Livingston and Leelili (1). As these formulas are less frequently used, we did not use them in our testing. *Table 1* shows the apps, the formulas they implement, and whether the formula is clearly specified in the app so the patient or provider knows which formula is used.

Table 1. BSA formulas used in apps

App ID	Du Bois	Mosteller	Other	Specifies Formula
appinventor.ai_medicina_android.Dialysis_Calculator	•			•
br.bsa	•	•		•
com.aci.itim.trauma				NO
com.hordev.doctor	•	•	•	•
com.rxdosecalc		•		•
co.mdapp.bsa	•	•	•	•
com.epocrates	•	•	•	•
com.grapp.kinderarztrechner	•	•		•
com.itmedicus.pdm	•	•		•
com.medcomis.device.android.egfr	•			•
com.mednt.drwidget_calcmed	•	•	•	NO
com.medpresso.android.ui	•	•		•
com.medtricks		•		NO
com.moaness.InfiniteDose.free		•		NO
com.mospeciol.bmiapp		•		•
com.niya.lijo.nursingcalculators		•		•
com.qxmd.calculate	•	•		•
com.redapps.egfrcalculator	•			•
com.tr.bsacalculator	•	•	•	•
com.wilcorp.IdealBodyStats			•	•
org.obstetrics.widgets	•			•
pe.com.codespace.nurse		•	•	•
Pedcall.Calculator	•	•		•

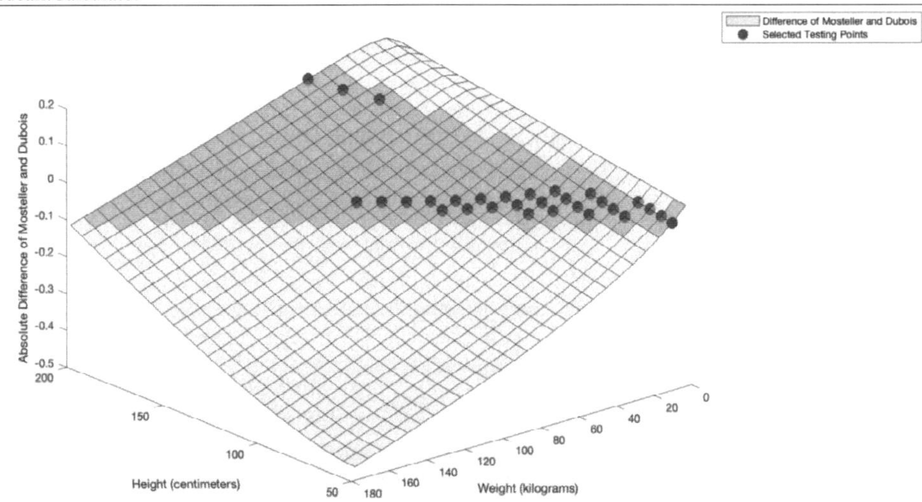

Figure 3. The absolute difference between Mosteller and Du Bois values. The green shaded area shows humanly possible Weight/Height combinations

3.1 Testing Points

To systematically test apps, we used 25 weight/height pairs. These 25 pairs are listed in the first two columns of *Table 2*. We used BMI to determine and eliminate pairs outside of what is possible for the human body (in *Figure 3*, we indicate the plausible combinations as the green area, and the implausible as the gray area). To establish ground truth, we calculated the BSA values for both Mosteller and BuBois.

Table 2. BSA value differences between Du Bois and Mosteller

Weight (kilograms)	Height (centimeters)	Du Bois (meters^2)	Mosteller (meters^2)	Absolute Difference	Relative Difference (%)
10	50	0.33	0.37	0.04676	14.35%
10	56	0.35	0.39	0.04057	11.47%
17	68	0.51	0.57	0.05632	11.04%
24	80	0.66	0.73	0.06550	9.85%
10	62	0.38	0.42	0.03407	8.94%
17	74	0.54	0.59	0.04852	8.94%
38	98	0.94	1.02	0.08080	8.63%
31	92	0.82	0.89	0.06985	8.52%
24	86	0.7	0.76	0.05660	8.08%
45	110	1.09	1.17	0.07870	7.19%
38	104	0.98	1.05	0.07028	7.19%
52	116	1.21	1.29	0.08560	7.08%
17	80	0.57	0.61	0.04047	7.05%
31	98	0.86	0.92	0.05996	6.98%
59	122	1.32	1.41	0.09090	6.87%
10	68	0.41	0.43	0.02730	6.70%
24	92	0.74	0.78	0.04747	6.45%
45	116	1.14	1.2	0.06730	5.92%
52	122	1.25	1.33	0.07360	5.87%
38	110	1.02	1.08	0.05950	5.84%
59	128	1.37	1.45	0.07840	5.72%
52	200	1.79	1.7	0.09470	5.57%
31	104	0.9	0.95	0.04987	5.56%
66	134	1.49	1.57	0.08210	5.53%
17	86	0.61	0.64	0.03219	5.32%
73	140	1.6	1.68	0.08450	5.28%
45	188	1.61	1.53	0.08040	5.24%
38	176	1.43	1.36	0.06840	5.02%
80	146	1.72	1.8	0.08600	5.01%
24	98	0.77	0.81	0.03813	4.95%

3.2 Finding: Formula Divergence

We then selected points (see *Table 2*) with the largest difference between the two formulas, to maximize the point that could reveal errors in the calculation — specifically, those apps that did not reference any formula in their implementations. Redlarski et al.'s work supports the points we selected for our testing as they represent individuals with abnormal bodies who are at higher risk for BSA calculation errors.

Table 2 reveals the first concerning finding: *the choice of formula matters, and simply using two different formulas can lead to BSA value differences of up to 14%.* Hence it is critical that apps state the formula used, so that apps and providers are on the same page regarding the formula used.

3.3 Finding: Incorrect Calculations

The result of the study revealed calculation errors in three apps, in both Mosteller and Bu Bois implementations. We now discuss each app in detail.

Paediatric Emergency-Dose-NICU. This app (Google Play package name "com.medtricks"; more than 100,000 downloads) states that it is using the Mosteller formula by displaying the mathematical formula but not its name. *Figure 6* shows the app's BSA calculation which *produces an incorrect value error of 34m^2 whereas the correct value is 1.21m^2*. In *Figure 4* we show the results of testing on the 25 points: the x-axis is the index for the testing points from *Table 2*, while the y-axis shows the BSA value. The app's results are shown in red, whereas the correct values, i.e., the ground truth, are shown in blue. The figure clearly shows the large discrepancy in the values. While BSA values are supposed to be within the range of 0.3 and 3, the app produces values that have an error of at least 8.63m2, growing to an error of $49.3m^2$. Overall, the app-calculated values were on average 2549% higher than the correct value. Having an app with more than 100,000 downloads and such a large margin of error is problematic and can have serious implications.

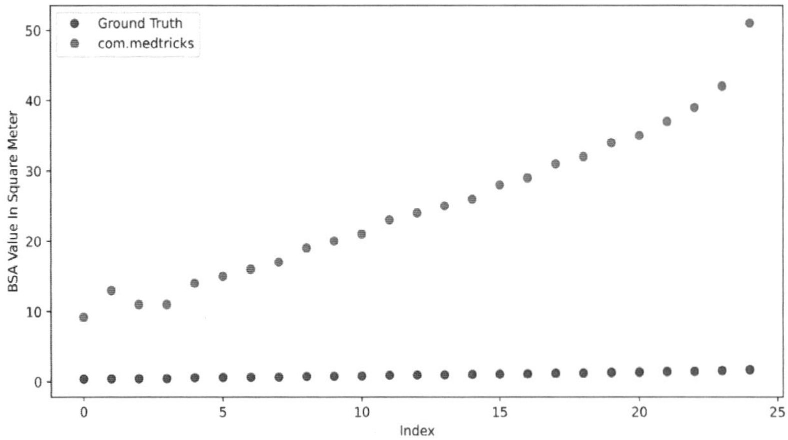

Figure 4. Incorrect values produced by app **com.medtricks**

GFR & BSA Calculator. This app (Google Play package name "com.medcomis.device.android.egfr"; more than 10,000 downloads) is using the Du Bois formula. In the calculation screenshot in *Figure 7,* we can see a test for *height=122cm* and *weight=59kg*. While the app displays a value of BSA=1.34m², the correct value is 1.32m². Such differences were present across most of the test points. In *Figure 5* we show the relative error on the y-axis, while the ground truth values are on the x-axis: only 4 out of 25 points had an accurate calculation, with the other 21 points' error being up to 2.86%. Again, this is problematic in care scenarios where accuracy is critical.

NSW Trauma. This app (Google Play package name "com.aci.itim.trauma"; more than 10,000 downloads) did not reference the formula used for the implementation, and due to the lack of precision in the result, we were unable to infer the formula. The calculation screenshot in *Figure 8,* we can see a test for *height=104cm* and *weight=38kg*. From the screenshot, notice how the result is displayed only in a single decimal point place, instead of the common practice of at least two decimal places. In *Figure 9* we show the relative error on the y-axis, in comparison to Mosteller's ground truth values on the x-axis. Only 2 out of 25 points have an accurate calculation, with the other 23 points' error being up to 8.66%. We also compared the app against Du Bois values and found errors of up to 22.7% and only 1 of the 25 tested points resulted in correct calculations.

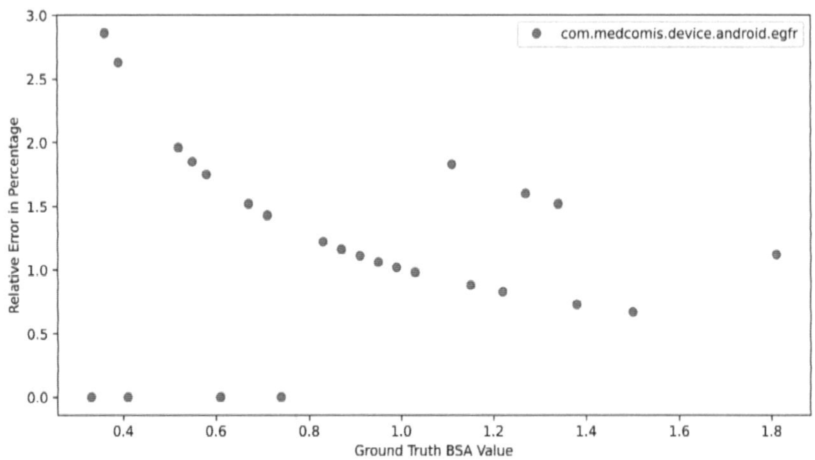

Figure 5. Relative error, in %, between ground truth and output of app **com.medcomis.device.android.egfr**

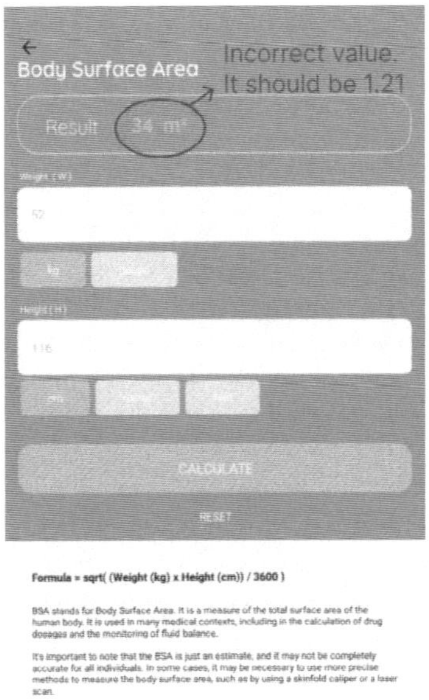

Figure 6. BSA calculation in app **com.medtricks**

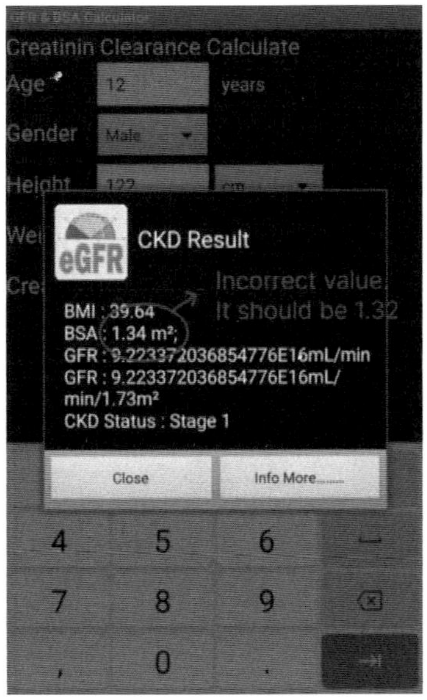

Figure 7. BSA calculation in app **com.medcomis.device.android.egfr**

Figure 8. BSA calculation in app **com.aci.itim.trauma**

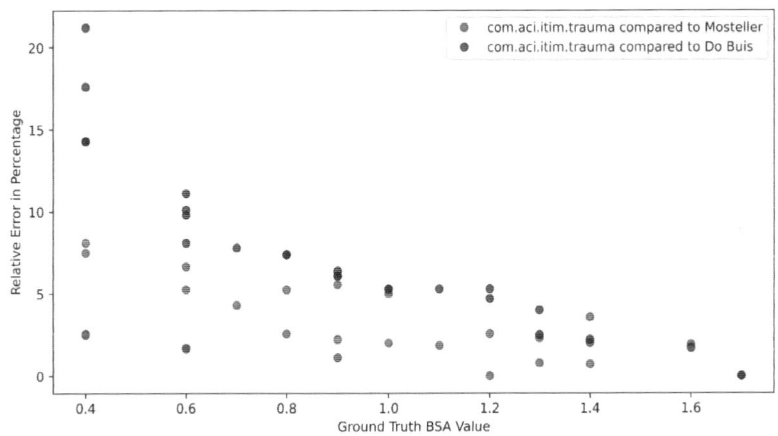

Figure 9. Relative error, in %, between ground truth and output of app **com.aci.itim.trauma**

BSA Calculator - Body Surface. After entering input values and pressing `Calculate', this app (Google Play package name "co.mdapp.bsa") crashed on our test platforms (a Google Pixel 1 phone running Android 8.1.0 and a Google Pixel 5 phone running Android 14) hence we could not conduct further testing.

Other findings. Although the degree of precision required for BSA-dosing is dependent on the type of treatment, we found variation between apps in terms of how many decimal points BSA values were presented, which can have an impact as well. Another area that highlights negligence by developers was input field validation: some apps allowed negative values, some apps did not allow decimal points, while other apps did not have any validation at all. All these design issues should be addressed.

4. DISCUSSION

Our study and findings shed light on the main causes of errors, and possible solutions, in the BSA space. First, BSA lacks a single coherent definition, leaving the developers and users guessing as to which definition to implement, and which definition an app uses, respectively. At the very least, as Table 1 suggests, apps should indicate the name of the formula prominently, and implement both of the two most popular formulas (Du Bois and Mosteller). Second, regulatory agencies (e.g., the FDA, FTC, CDC in the US) should provide reference input and output values for commonly used calculators, so developers have ground truth for validating their apps. Third, app stores could use a higher bar for publishing apps that implement critical calculators. Fourth, the research community should embark on a systematic effort of reviewing the correctness of medical formulas and apps that implement calculators for these formulas; (Rahaman et al., 2023) has a broader set of recommendations regarding medical calculator apps.

ACKNOWLEDGMENTS

We thank the anonymous reviewers for their feedback. This material is based upon work supported by the National Science Foundation under Grant No. CCF-2106710.

REFERENCES

Bois, D. D. & Bois, E. F. D. (1916) A height-weight formula to estimate the surface area of man. Experimental biology and medicine (Maywood, N.J.). [Online] 13 (4), 77–78.

Faisal, W. et al. (2016) Not All Body Surface Area Formulas Are the Same, but Does It Matter? Journal of global oncology. [Online] 2 (6), 436–437.

GitHub. (2023). apkeep - A command-line tool for downloading APK files from various sources. [online] Available at: https://github.com/EFForg/apkeep.

GitHub. (2023). DroidBot. [online] Available at: https://github.com/honeynet/droidbot.

Haffey, F. et al. (2013) A Comparison of the Reliability of Smartphone Apps for Opioid Conversion. Drug safety. [Online] 36 (2), 111–117.

Haycock, G. B. et al. (1978) Geometric method for measuring body surface area: A height-weight formula validated in infants, children, and adults. The Journal of pediatrics. [Online] 93 (1), 62–66.

Huckvale, K. et al. (2015) Smartphone apps for calculating insulin dose: A systematic assessment. BMC medicine. [Online] 13 (1), 106–106.

jq. (2023, December 21). GitHub. https://github.com/jqlang/jq

Moore, R. A., Waheed, A., & Burns, B. (2022). Rule of Nines. In StatPearls. StatPearls Publishing.

Mosteller, R. D. (1987) Simplified Calculation of Body-Surface Area. The New England journal of medicine. [Online] 317 (17), 1098–1098.

Pan, S. D. et al. (2016) Weight-based dosing in medication use: What should we know? Patient preference and adherence. [Online] 10 (Issue 1), 549–560.

Rahaman, S. et al. (2023) Diagnosing Medical Score Calculator Apps. Proceedings of ACM on interactive, mobile, wearable and ubiquitous technologies. [Online] 7 (3), 1–27.

Redlarski, G. et al. (2016) Body surface area formulae: An alarming ambiguity. Scientific reports. [Online] 6 (1), 27966–27966.

Samuel, R. et al. (2022) Could medical apps keep their promises? International Conference on eHealth. 173–181.

Statista. (n.d.). Google Play Store number of available apps by category 2022. [online] Available at: https://www.statista.com/statistics/1333698/google-play-store-number-of-available-apps-by-category/.

Tumbleson, C. (2021). iBotPeaches/Apktool. [online] GitHub. Available at: https://github.com/iBotPeaches/Apktool.

Vu, T. (2002) Standardization of body surface area calculations. Journal of oncology pharmacy practice. [Online] 8 (2–3), 49–54.

Yau, T. O. (2019) Precision treatment in colorectal cancer: Now and the future. JGH open. [Online] 3 (5), 361–369.

TOPOLOGICAL DATA ANALYSIS FOR SELECTION OF MACHINE LEARNING MODELS IN CEREBRAL STROKE DETECTION

Aleksandra Vatian, Natalia Gusarova, Ivan Tomilov, Vladimir Shilonosov, Julia Valitova, Vladimir Sashin, Maria Bobrova, Stanislav Mashechkov, Tatiana Polevaya and Alexey Zubanenko
ITMO University, Sankt-Petersburg 197101, Russia

ABSTRACT

Rapid and reliable diagnosis of cerebral stroke is a vital necessity, and among them, ischemic stroke is the most difficult to recognize on MRI images. Increasing the efficiency of stroke diagnosis is associated with the transition to increasingly "heavy" AI tools working in 3D mode, processing MRI images not pixel-by-pixel, but voxel-by-voxel and using complex multifactor information processing algorithms. The implementation of such products requires large computing resources, which are often unavailable outside large medical centers. In addition, existing explainable AI tools identify the affected area very roughly and generally, which reduces the doctor's confidence in the diagnostic result. On the other hand, one can use AI models operating in 2D mode - weaker, but faster and less demanding on computing resources.
The article uses a Siamese neural network as a base model. To improve classification efficiency, a model pretrain on "light" synthetic data based on Perlin noise is proposed. To objectify the choice of mode, the article uses the apparatus of topological data analysis, namely, changes in persistent entropy and Renyi entropy of data embeddings on fully connected layers of a neural network. It is theoretically justified and experimentally confirmed that using a 2D model, when trained on slices with maximum lesion visibility, produces ROC-AUC values no worse than using a full-scale 3D SOTA models, while allowing the clinician to selectively evaluate the individual slices he or she selects.

KEYWORDS

Cerebral Stroke, Topological Data Analysis, Persistent Entropy, Pre-Training, Perlin Noise

1. INTRODUCTION

The World Health Organization describes cerebral stroke as a clinical syndrome with rapidly increasing clinical signs of local or global impairment of brain function without an obvious cause other than vascular origin (Mosisa et al, 2023). The annual mortality rate from stroke is almost six million people worldwide. Stroke is the second most common cause of death among adults over 60 years of age, and the fifth most common among people aged 15 to 59 years, accounting for 11.8% of all deaths (National Collaborating Centre for Chronic Conditions (UK), 2008; Writing Group Members et al., 2016).

The predominant diagnostic tools for early detection of stroke in modern medicine is magnetic resonance imaging (MRI) (Valliani et al, 2019). As a result, a set of sections of brain tissue is formed, which the doctor can view individually (in 2D mode), as well as in the form of a pseudo-volume image (in 3D mode) and make his diagnostic conclusion. Although stroke is a disease caused by a disruption in the blood supply to brain tissue, there are two types of stroke - hemorrhagic and ischemic, which manifest differently on MRI images. In the case of a hemorrhagic stroke, a separate vessel ruptures, and the outflowing blood immediately forms a hematoma, which is visible on MRI images as a neoplasm with structure and characteristics differing from the surrounding tissues. In ischemic stroke, on the contrary, a spasm of the vessel occurs, and, as a result, the necrosis of the tissues being supplied before. MRI manifestations of this process increase gradually and may be different in structure and characteristics depending on the location and level of the lesion. Therefore, ischemic stroke is more difficult to recognize on MRI images, and at the same time, it is vitally important to carry out the correct diagnosis at the earliest possible stages of the disease.

To objectify the diagnostic process, as well as to help the radiologist, a constantly expanding arsenal of artificial intelligence (AI) tools is being developed. The main trend in research is associated with the transition to increasingly “heavy” machine learning models. As a rule, such models work in 3D mode, i.e. they process MRI images not pixel-by-pixel, but voxel-by-voxel, and implement complex multifactor information processing algorithms, which allows for increased automation of the diagnostic process, but comes at the cost of the expansion of information and computing resources used, especially when processing high resolution images. However, in real medical practice there are several factors that impede such a transition. First, as a rule, the deployment of such models requires computer equipment with sufficient software and hardware resources that allow you to quickly calculate a large amount of data typical for 3D models. Outside a specialized medical center, such equipment is most often not available.

Second, practitioners who are directly responsible for the patient’s life place high demands on the explainability of diagnostic AI tools, and for this a wide range of explainable AI tools (Speith, 2022; Saranya & Subhashini, 2003) have been proposed. However, as an analysis of the literature shows, XAI methods in the diagnosis of ischemic stroke make it possible to identify the affected area very roughly and generally, which is not enough to make a diagnostic decision. Therefore, to increase confidence in the diagnostic result, a combined intelligence procedure can be proposed: the doctor sequentially views slices of the MRI image and, relying on his expert knowledge and experience, identifies the most “suspicious” slices; on them, a selection of the affected area is carried out using AI and compared with an expert assessment. In this case, one can use AI models operating in 2D mode - weaker, but faster and less demanding on computing resources. Besides, the diagnostic process remains completely transparent to the doctor.

Obviously, the fundamental difference between 2D and 3D models lies in the amount of information simultaneously processed by the network, so it is reasonable to use an information approach to select a specific model. In this work, for this purpose, topological data analysis (Chazal & Michel, 2021) is used, specifically, the apparatus of persistent homologies.

The main results of our work are as follows:

- We have shown that in conditions of small datasets, which is typical for MRI images of ischemic stroke, the use of pre-training of difference models on “light” synthetic data allows to leverage efficiency of 2D –mode processing up to 3D- mode.
- We have shown that changes in topological characteristics (persistent entropy and Renyi entropy) of data embeddings on fully connected layers of a neural network can serve as an objective assessment of the effectiveness of the network architecture and, therefore, be used for its selection in conditions of limited resources.
- We have shown that using a 2D model, when trained on slices with maximum lesion visibility, produces ROC-AUC values no worse than using a full-scale 3D model, while allowing the clinician to selectively evaluate the individual slices he or she selects.

The rest of the article is organized as follows. The section Background and related works presents existing work on automating stroke diagnostics, briefly describes the use of topological data analysis and persistent homologies for it, and justifies the choice of the SOTA model. The Method and materials section describes the methods and datasets used, the Results and discussion section describes the experimental results obtained and compared with the existing SOTA model. The Conclusion section completes the work.

2. BACKGROUND AND RELATED WORKS

A plethora of AI solutions, primarily based on machine learning (ML), are used for automated recognition of stroke and its localization (Ruksakulpiwat, 2023). All researchers note that the most difficult problem is recognizing ischemic stroke. The most effective solutions include the use of deep convolutional neural networks (Zhang et al., 2021; Stib & Vasquez, 2020; Shinohara et al. 2020), simultaneous use of images from different modalities (Nishio et al., 2020; Lu et al., 2022), combining disparate algorithms into a single pipeline (Zhang et al., 2021), full-scale 3D image processing (Oman et al. 2019; Wang et al., 2021) etc. The efficiency of the best models is estimated as ROC-AUC = 0.93 (Oman et al., 2019) and ROC-AUC = 0.90±0.04 (Wang et al., 2021). To overcome the problem of small training samples, which is characteristic for the recognition of ischemic stroke, using generative models (Zhu et al., 2020) is proposed. In general, all above-mentioned models allow you to take into account many influencing factors, but they are "heavy" and do not fit well into the equipment available to practicing physicians.

An elegant way to "lighten" the model is based on physical features of medical artifacts, namely the symmetry of brain hemispheres. Symmetrical neural networks (Cui et al., 2021), first Siamese neural networks (Chen et al., 2020), are widely employed, and their combination with the few-shot learning (Barman et al., 2019; Shovkoplias et al., 2023) allows to work on small datasets. In determining of stroke, the best results are ROC-AUC = 0.914 with 0.88–0.95 confidence interval in three-dimensional mode (Barman et al., 2019) and ROC-AUC = 0.794–0.921 depending on the type of lesion in two-dimensional mode (Shovkoplias et al., 2023). As one can see, symmetric ML models are potentially not inferior in efficiency to heavier ones, but are more critical to the model settings, primarily to the choice of the analysis area. Therefore, for them the problem of choosing and setting model parameters is acute.

The dominant approach here is the heuristic one (see, for example, Thomas et al., 2020), which relies on the intuition and experience of developers and does not generalize well. More formalized methods use data statistics, like nested cross-validation (Larracy, Phinyomark, & Scheme, 2021) and Cohen's kappa metric (Castelo, 2022), but they are data-driven and single-criterion, which does not meet the real conditions of stroke detection. The choice of an ML model can also be considered as a decision-making process, which allows the use of widely known mathematical methods, such as linear and non-linear programming, genetic algorithms, AutoML etc. (Enns et al., 2015; Romero et al., 2022). However, the effectiveness of these approaches is limited by the completeness of considering all parameters of the problem, which is difficult to achieve. In real practice, generalized parameters is more in demand, which would allow, on a strictly formal basis, to obtain comparative assessments of the prospects of specific architectural and parametric solutions. For example, in the work (Shovkoplias et al., 2023) the theory of Probably Approximately Correct learning (Valiant, 2013) is used as a methodological basis. Accordingly, the generalized parameter here is the Vapnik-Chervonenkis dimension (Vapnik & Chervonenkis, 2015), for which it is difficult to offer a meaningful interpretation that is meaningful for a practitioner.

It seems more organic to use topological data analysis (Chazal & Michel., 2021; Wasserman, 2018), which allows us to directly consider the topological features of medical artifacts involved in the construction of Siamese neural networks. In recent years, TAD has established itself as one of the most advanced means of studying the internal structure of the feature space of a neural network (Carlsson & Gabrielsson, 2020; Hensel, Moor, & Rieck, 2021; Moroni & Pascali, 2021; Zia et al., 2023). The main tools in TDA are simplicial complexes, forming a kind of multidimensional clusters in input data structure or feature space of ML model. To visualize it, persistent homologies (PH) (Zomorodian & Carlsson, 2005) are used - a set of diagrams in the form of segments (barcodes), which describe the relative position of points in feature space through the distance parameter between points (filtering) ε. At the initial moment ($birth_t$) all points are separated from each other, which corresponds to the birth of individual homologies. As ε increases, some points unite into clusters, which is considered as the moment of disappearance of homology ($death_t$).

Formation and evaluation of PHs are the most widely used applications of the TDA apparatus to the analysis of neural networks. In work (Goibert, Ricatte, & Dohmatob, 2022), the topology of the induced graph describing the propagation of the computational process from the input to the output of the NN is studied, in works (Chen et al. 2018; Ramamurthy, Varshney, & Mody, 2018) - the topology of the interface of classifiers based on NN, in works (Gabrielsson & Carlsson, 2019; Rieck et al., 2019) - the topology of NN inner space. In the last case, the subject of consideration is the spatial-topological configuration of embeddings formed by the NN on its intermediate layers.

Most well-known works study the actual topological changes performed by NNs. The connection between the topological characteristics of a neural network and its typical metrics is considered only in a few works. Namely, in Gabrielsson & Carlsson (2019) the connection between accuracy on test data (accuracy) and the lifetime of the most persistent one-dimensional homology of the first layer of the neural network is shown. In Rieck et al. (2019) estimates of the minimum achievable error in NN training were obtained depending on the topological characteristics of the input data and the first layers of fully connected NNs. The use of TDA for comparative assessment of various NN architectures is not presented in the literature available to the authors.

3. METHOD AND MATERIALS

3.1 Neural Network Configuration

For the study, we used a Siamese neural network (SNN) of typical architecture (Barman et al., 2019) (Figure 1), which is designed to compare the symmetry of MRI images of 2D and 3D dimensions. SNN has two parallel branches, which include four inception modules with identical weights, as well as a common branch containing convolutional and pooling modules. An MRI image of the brain is supplied to the network input in the form of a set of slices, potentially containing the affected area. Sections of the affected area falling within the slices are highlighted in red.

During preprocessing, the input image is divided into two symmetrical halves (left and right hemispheres), each of which is fed into its own channel. Thus, two branches of the network process the left and right images independently, and in the common branch the difference between the feature representations of the input data is calculated. The architectures of both branches are identical, they include four InceptionModule (IM) layers, each of which extracts features from the input data x_1, x_2 through parallel paths $f_1(x_1)$ and $f_2(x_2$. Both branches have the same sets of weights $W * f_1(x_1)$ and $W * f_2(x_2)$ and identical convolution and pooling operations. The common branch includes multiple IM layers, as well as pooling and fully connected layers for the final output. When solving a classification problem, the network generates an output of 1 if a pathology is detected, and 0 otherwise.

The network was trained in two modes: with and without preliminary training on a simulating dataset.

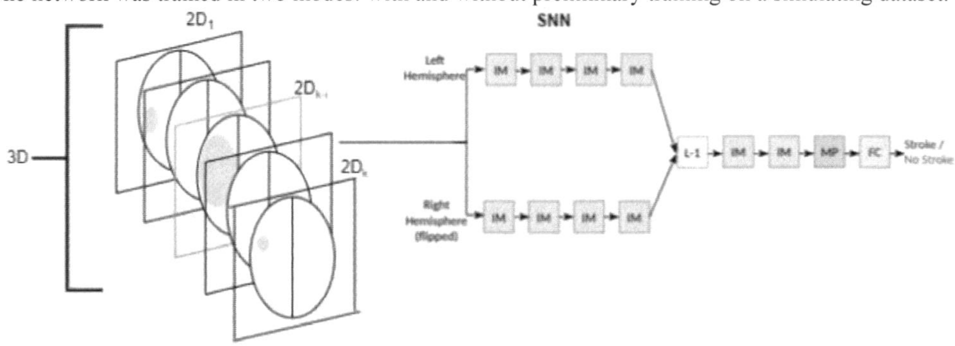

Figure 1. Basic model architecture: IM – Inception Module, L-1 – Merge Layer with L-1 difference, MP –Max Pooling, FC – Fully Connected Layer. Basic inputs: 3D – all slices per patient; 2D – one slice per patient

3.2 Metrics Used

To assess the classification properties of SNN, we used typical metric, namely ROC-AUC score:

$$roc_{auc} = \int_0^1 TPR\big(FPR^{-1}(t)\big)dt \quad (1)$$

where TPR – true positive rate, FPR – false positive rate.

To assess the topological properties of SNN, we built zero PHs for embeddings generated by SNN on each of the last four layers, and then we calculated the following metrics for each layer:

- Persistent entropy:

$$entropy = -\Sigma_{i=1}^{n} p_i \log p_i \, , p_i = \frac{death_i - birth_i}{\Sigma_{i=1}^{n}[death_i - birth_i]} \quad (2)$$

- Rényi Entropy for α=0.5, 2, 3:

$$renyi\_entropy_\alpha = \frac{1}{1-\alpha} log \Sigma_{i=1}^{n} p_i^\alpha \, , p_i = \frac{death_i - birth_i}{\Sigma_{i=1}^{n}[death_i - birth_i]} \quad (3)$$

- The difference in persistent entropies between adjacent layers

$$delta_{entr} = |entropy_l - entropy_{l-1}|, l - layer \ (4)$$

- The difference in Rényi entropies between adjacent layers

$$delta_{renyi_\alpha} = \left|renyi\_entropy_{\alpha_l} - renyi\_entropy_{\alpha_{l-1}}\right|, l - layer \ (5)$$

3.3 Experimental Data and Scenarios

Two datasets were used in the experiments. The native dataset consisted of MRI images of the brain tissue in FLAIR format; the images were desculled, normalized and divided into left and right hemispheres; the total size of the dataset is 200 images, of which 100 are normal and 100 with pathologies; image size was 64×92×26 for 3D variant and 64×92 for 2D variant.

Figure 2 shows slices of a real MRI image of the brain, including areas of ischemic stroke (highlighted in ovals). It is easy to see that the image contains other hyperintense zones of a similar configuration, which, however, are not affected. This makes it difficult to train a neural network and reduces its potential efficiency, especially when working on individual slices, i.e. in 2D mode. To overcome this difficulty, we used model pretraining. To do this, we generated a synthetic dataset based on Perlin noise (Etherington, 2022). Its advantages over generative models for simulating brain tissue are its low computational and memory requirements. Moreover, using Perlin noise allows to simulate healthy and affected fragments of brain tissue separately and to combine them in a controlled manner, i.e. create more realistic conditions for preliminary training of the model.

Examples of images from the synthetic dataset are shown in Figure 3. Simulation parameters were selected with the participation of a radiologist to reproduce the typical problems encountered in distinguishing between hyperintense areas, normal and stroke-affected. In Figure 3a shows a simulation of healthy brain tissue, Figure 3b – tissues with normal hyperintense areas, in Figure 3c – tissues with areas of ischemic stroke. The dataset for pretraining was composed of randomly selected images of the second and third groups (Figure 3b and Figure 3c), taken in equal proportions, the image sizes were normalized identically to the native dataset, the total volume of the dataset was 200 images.

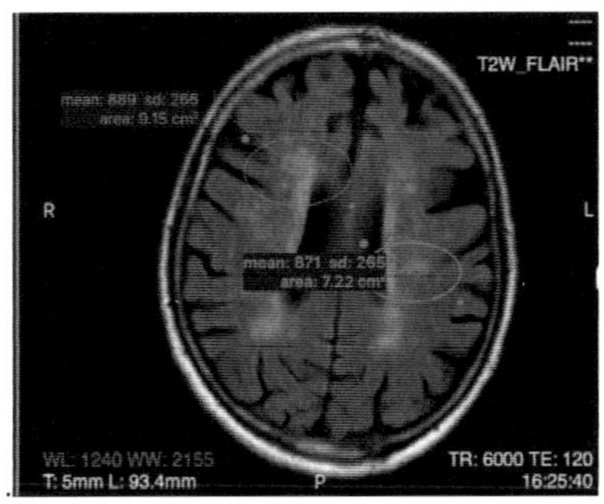

Figure 2. Slice of MRI image of a brain with ischemic stroke. Damage zones are marked with ovals; the remaining hyperintense zones are not stroke zones

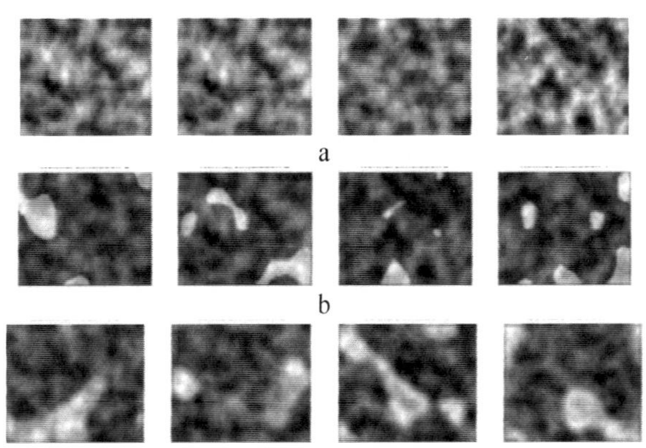

c

Figure 3. Examples of simulated images from a synthetic dataset: a - healthy brain tissue, b – tissue with normal hyperintense areas, c – tissue with areas of ischemic stroke

The experiment was carried out according to three scenarios:

I. SNN was trained without pretraining on a native dataset and was used to classify stroke in 2D and in 3D modes;

II. SNN was pre-trained on a synthetic dataset and trained on a native dataset afterwards; then the wholly trained SNN was used to classify stroke in 2D mode, and the central frame from the MRI package was used as the analyzed frame;

III. SNN was pre-trained on a synthetic dataset and trained on a native dataset afterwards; then the wholly trained SNN was used to classify stroke in 2D mode, and the analyzed frame was a frame from the MRI package, selected by the radiologist as the most informative.

Then, for each scenario, the differences between persistent entropirs (4) and Renyi entropies (5) were calculated for four adjacent layers of the common branch of the network (Figure 1). In addition, for each scenario the classification efficiency was assessed using the ROC-AUC score metric (1).

4. RESULTS AND DISCUSSION

An example of the ratio of PH lifetimes for 2D and 3D SNN operating modes is shown in Figure 4. It is easy to see that the lifetime of homologies in the first case is significantly greater than in the second, and, as our experiments have shown, this ratio persists for all SNN layers, regardless of the preprocessing of the dataset. This confirms that, all other things being equal, 3D SNN operating mode provides greater data separability than 2D one. A more detailed comparison of topological characteristics, carried out using metrics 8 and 9, is graphically presented in Figure 5.

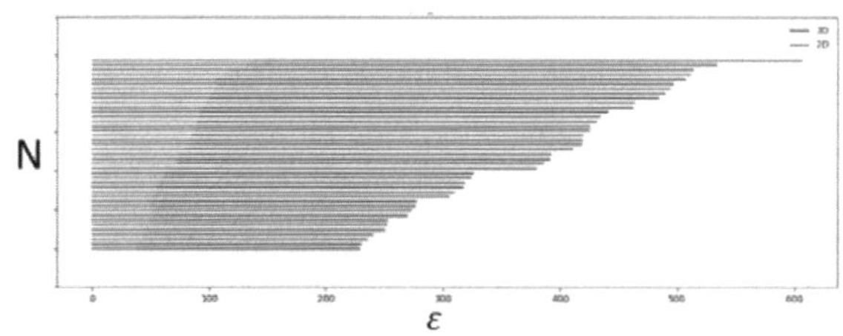

Figure 4. An example of the ratio of zero GHGs on one layer for 2D (blue lines) and 3D (red lines) SNN operating modes: ε – filtration parameter, N – homology number in birth order

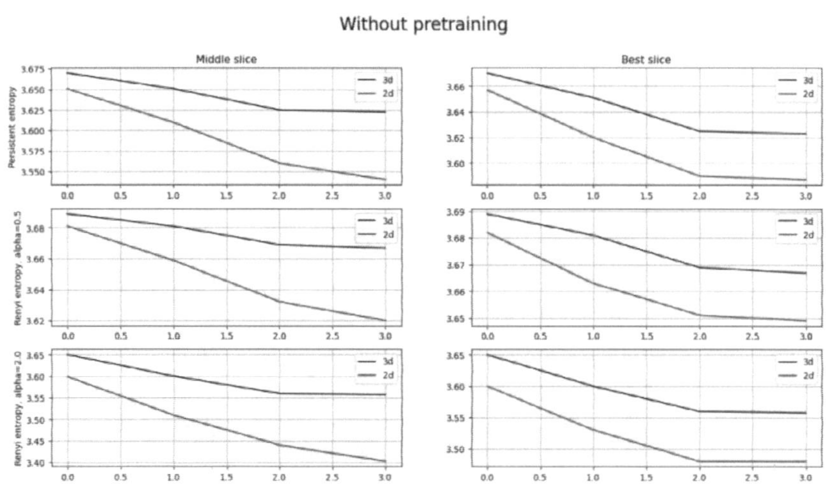

(a)

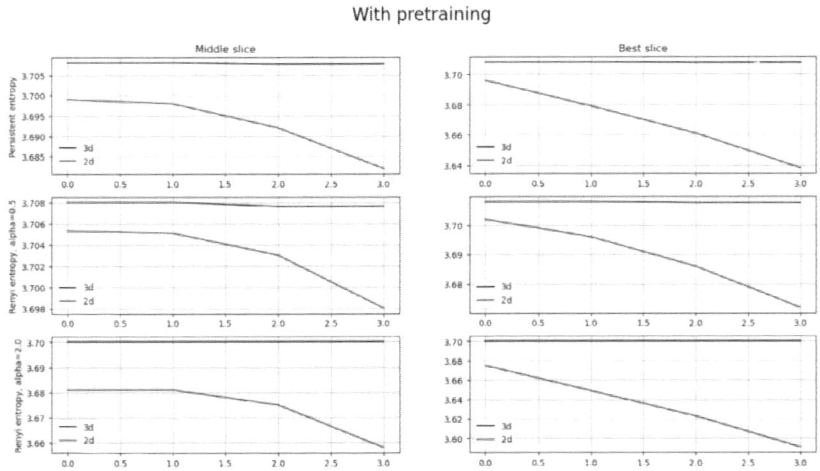

(b)

Figure 5. a) b) Comparison of persistent entropies between adjacent layers for different experimental scenarios

A comparison of the graphs shows that when using a pretrain in 3D mode, the persistent entropy is practically unchanged across layers, while in 2D mode such a change takes place. Moreover, in 2D mode without pretrain, the use of the best slices provides a smaller drop in entropy than in the middle ones, and in 2D mode with pretrain the opposite effect is observed. It means that pretraining on synthetic data allows the model to extract more information from the data during additional training. In other words, with the same architectural solutions, the 3D model extracts redundant information, which can reduce the efficiency of its configuration, while a simpler 2D model works with less information, but uses it more efficiently. At the same time, an inverse relationship is observed on the 3D model, which means that the effect of pretraining on the 3D model will have less effect. The conclusions obtained are confirmed by comparing ROC-AUC metrics for 2D and 3D models (Table 1).

Table 1. Experimental assessment of the ROC-AUC score for the classification of ischemic stroke during operation of the SNN in different scenarios

Mode type	SOTA	SNN, not pre-trained		SNN, pre-trained	
		middle slice	best slice	middle slice	best slice
2D	0, 794	0,851	0,969	0,890	0.984
3D	0,899	0,973		0.995	

5. CONCLUSION

In this study, the use of Siamese neural networks (SNNs) was shown to be effective for ischemic stroke classification in resource-limited settings.

The paper proposes a method for assessing the degree of informativeness of SNNs using topological data analysis, which allows us to objectify the differences between 2D and 3D modalities and thereby optimize the choice of neural network architecture when classifying medical images depending on specific conditions.

It is theoretically justified and experimentally confirmed that the use of pre-training of difference models on "light" synthetic data allows leveraging the efficiency of 2D-mode processing up to 3D-mode.

In addition, allowing the clinician to selectively select slices of greatest interest and perform automated classification based on them increases the explainability of AI systems in medicine and staff confidence in their work.

ACKNOWLEDGEMENT

This work was supported by Russian Science Foundation, Grant 23-11-00346.

REFERENCES

Barman, A. et al. (2019). Determining ischemic stroke from ct-angiography imaging using symmetry-sensitive convolutional networks. *2019 IEEE 16th International Symposium on Biomedical Imaging* (ISBI 2019) Venice, Italy, April 8-11, 2019

Carlsson, G., & Gabrielsson, R.B. (2020). *Topological Approaches to Deep Learning.* Topological Data Analysis (Springer), 119–146. 10.1007/978-3-030-43408-3_5

Castelo, G.P. (2022). Minimum Sample Size Estimation in Machine Learning. https://openaccess.uoc.edu/bitstream/10609/146007/8/gprolcasteloTFM0622report.pdf

Chazal, F., & Michel, B. (2021). An Introduction to Topological Data Analysis: Fundamental and Practical Aspects for Data Scientists. *Front. Artif. Intell. Sec. Machine Learning and Artificial Intelligence, 4.* https://doi.org/10.3389/frai.2021.667963

Chazal, F., & Michel, B. (2021). An Introduction to Topological Data Analysis: Fundamental and Practical Aspects for Data Scientists. *Front. Artif. Intell.*, 29 September 2021. *Sec. Machine Learning and Artificial Intelligence.* V.4 https://doi.org/10.3389/frai.2021.667963

Chen, C. et al. (2018). A Topological Regularizer for Classifiers via Persistent Homology. arXiv:1806.10714v3 [cs.LG] 16 Oct 2018

Chen, H. et al. (2020). Anatomy-Aware Siamese Network: Exploiting Semantic Asymmetry for Accurate Pelvic Fracture Detection in X-ray Images. *Computer Vision – ECCV 2020.* ECCV 2020. LNCS, vol 12368. Springer, Cham. doi.org/10.1007/978-3-030-58592-1_15

Cui, L. et al. (2021). Deep symmetric three-dimensional convolutional neural networks for identifying acute ischemic stroke via diffusion-weighted images. *Journal of X-Ray Science and Technology, 29*(2021), 551–566. doi:10.3233/XST-210861

Enns, E.A. et al. (2015). Identifying best-fitting inputs in health-economic model calibration: a Pareto frontier approach. *Med Decis Making.* 2015 Feb; 35(2): 170–182.

Etherington, T.R. (2022). Perlin noise as a hierarchical neutral landscape model. *Web Ecology, 22*(1): 1–6. doi:10.5194/we-22-1-2022.

Gabrielsson, R.B., & Carlsson, G. (2019). Exposition and Interpretation of the Topology of Neural Networks. arXiv:1810.03234v3 [cs.CV] 18 Oct 2019

Goibert, M., Ricatte, T., & Dohmatob, E. (2022). An Adversarial Robustness Perspective on the Topology of Neural Networks https://arxiv.org/abs/2211.02675

Hensel, F., Moor, M., & Rieck, B. (2021). A Survey of Topological Machine Learning Methods. *Front. Artif. Intell.*, 26 May 2021. *Sec. Machine Learning and Artificial Intelligence.* Volume 4 – 2021. https://doi.org/10.3389/frai.2021.681108

Larracy, R., Phinyomark, A., & Scheme, E. (2021). Machine Learning Model Validation for Early Stage Studies with Small Sample Sizes. *IEEE Eng Med Biol Soc.* Nov 2021:2314-2319. doi: 10.1109/EMBC46164.2021.9629697.

Lu, J. et al. (2022). Identification of early invisible acute ischemic stroke in non-contrast computed tomography using two-stage deep-learning model. *Theranostics*, *12*(12), 5564–5573. doi:10.7150/thno.74125

Moroni, D., & Pascali, M.A. (2021). Learning topology: bridging computational topology and machine learning. *Pattern Recognition and Image Analysis: Advances in Mathematical Theory and Applications*, 2021, 443, pp. 443-453. ffhal-03351647f

Mosisa, W. et al (2023). Survival status and predictors of mortality among adult Stroke patients admitted to Jimma University Medical Center, South west Ethiopia: A retrospective Cohort study. *Vasc Health Risk Manag*, *19*, 527-541.

National Collaborating Centre for Chronic Conditions (UK). (2008). *Stroke: National Clinical Guideline for Diagnosis and Initial Management of Acute Stroke and Transient Ischaemic Attack (TIA).* Royal College of Physicians (UK), London, UK.

Nishio, M. et al. (2020). Automatic detection of acute ischemic stroke using non-contrast computed tomography and two-stage deep learning model. *Comput Methods Programs Biomed*, *196*:105711. doi:10.1016/j.cmpb.2020.105711

Oman, O. et. al. (2019). 3D convolutional neural networks applied to CT angiography in the detection of acute ischemic stroke. *Eur Radiol Exp*, *3*(1)8. doi:10.1186/s41747-019-0085-6

Ramamurthy, K.N., Varshney, K.R., & Mody, K. (2018). Topological Data Analysis of Decision Boundaries with Application to Model Selection. arXiv:1805.09949v1 [stat.ML] 25 May 2018

Rieck, B., et al. (2019). Neural persistence: A complexity measure for deep neural networks using algebraic topology. *International Conference on Learning Representations* (ICLR) (2019)

Romero, R.A.A, et al. (2022). Benchmarking AutoML frameworks for disease prediction using medical claims. *BioData Mining* V. 15, 15. https://doi.org/10.1186/s13040-022-00300-2

Ruksakulpiwat, S. et al. (2023). Using Neural Networks Algorithm in Ischemic Stroke Diagnosis: A Systematic Review. *J Multidiscip Healthc, 16*, 2593-2602. https://doi.org/10.2147/JMDH.S421280

Saranya, A. & Subhashini (2023). Systematic review of Explainable Artificial Intelligence models and applications: Recent developments and future trends. *Decision Analytics Journal*, 7, 100230.

Shinohara, Y. et al. (2020). Development of a deep learning model to identify hyperdense MCA sign in patients with acute ischemic stroke. *Jpn J Radiol*. 2020;*38*(2), 112–117. doi:10.1007/s11604-019-0089

Shovkoplias, G. et al. (2023). Proactive selection of machine learning models for small sample sizes in cerebral stroke detection based on PAC-learning theory. *12th International Young Scientist Conference on Computational Science* (YSC 2023)

Speith, T. (2022). A Review of Taxonomies of Explainable Artificial Intelligence (XAI) Methods. *2022 ACM Conference on Fairness, Accountability, and Transparency* (FAccT '22), June 21–24, 2022, Seoul, Republic of Korea. 12p. https://doi.org/10.1145/3531146.3534639

Stib, MT, & Vasquez, J. (2020). Detecting large vessel occlusion at multiphase CT angiography by using a deep convolutional neural network. *Radiology* 2020;*297*(3), 640–649. doi:10.1148/radiol.2020200334

Thomas, R.M. et al. (2020). Dealing with missing data, small sample sizes, and heterogeneity in machine learning studies of brain disorders. *Machine Learning. Methods and Applications to Brain Disorders*. Chapter 14. 2020, pp. 249-266.

Valiant, L. (2013). *Probably Approximately Correct: Nature's Algorithms for Learning and Prospering in a Complex World*. Basic Books, 2013. 208 p.

Valliani, A. A. A. et al. (2019). Deep learning and neurology: a systematic review. *Neurology and Therapy*, *8*(2), 351–365.

Vapnik, V.N., & Chervonenkis, A.Ya. (2015). On the Uniform Convergence of Relative Frequencies of Events to Their Probabilities. *Measures of Complexity*. p. 11. doi:10.1007/978-3-319-21852-6_3. ISBN 978-3-319-21851-9.

Wang, CY. et al. (2021). Deep learning-based identification of acute ischemic core and deficit from non-contrast CT and CTA. *J Cereb Blood Flow Metab*, *41*(11), 3028–3038. doi:10.1177/0271678x211023660

Wasserman, L. (2018). Topological data analysis. *Annual Rev. Stat. Appl.* 5, 501–532 (2018)

Writing Group Members et al. (2016). Heart Disease and Stroke Statistics-2016 Update: A Report from the American Heart Association. *Circulation*, *133*(4), 38-360.

Zhang, SJ et al. (2021). Stroke lesion detection and analysis in MRI images based on deep learning. *J Healthc Eng*. 2021:5524769. doi:10.1155/2021/5524769

Zhu, L. et al. (2020). Make-A-Volume: Leveraging Latent Diffusion Models for Cross-Modality 3D Brain MRI Synthesis. arXiv: 07.10094v1 [eess.IV] 19 Jul 20

Zia, A. et al. (2023) Topological Deep Learning: A Review of an Emerging Paradigm. arXiv:2302.03836v1 [cs.LG] 8 Feb 2023.

Zomorodian, A., & Carlsson, G., 2005. Computing persistent homology. *Discrete Comput. Geom, 33*(2), 249–274. https://geometry.stanford.edu/papers/zc-cph-05/zc-cph-05.pdf 17

AI-BASED APPROACH FOR EARLY DIAGNOSIS SUPPORT IN HEMORRHAGIC STROKE

Athanase Sawadogo[1], Lydie Simone Tapsoba[2], Rodrique Kafando[3], Abdoul Kader Kabore[3], Aminata Sabane[1] and Tegawende François d'Assise Bisyande[1]
[1]*Computer science department, UFR/SEA, Joseph KI-ZERBO University, Burkina Faso*
[2]*Mathematics and Informatics laboratory (LAMI), Ouagadougou, Burkina Faso*
[3]*Centre d'Excellence CITADEL, Université Virtuelle of Burkina Faso, Burkina Faso*

ABSTRACT

A hemorrhagic stroke is a life-threatening medical condition that happens when a blood vessel in your brain ruptures and bleeds. It constitutes a burden on health services and the victim's family. The current definitive diagnosis of stroke is based on brain scanning. However, the clinical diagnosis of hemorrhagic stroke is complex and depends on the skills and experience of the practitioner. Human diagnostic errors lead to delays in treatment and thus compromise clinical outcomes. Our vision is to propose an artificial intelligence approach for medical assistance in the early clinical diagnosis of hemorrhagic strokes. We studied and compared three machine learning models, namely logistic regression, Random Forest and artificial neural networks, to choose the best one after setting up a stroke dataset and identifying the most important characteristics. We can conclude that our system designed with artificial intelligence is important with satisfactory results to help health workers make the rapid diagnosis of hemorrhagic stroke and promote rapid treatment of suspected patients.

KEYWORDS

Hemorrhagic Stroke, Machine Learning, Clinical Data

1. INTRODUCTION

Stroke, a leading cause of global mortality, remains a significant public health concern (Incidence & Collaborators, 2018; Luft, Andrea & Katan, 2018). According to World Health Organization (WHO) reports cited by Mendis, Puska, & Norrving (2011), strokes account for 6.2 million deaths annually, with projections indicating sustained high mortality rates through 2030. Ischemic strokes, caused by blockages in cerebral blood vessels, and hemorrhagic strokes, resulting from ruptured arteries, are the two primary types (Types of stroke, 2023). Early diagnosis of hemorrhagic stroke is crucial for prompt medical intervention, as clinical symptoms often overlap with ischemic strokes, complicating accurate diagnosis (Shehab, et al., 2022). The complexity of hemorrhagic stroke diagnosis underscores the potential of artificial intelligence (AI) and data analysis techniques in medical applications (Shin & Lee, 2020; Shatte, Hutchinson, & Teague, 2019). Misdiagnosis can lead to treatment delays and adverse outcomes, emphasizing the need for reliable diagnostic tools. Our research aims to leverage AI to enhance clinical diagnosis by identifying pertinent risk factors and characteristics from stroke registries. This involves developing an AI model tailored to classify hemorrhagic strokes effectively, structured into literature review, methodology, and results sections to achieve our objectives.

2. STAT OF THE ART

The research we studied spans the period from 2018 to 2023. In most studies, the authors compared methods and algorithms to identify the best ones. Several studies focused on predicting pre-hospital outcomes (Zhao, et al., 2021; Qu, et al., 2022; Heo, JoonNyung, et al., 2019; Dr. V. Jyothsna & Dr. M. Rajkumar, 2023; Sailasya & Kumari, 2021; S K Uma & Rakshith S R, 2022; Bandi, Bhattacharyya, & Midhunchakkravarthy, 2020). They used some of machine learning algorithms such as Random Forest (RF), Logistic Regression(LR), Generalized Regression Neural Network (GRNN), Deep Neural Network (DNN), K-Nearest Neighbors

(KNN), Support Vector Machine (SVM), Decision Trees (DT), Naïve Bayes(NB), Multi-Layer Perceptron (MLP), Gaussian Naive Bayes (GNB), Bernoulli Naive Bayes (BNB), Radial Basis Function Kernel Support Vector Machine (RBG-SVM), Gradient Boosting Classifier (GB Classifier), Stochastic Gradient Descent (SGD), AdaBoost, AdaBoost with Stochastic Gradient Descent (SGD) to train models to classify and predict stroke cases in general. Their results were interesting, ranging from 80% to 98.54%. Unfortunately, they did not use the same data sizes, characteristics, algorithms, or evaluation metrics in their comparisons.

Classification models have been used to predict post-stroke functional outcomes (Chang et al., 2021; Kim, Choo, & Chang, 2021; Ashrafuzzaman, Saha & Nur, 2022). These authors compared various algorithms such as DT, NB, KNN, Linear Discriminant Analysis, AdaBoost (AB), SVM, LR, RF, and Deep Neural Networks (DNN). They focused on combining and comparing methods for predicting motor function outcomes after stroke (Chang et al., 2021; Kim, Choo, & Chang, 2021). Chang et al. (2021) used a stacking model that combines predictions from several ML models, demonstrating that this method does not improve outcomes. As their studies were based on determining functional outcomes, they used features that provided good results but are not accessible at the pre-hospital level.

Other authors specifically focused on hemorrhagic strokes, but rather on predicting the risk of intracranial hemorrhage (ICH) in patients on hemodialysis (Fengda Li, et al., 2023), as well as evaluating and comparing the performance of machine learning models to predict mortality 90 days post-discharge (Tang, et al., 2022). These last two studies also did not take into account patients in general clinical situations.

Studies show that the Logistic Regression algorithm is used in all stroke classification studies, followed by Random Forest. Random Forest achieves the highest precision in three studies (Zhao, et al., 2021; Dr. V. Jyothsna & Dr. M. Rajkumar, 2023; Bandi, Bhattacharyya, & Midhunchakkravarthy, 2020) with 83%, 95.56%, and 94.32% respectively. Neural networks are less frequently used but demonstrate greater efficiency in their applications compared to other methods (Qu et al., 2022; Tang et al., 2022; Ashrafuzzaman, Saha, & Nur, 2022; Kim, Choo, & Chang, 2021; Sailasya & Kumari, 2021; and S K Uma & Rakshith S R, 2022). From these studies, 13 critical clinical characteristics of hemorrhagic stroke have been identified (Qu et al., 2022; Tang et al., 2022; Li et al., 2023; Ashrafuzzaman, Saha, & Nur, 2022; Kim, Choo, & Chang, 2021; Bandi, Bhattacharyya, & Midhunchakkravarthy, 2020; Sailasya & Kumari, 2021; and Heo et al., 2019). In some studies, resampling methods such as SMOTE, RUS, ADASYN, Borderline, and SMOTEENN were applied to the study data, and the authors concluded that these methods do not improve model results (Tang et al., 2022; Sailasya & Kumari, 2021). Regarding evaluation metrics, ten studies used AUC to compare models (Qu, et al., 2022; Heo, JoonNyung, et al., 2019; Dr. V. Jyothsna & Dr. M. Rajkumar, 2023; Sailasya & Kumari, 2021; Bandi, Bhattacharyya, & Midhunchakkravarthy, 2020; Chang et al., 2021; Kim, Choo, & Chang, 2021; Ashrafuzzaman, Saha & Nur, 2022; Fengda Li, et al., 2023; Tang, et al., 2022). This literature review allows for a comparison of models such as Logistic Regression, Random Forest, and Artificial Neural Networks, guiding the selection of the best approach for an AI-based medical decision support system for the clinical diagnosis of hemorrhagic stroke, which will be presented in the following chapter.

3. METHODOLOGY

3.1 Proposal Approach

The approach to our work is a process that extends from setting up the dataset to validating the solution. The main steps of this approach are: establishment of a dataset, exploration and preprocessing of data, selection of independent characteristics, division of preprocessed data into sub-bases, construction of models, evaluation and validation of models. The Figure 1 of the architecture clearly specifies this process.

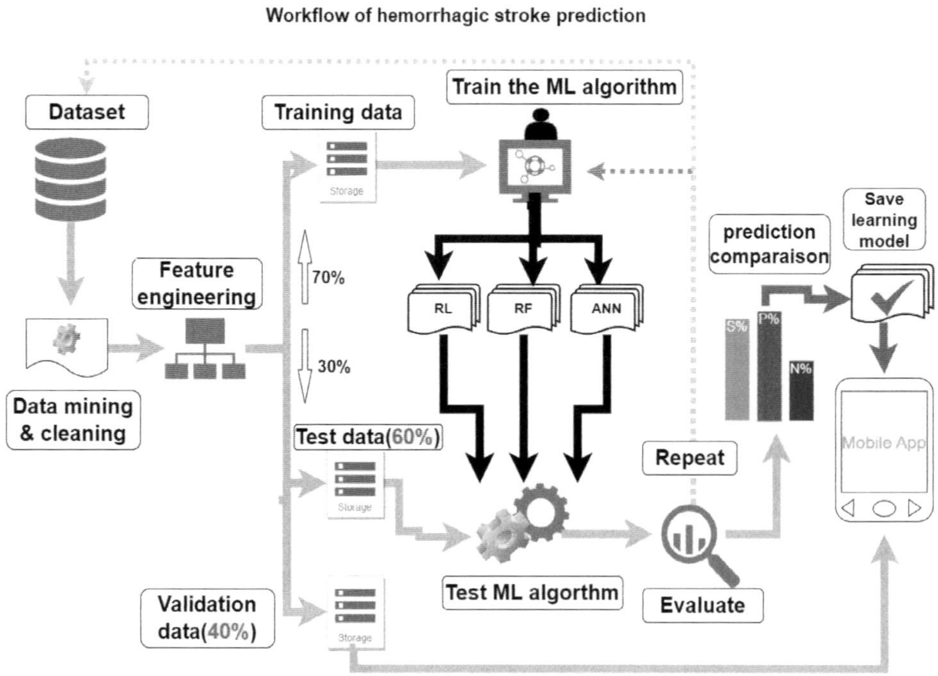

Figure 1. The architecture of the methodology. ML: Machine Learning, RL: Logistic Regression, RF: Random Forest, ANN: Artificial Neural Network

3.2 Implementation

3.2.1 Dataset

We used anonymized secondary data from a hospital in Burkina Faso, consisting of 148 records of patients who experienced stroke confirmed by CT and/or MRI, including only those with hemorrhagic or ischemic strokes. The dataset provided information on clinical symptoms, etiological assessment, treatment, and clinical course. Inclusion criteria focused on relevant clinical information for stroke patients with available radiographic results, while exclusion criteria omitted laboratory tests and scans except for stroke results used as the diagnostic target. Basic analysis and preprocessing involved extracting data from mdb and prj formats into csv and xlsx using Epi Info 7.2.5.0, followed by manual analysis in Excel to select clinical characteristics for AI tools. Initially, the dataset included 148 rows and 129 columns, covering sociodemographic, clinical, paraclinical, etiological, patient care, and pathology evolution data. Five additional columns were added to separate the patient's clinical history into individual characteristics, resulting in 28 selected variables for the study.

3.2.2 Exploratory Analysis

We used Python and the Pandas library to visualize the dataset, which contains 28 columns and 148 rows. The columns include age, sex, profession, origin, sickle cell anemia, blood pressure, stroke history, diabetes, kidney failure, chronic brain disease, family medical history, swallowing disorder, vomiting, nausea, Glasgow Coma Scale, consciousness disorder, HIC syndrome, paralysis, paresis, CKD, motor deficiency, aphasia, commitment, alcohol consumption, stress, tobacco consumption, obesity, and stroke status (hemorrhage). Initially, the dataset columns contained strings, booleans, and integers, with no duplicated data. We identified zero values and calculated the number of null values, finding that some columns had missing values, but the present values exceeded the missing ones. Only the 'Age' attribute contained random numerical values. Using skewness-symmetry, we confirmed the data was not skewed. The 'AVCH' variable was the target for classification: 'False' indicated no risk of hemorrhagic stroke (indicating ischemic stroke), and 'True' indicated a positive case for hemorrhagic stroke. The dataset was almost balanced.

3.2.3 Data Preprocessing

Data preprocessing is essential to remove noise and outliers, ensuring the model trains efficiently by addressing insignificant variables. The steps taken for data cleaning included balancing the data by aligning the number of observations for each class of the target variable. Initially, there were 78 positive AVCH cases and 70 negative cases; the majority class ("NOT AVCH") was subsampled to match the number of positive cases. Value conversion transformed column values into different types for consistency, and data scaling normalized the numeric variable "Age" to aid in studying variable correlations and preparing for independent variable selection. Correlation analysis determined the relationships between variables and assessed their importance concerning the target variable, aiding in selecting the most informative features. Data augmentation was then used to multiply data without duplicates based on specific criteria, resulting in a simulation dataset. After initial processing, 140 rows of data were obtained, and augmentation increased this to 50,000 samples, creating a large "simulation data" set for training effective models. Data augmentation is a strategy aimed at multiplying data without creating duplicates by using various criteria to conduct simulations. We used the "RandomOverSampler" class from the "imblearn.over_sampling" module of the "Imbalanced-Learn" (imblearn) library to balance our dataset. This class performs random oversampling of the minority classes to achieve a more balanced dataset. After instantiating "RandomOverSampler", we applied the "fit_resample()" method from the same library. This method takes the imbalanced data as input and returns a balanced sample in terms of classes. However, this augmented data is far from real-world data, potentially leading to poorly performing models when deployed in real situations.

3.2.4 Feature Engineering

We used three feature selection methods to identify the most important features for training our models and built a new database with these influential features. These methods included the use of the ExtraTreeClassifier algorithm with feature_importances, SelectKBest from scikit-learn's sklearn.feature_selection module, and logistic regression employing the ExhaustiveFeatureSelector (EFS) from the mlxtend.feature_selection module. We summarized the results of these three methods, retaining variables that appeared twice in the results and those ranked in the top three that were not already selected. The final features selected for model training were: "Age", "Stroke", "ATH", "HIC Syndrome", "CKD", "Smoking", "Diabetes" and "Occupation".

3.2.5 Division of Data

We considered both the original and augmented data. Each dataset was divided into sub-datasets with 70% for training and 30% for testing and validation, as shown in Table 1.

Table 1. Dividing data based on original data and augmented data

	Total of data	Training data	Test data	Validation data
Pourcentages	100%	70%	60% from 30%	40% from 30%
Number of original data	140	98	29	13
Number of augmentation data	50000	35000	9000	6000

3.2.6 Models Building

Based on the literature review, we opted to construct three classification models: logistic regression, Random Forest, and artificial neural networks, aiming to compare their efficiencies subsequently. **Logistic Regression (LR):** Utilizing LogisticRegressionCV from sklearn's model_selection module, which integrates cross-validation for automatic parameter selection. We employed GridSearchCV to explore hyperparameter combinations such as regularization parameter Cs= [2, 4, 5, 6], cross-validation settings cv= [4], penalty ['l1', 'l2'], and solver ['saga', 'liblinear']. Parameters were automatically chosen based on data size and "refit" was set to True for automatic model adjustment after parameter selection. **Random Forest (RF):** Constructed using RandomForestClassifier from sklearn.ensemble, with parameters set as oob_score=True, random_state=42, warm_start=True, and n_jobs=-1. Hyperparameter tuning via GridSearchCV determined the optimal number of estimators (n_estimators=15 with original data, n_estimators=30 with augmented data), considering data size variations and possible values like 15, 30, 40, and 50. **Artificial Neural Networks (ANN):** Developed using Keras from TensorFlow, configured with a sequential model. The network included an input layer (units=50, activation='relu', input_dim=8), three hidden layers (units=100, activation='relu'), and an output layer (unit=1, activation='sigmoid'). The model was compiled with optimizer='adam', loss='binary_crossentropy', and metrics=['accuracy']. Hyperparameters for layers and neurons were manually adjusted to optimize performance, bypassing formal hyperparameter selection techniques.

3.2.7 Evaluation and Validation Methods

In the first step, we assessed and compared the models' performance using evaluation metrics such as accuracy_score, f1_score, precision_score, recall_score, and roc_auc_score from the Sklearn.metrics module in Python. Additionally, we utilized confusion matrices. In the second step, we employed validation data to evaluate the best-performing model saved from the first step.

3.2.8 Development Environment and Tools

We developed our project using Windows 10 on a Lenovo PC with 500GO of disk and Google Colab for Python execution. Our toolkit included Epi Info 7 for data conversion, Excel for preprocessing, Draw.io for diagrams, overleaf for writing, PowerPoint for presentations.

4. OUTCOMING

4.1 Results Obtained

4.1.1 Comparison of the Results obtained with the Original Data and Simulation Data

We compared the outcomes of the three trained models with both real data and simulated data, presenting the results in Table 2.

Table 2. Comparison of models built on the basis of original data et simulation data

Data Category	Models	Accuracy	F1-score	Precision	Recall	AUC	Confusion Matrix
Original data (140)	**Random Forest**	**73.07%**	**69.56%**	**66.66%**	**72.72%**	**73.03%**	**73.07%**
	ANN Model	50%	51.85%	43.75%	63.63%	51.81%	50%
	Logistic Regression	46.15%	22.22%	28.57%	18.18%	42.42%	53.84%
Simulation data (50000)	**Random Forest**	**99.33%**	**99.33%**	**98.32%**	**100%**	**99.15%**	**99.33%**
	ANN Model	98.61%	98.58%	100%	97.21%	98.60%	98.61%
	Logistic Regression	68.66%	66.61%	68.27%	68.95%	68.66%	99.15%

4.1.2 Saving and Validating Best Model

When validating the saved model, the Random Forest model achieved a prediction accuracy of 99.05% using the validation data.

4.2 Discussion

4.2.1 Interpretation of Results

We found that the Random Forest model outperforms other models with an accuracy of 99.33% on test results, compared to 98.61% for ANN and 68.66% for logistic regression. Due to its superior performance, we chose to save the Random Forest model and interpret its results. Considering accuracy, we notice that the correct predictions among all the predictions made were 99.33%. This shows that the model is effective in classifying hemorrhagic strokes using clinical data. This performance is crucial in the context of diagnosing hemorrhagic stroke, as it minimizes the risk of misclassifying healthy individuals as sick. Moreover, the model achieved a validation accuracy of 99.05% on new data, demonstrating its robustness and reliability. About the model that is strong classification performance, it can be attributed to several factors: algorithmically, the Random Forest model excels in mitigating overfitting by aggregating predictions from multiple decision trees, enhancing generalization to new data. Its capability to handle large and complex datasets, along with robustness in managing nonlinear characteristics, further improves its effectiveness. Automatic hyperparameter selection tailored to the dataset size optimized model performance. Additionally, the high data quality, ensured through rigorous preprocessing, and the positive impact of feature selection techniques collectively contributed to the model's success. In conclusion, we are pleased with the results, which align well with our expectations.

4.2.2 Put into Perspective with the State of the Art

As we did not use the same dataset or data size as those in the state-of-the-art studies, direct comparison of our results is not feasible. However, Table 3 , we attempt to compare the numerical outcomes of our Random Forest models trained on the original reduced dataset and simulated data against those from the state-of-the-art studies we deemed superior.

Table 3. Put into perspective with the results of the state of the art

	High-performance model	**Data size**	**Accuracy**
(Zhao, et al., 2021)	Random Forest	4914	83%
(Dr. V. Jyothsna & Dr.M. Rajkumar, 2023)	Random Forest	-	95%
(Bandi, Bhattacharyya, & Midhunchakkravarthy, 2020)	Random Forest	4799	94%
Our study	Random Forest	140 for original data	73.07%
		50000 for simulation data	99.33%

The table illustrates that Random Forest accuracy rates reported in the state-of-the-art studies range from 83% to 95%. Despite having limited original data, we achieved 73.07% accuracy, which improved to 99.33% with simulation data. This suggests that these accuracy rates can improve significantly with sufficient data.

4.2.3 The Difficulty Encountered and the Research Limit

Our main challenge lies in the inadequate quantity of data available for model training. Despite initiating a data request procedure during the internship, delays in response and ethical considerations related to data usage hindered its success. Due to insufficient data, the mobile application we developed is not suitable for real-world deployment, as the integrated model was trained using simulation data.

5. CONCLUSION

Our research focused on developing an AI-based approach to aid in the early diagnosis of hemorrhagic strokes, recognizing their complexity and distinct clinical characteristics. We identified key stroke features using AI techniques and explored advanced algorithms suitable for diagnosis. Our work resulted in a high-performance Random Forest model achieving 99.33% accuracy, albeit trained on simulated data not directly applicable in real scenarios. Future steps involve adapting the model with real data to enhance its practical utility as a supportive tool for healthcare professionals, facilitating faster diagnosis and decision-making without replacing medical expertise.

ACKNOWLEDGEMENT

We extend our gratitude to Dr. R. Aristide YAMEOGO, cardiologist and faculty member at Joseph KI-ZERBO University, and the Director of the Interdisciplinary Center of Excellence, for generously providing us with anonymous data for our research.

REFERENCES

Ashrafuzzaman, M., Saha, S., & Nur, K. (2022). Prediction of Stroke Disease Using Deep CNN Based Approach. *Journal of Advances in Information Technology Vol, 13*(6). From http://www.jait.us/issues/JAIT-V13N6-604.pdf

Bandi, V., Bhattacharyya, D., & Midhunchakkravarthy, D. (2020). Prediction of Brain Stroke Severity Using Machine Learning. *Revue d'Intelligence Artificielle, 34*(6). From https://scholar.archive.org/work/e3ff3d4fwbdbtfwb2ntynco6pq/access/wayback_/http://www.iieta.org/download/file/fid/48077

Chang, S.-C., Chu, C.-L., Chen, C.-K., Chang, H.-N., Wong, A., Chen, Y.-P., & Pei, Y.-C. (2021). The comparison and interpretation of machine-learning models in post-stroke functional outcome prediction. *Diagnostics, 11*(10), 1784. From https://www.mdpi.com/2075-4418/11/10/1784/pdf

Dr. V. Jyothsna, & Dr.M. Rajkumar. (2023). Application Of Supervised Machine Learning Techniques For Classification Of Brain Stroke. *Journal of Clinical Otorhinolaryngology, Head, and Neck Surgery, Associate Dean (Academic A fairs) and Associate Professor Department of CSIT, School of Computing, MOHAN BABU UNIVERSITYl vol, 27*(1), 1001-1781. From https://www.lcebyhkzz.cn/article/view/2023/377.pdf

Heo, JoonNyung, Yoon, Jihoon G, Park, H., Kim, Y., Nam, H., & Heo, J. (2019). Machine learning--based model for prediction of outcomes in acute stroke. *Stroke, 50*(5), 1263-1265. From https://www.ahajournals.org/doi/full/10.1161/STROKEAHA.118.024293

Incidence, G. 2., & Collaborators, P. (2018). Global, regional, and national incidence, prevalence, and years lived with disability for 354 diseases and injuries for 195 countries and territories, 1990-2017: a systematic analysis for the Global Burden of Disease Study 2017. *Lancet (London, England)*(392(10159)), 1789–1858. doi:10.1016/S0140-6736(18)32279-7

Kim, J., Choo, Y., & Chang, M. (2021). Prediction of motor function in stroke patients using machine learning algorithm: Development of practical models. *Journal of Stroke and Cerebrovascular Diseases, 30*(8), 105856. From https://doi.org/10.1016/j.jstrokecerebrovasdis.2021.105856

Types of stroke. (2023, 11 02). From coeuretavc: https://www.coeuretavc.ca/avc/questce-quun-avc/les-types-d-avc

Li, F., Chen, A., Li, Z., Gu, L., Pan, Q., Wang, P., ... Feng, J. (2023). Machine learning-based prediction of cerebral hemorrhage in patients with hemodialysis: A multicenter, retrospective study. *Frontiers in Neurology, 14*, 1139096. From https://www.frontiersin.org/articles/10.3389/fneur.2023.1139096/full

Luft, Andrea, & Katan, M. (2018, 05 23). Global Burden of Stroke. *Semin Neurol, 38*, 208-211. doi:10.1055/s-0038-1649503

Mendis, S., Puska, P., & Norrving, B. (2011, 01 01). Global atlas on cardiovascular disease prevention and control. *World Heart Federation and World Stroke Organization*. From https://www.researchgate.net/publication/311885270_Global_atlas_on_cardiovascular_disease_prevention_and_control_WHO

Qu, S., Zhou, M., Jiao, S., Zhang, Z., Xue, K., Long, J., ... others. (2022). Optimizing acute stroke outcome prediction models: Comparison of generalized regression neural networks and logistic regressions. *Plos one, 17*(5), e0267747. From https://journals.plos.org/plosone/article?id=10.1371/journal.pone.0267747

S K Uma, & Rakshith S R. (2022). Stroke Analysis Using 10 ML Comparison. *International Journal for Research in Applied Science & Engineering Technology (IJRASET) Vol, 10*(10), 2321-9653. From https://doi.org/10.22214/ijraset.2022.45895

Sailasya, G., & Kumari, G. (2021). Analyzing the performance of stroke prediction using ML classification algorithms. *International Journal of Advanced Computer Science and Applications, 12*(6). From https://pdfs.semanticscholar.org/df5c/7d1bd7a59009dc51b9db903aa7f144241879.pd

Shatte, A., Hutchinson, D., & Teague, S. (2019). Machine learning in mental health: a scoping review of methods and applications. *Psychological medicine, 49*(9), 1426-1448. From https://doi.org/10.1017/S0033291719000151

Shehab, Mohammad and Abualigah, Laith and Shambour, Qusai and Abu-Hashem, Muhannad A and Shambour, Mohd Khaled Yousef and Alsalibi, ... Amir H. (2022). Machine learning in medical applications: A review of state-of-the-art methods. *Computers in Biology and Medicine, 145*, 105458. From https://doi.org/10.1016/j.compbiomed.2022.105458

Shin, Y., & Lee, I. (2020). Machine learning for enterprises: Applications, algorithm selection, and challenges. *Business Horizons, 63*(2), 157-170.

Tang, J., Wang, X., Wan, H., Lin, C., Shao, Z., Chang, Y., ... Du, Y. (2022). Joint modeling strategy for using electronic medical records data to build machine learning models: an example of intracerebral hemorrhage. *BMC Medical Informatics and Decision Making, 22*(1), 278. From https://link.springer.com/article/10.1186/s12911-022-02018-x

Zhao, Y., Fu, S., Bielinski, S., Decker, P., Chamberlain, A., Roger, V., ... Larson, N. (2021). Natural language processing and machine learning for identifying incident stroke from electronic health records: algorithm development and validation. *Journal of medical Internet research, 23*(3), e22951. From https://www.jmir.org/2021/3/e22951/

TEST AMPLIFICATION FOR MEDICAL APPS IMPLEMENTING LINEARLY-APPROXIMATABLE FUNCTIONS

Ian Almutawa, Muyeed Ahmed, Iulian Neamtiu and Peter Flis
New Jersey Institute of Technology, USA

ABSTRACT

Mobile (smartphone) apps are increasingly used in medical, health, and fitness settings, but their reliability can be poor, which has negative healthcare consequences. Researchers have started to scrutinize the correctness and reliability of such apps, but the techniques used are unsatisfactory. For example, prior work has used ad-hoc techniques to discover some errors and incorrect calculations, but we still lack automated procedures that systematically look for errors. We introduce a novel approach that provides test "amplification": given a few test cases for an app, we use a linear approximation to model app behavior, compute an error function, and then look for its maxima. The error function maxima are further test cases that exhibit higher errors. Our approach can generally be applied to any calculations based on linearly-approximatable functions. We have applied our approach to 54 Android apps. We have primarily focused on apps computing the Basal Metabolic Rate (BMR). While manual testing discovered errors of at most 0.3%, amplification found errors as large as 5.4%. We also show how our approach can perform effective test amplification, from 2.4% to 8%, for recently-found issues in Android apps that calculate the Body-Surface Area.

KEYWORDS

Medical Apps, Health Apps, Android, Software Reliability, Software Testing, Linear Approximation

1. INTRODUCTION

Mobile (smartphone) apps are used in a variety of health-related scenarios where calculation accuracy is paramount. For example, medical apps can compute drug dosages or diagnostic scores, fitness apps can keep track of, and recommend, exercise or dietary regimens. However, the low barrier to entry (ease of publishing apps onto Google Play or Apple's App Store) has led to a proliferation of medical, health, and fitness apps that are under-scrutinized in terms of correctness, reliability, or security (Samuel et al., 2022). Researchers have started to address these issues using a standard approach: first, identifying a problem domain, e.g., insulin, opioid, body-surface area calculations (Huckvale et al., 2015, Haffey et al., 2013, Almutawa et al., 2024); second, within this domain, developing test cases with known inputs and outputs. While these approaches have managed to expose certain errors, they have several issues. First, coverage is limited to the manually-constructed inputs: developing new test cases takes additional manual effort. Second, the chosen inputs might not cover the input domain effectively, hence the researchers/testers do not know if the error found actually represents the most severe violations for the specification. For example, if a test case has revealed a 0.5% error, it is unclear if the error can be even higher for other inputs. Third, for many medical calculations, several alternative formulas can be used; while some apps do specify the formula they use, other apps do not, which can lead to incorrect results or user confusion. We introduce an approach that addresses these problems. Given a "ground truth" function GT, we construct a function A that models app behavior, as well as an error function $\Delta = GT - A$. Our approach is centered on finding maxima of Δ.

To evaluate our approach, we focus on two widely-used health calculations, First, Basal Metabolic Rate (BMR), a linear function, as implemented in 31 apps. We also examine 23 Body-Surface Area (BSA) apps, to show how our approach can effectively improve upon recent work in this area. While BSA is not a linear function itself, we can construct a linear approximation via a log transformation.

Our results show that (1) we can effectively infer the formulas used by apps; (2) we can amplify test seeds effectively, automatically finding errors that are 3–18 times greater than those found manually in the seeds; (3) as few as 10–20 seeds can suffice to allow effective amplification.

Prior work in this area has used manually constructed inputs. (Huckvale et al., 2015) used minimum, maximum, and typical values for insulin calculators. (Haffey et al., 2013) used 1mg, 100mg, and range median for opioid conversion. (Almutawa et al., 2024) have focused on BSA, which can be computed using two main formulas, and used input points that aimed to maximize the difference between the two formulas.

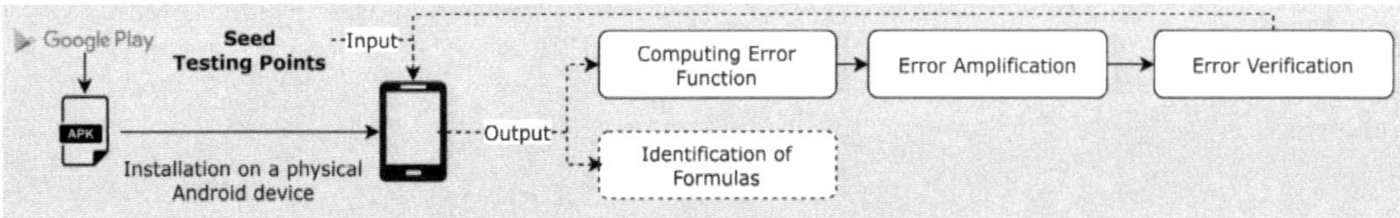

Figure 1. Overview of the amplification process

2. APPROACH

The architecture of our approach is shown in Figure 1: given an app implementing a linearly-approximatable function, we manually construct a few inputs, and record the app output. These input-output pairs (seeds) will then be used toward several purposes. First, we can identify the formula used (for apps that do not specify the formulas they implement). Second, we compute the error function and its maxima over the input range, to amplify any potential error; these maxima are then used as test cases.

Definitions. We denote $y=GT(\underline{x})$ as a multivariable "ground truth" function, i.e., the formula that the app is supposed to implement. This formula maps multivariable inputs $\bar{x}_i$ to output y_i. While GT does not need to be linear, it needs to be linearly-approximatable. Our approach starts with a few test "seeds", $<(\underline{x}_1,y_1),...,(\underline{x}_n,y_n)>$ where again y_i is the output produced by the app for the given input vector $\bar{x}_i$; our experiments show that as little as 10–20 seeds suffice. From the seeds, we model app behavior by constructing a multivariable linear function $A(\underline{x})$, i.e., a linear approximation of GT, constructed via linear regression from the given test seeds. Next, we compute the error function $\Delta = GT - A$. Then, we find values $\overline{x^a}$ of Δ across the input range (minima and/or maxima), that maximize the absolute value of GT - A. For simplicity let us denote $\Delta_i = GT(\bar{x}_i) - A(\bar{x}_i)$, i.e., the error at point i. Let a be a maximizing point for the error, i.e., $\Delta_a = GT(\overline{x^a}) - A(\overline{x^a})$. Let m be the point (among the seeds) with the maximum error, in absolute value, i.e., $\Delta_m = GT(\bar{x}_m) - A(\bar{x}_m)$. We define *amplification* as the ratio Δ_a/Δ_m. If our approach is successful, $\Delta_a/\Delta_m > 1$.

3. AMPLIFICATION: BASAL METABOLIC RATE

Basal Metabolic Rate (BMR). The BMR is a person's rate of burning calories per unit of time at rest; it typically represents 60–80% of a person's daily caloric needs (Manini, 2010); for context, examples of BMR values are 500 kcal/day for a toddler or 1500 kcal/day for an adult. BMR is widely used in health and fitness scenarios to prescribe nutrition regimens, e.g., to gain or lose weight. While several BMR formulas exist, and the formula to use depends on gender, without loss of generality in this paper we focus on three popular formulas that compute a male's BMR given their *weight*, *height*, and *age*. The formulas (i.e., multivariate functions), named after their authors, *Original Harris-Benedict* (Harris et al., 1918), *Revised Harris-Benedict* (Roza et al., 1984), and *Mifflin St. Jeor* (Mifflin et al., 1990), are provided below:

$$BMR_{Original\ Harris-Benedict} = 13.751 * Weight + 5.0033 * Height - 6.755 * Age + 66.473 \quad (1)$$

$$BMR_{Revised\ Harris-Benedict} = 13.397 * Weight + 4.799 * Height - 5.677 * Age + 88.362 \quad (2)$$

$$BMR_{Mifflin\ St\ Jeor} = 10 * Weight + 6.25 * Height - 5 * Age + 5 \quad (3)$$

To be clear, in this scenario, the formulas are the ground truth functions *GT*; $\underline{x}_i$ are *<Weight, Height, Age>* combinations; and y_i is the BMR computed by the app.

3.1 Motivating Examples: Errors in BMR Calculation

We now present two apps where we discovered errors. First, app *com.roliosolutions.bmr.bmi.calculator*, shown in Figure 2, which uses the *Original Harris-Benedict* formula. Upon entering the height, weight, and age as 173 cm, 86 kg, and 65 years, respectively, the app calculated the BMR to be 1667 kcals/day. However, based on equation (1), the BMR value should be 1675.61 kcals/day. Figure 3 shows a similar discrepancy in another app, *com.microhealthllc.bmr_calculator*, which uses the same formula. Given the same input as before (though note that the app displays the height and weight in the imperial, rather than metric system), the app reported a BMR of 1671.3 kcals/day, whereas the correct value is 1675.61. These errors suggest that the apps may either have incorrectly implemented the formula or used a different equation entirely.

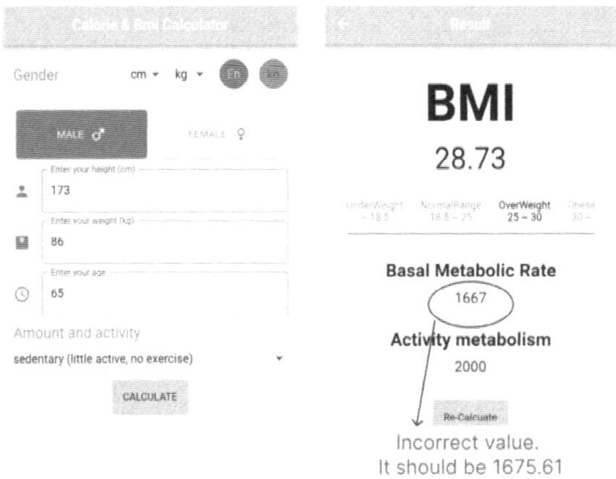

Figure 2. BMR calculation in app
com.roliosolutions.bmr.bmi.calculator

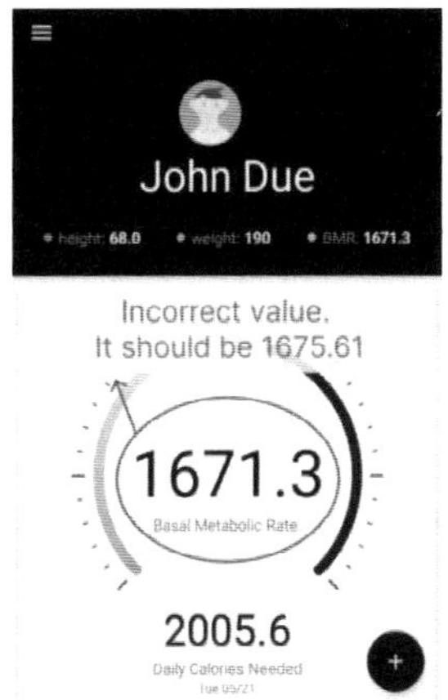

Figure 3. BMR calculation in app
com.microhealthllc.bmr_calculator

3.2 Results: Inferred Formulas

We now discuss how we compute the function $A(\underline{x})$. Given the seeds $\bar{x}_i$ in the form *<Weight, Height, Age>* combinations and the BMR y_i calculated by the app, we use linear regression analysis as follows:

$$A(\bar{x}) = \beta_1 * Weight + \beta_2 * Height + \beta_3 * Age + \beta_0 \quad (4)$$

Here, β_1, β_2, and β_3 are the coefficients for weight, height, and age, while β_0 is the intercept. Table 1 shows three apps with their implemented formula that we derived using the regression analysis: the coefficients are shown under β's in each row, while the *Δ*, i.e., the absolute difference between the coefficients or intercepts implemented by the apps, and GT, are shown in parentheses, in italics.

Table 1. BMR: derived formulas used in apps and the Δ vs GT [MSJ: Mifflin St Jeor, OHB: Original Harris-Benedict, RHB: Revised Harris-Benedict]

App ID	Formula	Coefficient (*Δ*)			Intercept (*Δ*)
		β_1	β_2	β_3	β_0
com.calculatorio.fitness_health.bmr	MSJ	10.01 (*0.01*)	6.246 (*0.004*)	-5.011 (*0.011*)	5.457 (*0.457*)
com.firstcenturythinking.bmrcalculator	OHB	13.783 (*0.031*)	4.986 (*0.017*)	-6.767 (*0.012*)	66.827 (*0.354*)
com.chowdhuryelab.bmiandbmrcalculator	RHB	13.397 (*0*)	4.799 (*0*)	-5.677 (*0*)	88.365 (*0.003*)

Therefore, for app *com.calculatorio.fitness_health.bmr* we have:

$A(\bar{x}) = 10.01 * Weight + 6.246 * Height - 5.011 * Age + 5.457$

$\Delta(\bar{x}) = -0.01 * Weight + 0.004 * Height + 0.011 * Age - 0.457$

$GT(\bar{x}) = 10 * Weight + 6.25 * Height - 5 * Age + 5$

Section 3.3 will show how we look for maxima of Δ to potentially amplify the error. In order to evaluate our regression models, i.e., inferred formulas, for each app we computed the R^2 score between the actual outputs from the app and the predicted outputs by the model across 25 test cases. Table 2 shows the average R^2 scores of all inferred formulas. The R^2 values ranged between 0.99–1, indicating a high level of model accuracy, thus proving the derived formulas to be a very close approximation of the app's formula.

Table 2. BMR: Average R^2 score for the 3 formulas

Formula	Apps	Avg. R^2
Mifflin St. Jeor	10	0.99
Original Harris-Benedict	1	1
Revised Harris-Benedict	7	0.99

3.3 Error Amplification using the Derived Formulas

The amplification is performed by looking for the maxima of Δ given using equation (5). Here, *a, b, c,* and *i* are the coefficients and the intercept of the ground truth formula mentioned by the app.

$$\Delta(\bar{x}) = (a - \beta_1) * Weight + (b - \beta_2) * Height + (c - \beta_3) * Age + (i - \beta_0) \quad (5)$$

In other words, the problem reduces to finding point(s) $\bar{x} = \langle Weight, Height, Age \rangle$ where $\Delta(\bar{x})$ is maximum (in absolute value). We solve this numerically using Python over ranges defined as follows. We used two different possible ranges for *Age*: 2–17 years and 18–100 years, respectively. These age ranges induce weight and height ranges which we derived from World Health Organization percentile charts for height and age, given the age (WHO, 2021).

Table 3. BMR: Amplification across different apps.

App	Error Achieved with Seeds (%)		Maximum Error After Amplification (%)	Amplification Factor (x)	
	Mean	Max		Mean	Max
com.nacdevstudios.basalmetabolicrate	0.146	0.293	5.421	37.13	18.5
com.firstcenturythinking.bmrcalculator	0.079	0.099	0.964	12.15	9.69
com.chowdhuryelab.bmiandbmrcalculator	0.000	0.000	0.001	7.68	5.31
com.cit.bmi.bmr.bodyfatpercentage.calculator	0.017	0.041	0.124	7.39	3.06
com.calculators.tech.tdee	0.017	0.041	0.124	7.39	3.06
com.agochar.caloriecalculator	0.017	0.041	0.124	7.39	3.06
com.microhealthllc.bmr_calculator	0.232	0.313	1.627	7.01	5.20
com.sadeqe.bmr_calculator	0.017	0.041	0.082	4.90	2.03
com.ca.bmr.calclator	0.015	0.036	0.072	4.83	1.99
com.rickardapps.bmrcal	0.019	0.037	0.072	3.82	1.93
com.roliosolutions.bmr.bmi.calculator	0.321	0.431	1.183	3.68	2.75
com.developerJamey.maintain_calculator	0.425	0.514	1.267	2.98	2.47
com.calculatorio.fitness_health.bmr	0.025	0.046	0.059	2.32	1.28

Table 3 shows our results. For each app, the second and third columns show the mean and maximum error achieved with seeds (manual test inputs), in percent, relative to *GT*. Notice that the error is low, typically less than 0.5%. After amplification, we found one case where the error *exceeded 5%* (first row in the table) and several cases where the error exceeded 1%. *Overall, the error increased by a factor of 1.28x–18.5x.* This indicates that our approach is effective at amplifying the seed test cases.

3.4 Inferring Formulas Implemented by Apps

Many apps fail to disclose the BMR formula used (a problem also noted for insulin calculators (Huckvale et al., 2015), as illustrated next. Notice how app *com.bmi.bmr.calculator.tracker.health.tdee* (Figure 4) fails to specify which formula it uses, whereas app *com.agochar.caloriecalculator* (Figure 5) indicates the formula. Failing to specify the formula is problematic because it introduces user confusion and discrepancies in expectations, e.g., if the user switches apps or providers, they will get different outcomes for the same input.

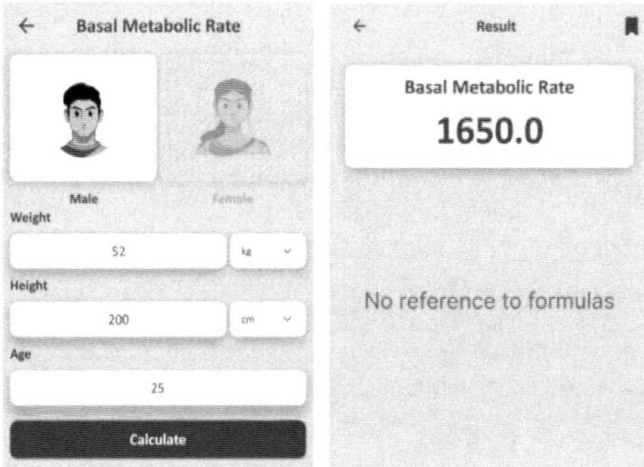

Figure 4. BMR calculation in app
com.bmi.bmr.calculator.tracker.health.tdee

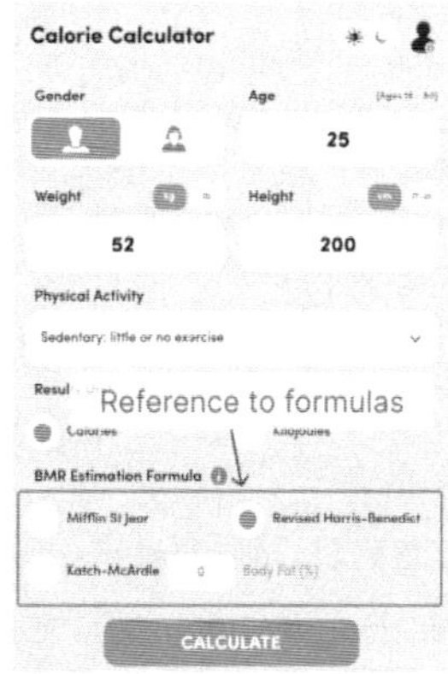

Figure 5. BMR calculation in app
com.agochar.caloriecalculator

Table 4. Apps that do not specify formula: Inferred BMR formulas.

	App ID	Inferred Formula	R^2	p-value
No Error Identified	org.fredriks.bmrcalc	Mifflin St Jeor		
	com.technopaths.health.bmr.weight.water.calorie.bmi.calculator	Mifflin St Jeor		
	com.iw.bmr	Mifflin St Jeor		
	webemobile.com.br.taxametablicabasal	Original Harris–Benedict		
	com.droidinfinity.healthcalculator	Mifflin St Jeor		
	body.health.check.tdee.calculator	Mifflin St Jeor		
	com.nacdevstudios.basalmetabolicrate	Mifflin St Jeor		
Error Identified	com.bmi.bmr.calculator.tracker.health.tdee	Mifflin St Jeor	1	0
	com.cit.bmi.bmr.bodyfatpercentage.calculator	Mifflin St Jeor	0.99	0
	com.d.bmrhsb	Mifflin St Jeor	1	0
	air.com.intemodino.tdeecalculator	Mifflin St Jeor	1	0
	com.tdee.calculator.tracker.bmi.bmr.body.fat.calc	Mifflin St Jeor	1	0
	com.sadeqe.bmr_calculator	Mifflin St Jeor	0.99	0
	com.calculators.tech.tdee	Mifflin St Jeor	0.99	0
	com.ca.bmr.calclator	Original Harris–Benedict	0.99	0
	com.jenos.bmicalculator	Original Harris–Benedict	0.99	0
	com.chowdhuryelab.bmiandbmrcalculator	Revised Harris–Benedict	0.99	0
	com.roliosolutions.bmr.bmi.calculator	Original Harris–Benedict	0.99	0
	com.microhealthllc.bmr_calculator	Original Harris–Benedict	0.99	0
	com.developerJamey.maintain_calculator	Original Harris–Benedict	0.99	0

For cases such as those in Figure 4, to infer the formula used by an app, we first calculated the error for 25 app outputs with the ground truth of the three formulas, giving us, E_{OHB}, E_{RHB} and E_{MSJ}. Next, we conducted three independent two-sample t-tests (SciPy, 2024), *t-test*(E_{OHB}, E_{RHB}), *t-test*(E_{OHB}, E_{MSJ}) and *t-test*(E_{RHB}, E_{MSJ}), each producing three t-statistic values (t_{O_R}, t_{O_M}, t_{R_M}) and three p-values (p_{O_R}, p_{O_M}, p_{R_M}). We then select the largest absolute t-statistic value, $max(abs(t_{O\_R}), abs(t_{O\_M}), abs(t_{R\_M}))$, and depending on that t-statistic value we can determine which of the three formulas is closest to the app's output. Our experiments demonstrate the effectiveness of the method, as we were able to infer the intended formula with very low error, substantially less than 1% for each app. Table 4 contains a list of 20 out of 30 apps that did not state or reference the formula used in their implementation. In the table, we list the formulas we inferred. Note that the first 7 apps did not have errors, so computing R^2 values (goodness of fit) would be moot. The bottom 11 apps did have errors. We show the inferred formulas along with R^2 values and p-values. The R^2 and p-values show that our approach for inferring formulas is effective.

4. AMPLIFICATION: BODY-SURFACE AREA

Body-Surface Area (BSA). The BSA is a person's skin surface area. The BSA is used in several safety-critical scenarios, such as computing chemotherapy dosages (and several drug dosages in general) or assessing burns for burn victims (Moore, 2022). Recent work has discovered that BSA-calculating apps can have calculation errors (Almutawa et al., 2024). However, as with all mobile medical calculating apps, it is unclear if their examined BSA apps can have *larger, undiscovered* errors beyond those discovered via manual testing. Therefore, we apply our approach to the BSA apps examined by prior work. The most popular BSA formulas (named after their authors: *Mosteller* and *DuBois*) use *weight* and *height* as inputs:

$$BSA_{Mosteller} = 0.016667 * Weight^{0.5} * Height^{0.5} \tag{6}$$

$$BSA_{Du\ Bois} = 0.007184 * Weight^{0.425} * Height^{0.725} \tag{7}$$

Note that these definitions are *not* linear functions. Nevertheless, we can use a log transformation to make the function linearly-approximatable hence amenable to our approach, e.g., we have:

$$log(BSA_{Mosteller}) = log(0.016667) + 0.5 * log(Weight) + 0.5 * log(Height)$$

4.1 Results: Inferred Formulas

We now discuss how we compute the formula for BSA for each app, $A(\bar{x})$. In order to facilitate linear regression analysis, we applied a logarithmic transformation to the equation. Specifically, for each test case, we calculated the logarithms of weight, height and BSA output from the app. Given the seeds $\bar{x}_i$ in the form *<log(Weight), log(Height)>*, and the *log(BSA)* as y_i, we performed linear regression to infer the formula:

$$R = \beta_1 * log(Weight) + \beta_2 * log(Height) + \beta_0 \tag{8}$$

$$A(\bar{x}) = e^R \tag{9}$$

Here, β_1 and β_2 are the coefficients for the variable *log(weight)* and *log(height)*, respectively, and β_0 is the intercept. Table 5 shows the coefficients and intercept of three apps that we derived using the regression analysis. The coefficients are shown under β's in each row, while the Δ, i.e., absolute difference between the coefficients or intercepts implemented by the apps, and GT, are shown in parentheses, in italics.

Table 5. BSA: derived formulas used in apps and the Δ vs. GT

App ID	Formula	Coefficient (Δ)		Intercept (Δ)
		β_1	β_2	β_0
br.bsa	Mosteller	0.5 (*0*)	0.499 (*0.001*)	-4.094 (*0*)
com.aci.itim.trauma	Mosteller	0.548 (*0.048*)	0.396 (*0.104*)	-3.785 (*0.309*)
com.medcomis.device.android.egfr	Du Bois	0.427 (*0.002*)	0.716 (*0.009*)	-4.89 (*0.046*)

Therefore, for app *com.aci.itim.trauma* we have:

$$A(\bar{x}) = e^{0.548* log\ (Weight)+0.396*log\ (Height)-3.785}$$

$$\Delta(\bar{x}) = e^{-0.048*log\ (Weight)+0.104*log\ (Height)-0.309}$$

$$GT(\bar{x}) = e^{0.5*log\ (Weight)+0.5*log\ (Height)-4.094}$$

Table 6 illustrates the average R^2 scores of all inferred formulas grouped by the two intended formulas. The average R^2 values for both Mosteller and Du Bois is 0.99, indicating a high level of model accuracy. This suggests the derived formulas are a close approximation of the formula used by the apps.

Table 6. BSA: Average R^2 score for the apps using the BSA formulas

Formula	Apps	Avg. R^2
Mosteller	15	0.99
Du Bois	14	0.99

4.2 Error Amplification using the Derived Formulas

The amplification is performed by looking for maxima of Δ given using equation (10). Here, *a, b* and *i* are the coefficient and the intercept of the ground truth formula mentioned by the app.

$$\Delta(\bar{x}) = e^{(a-\beta_1)*\log(\text{Weight})+(b-\beta_2)*\log(\text{Height})+(i-\beta_0)} \quad (10)$$

Table 7 shows the results demonstrating that amplification is effective. First, app *com.medtricks* was found to have a 2745% error (*off by a factor of 27*) by prior work; our amplification further increased this to 3196%. For app *com.aci.itim.trauma* we found a 7.96% error after amplification (i.e., a 3.27x amplification).

Table 7. BSA: Amplification across different apps

App	Error With Regular Test Cases (%)		Maximum Error (%)	Amplification (x)	
	Mean	Max		Mean	Max
com.medtricks	2726.59	2745.53	3196.06	1.17	1.16
com.aci.itim.trauma	1.669	2.439	7.966	4.77	3.27

5. BALANCING TEST EFFORT AND RESULTING ERROR

Testing many seeds, i.e., a large number of manually-constructed inputs, can be costly and time-consuming. Moreover, our experience from the course of this research indicates that seed "quality" matters: developers need to pick testing points outside the most common values for inputs (e.g., age, weight, height) in order to "stumble" upon large errors. Our test amplification approach alleviates this issue.

Therefore, we performed an analysis to find a number of seeds that balance effort with error, in other words, are sufficient to support effective error discovery and subsequently, amplification. By systematically increasing the number of seeds and analyzing the impact on model accuracy, we can identify a threshold where additional data points no longer significantly enhance the model's performance. We use mean absolute percentage error (MAPE) as a metric, as follows: to calculate the MAPE, we randomly sampled n seeds (where n=[5,10,15,20,25]) from the test cases for each dataset. We then applied linear regression, utilized the resulting model to predict outcomes for the 25 test cases, and subsequently computed the MAPE.

Individual apps. Figure 6 shows this balance for BMR app *com.microhealthllc.bmr_calculator*. The x-axis shows the number of seeds, while the y-axis has the MAPE. Notice how 10 seeds drastically improve model performance, and beyond 10 seeds, the gains are marginal. In contrast, in Figure 7 we show this balance for a BSA app, *com.medpresso.android.ui*. In this case, we need 20 seeds to achieve a stable MAPE.

All apps. We now compute this trade-off across all apps. Figure 8 shows the average MAPE across all BMR apps with differing numbers of seeds: note the drastic, 8x decrease in MAPE, from 0.42% to 0.05%, when moving from 5 to 10 seeds. For BSA apps (Figure 9), the increase from 5 to 10 seeds only yields a MAPE reduction from 0.71% to 0.61% and it takes 20 test cases to stabilize the error to around 0.56%.

These findings allow us to conclude that for apps modeling linear formulas, such as BMR, 10 seeds might be sufficient, whereas for non-linear formulas, such as BSA, 20 points would be needed. Nevertheless, even 20 input points is not a substantial burden for developers, users, or researchers.

6. DISCUSSION

We developed a principled approach that improves the state-of-research in testing medical mobile apps by inferring formulas they use, and then amplifying manually-constructed test inputs. Our study shows that even as little as 10–20 test inputs are sufficient for allowing amplification to "get traction" and discover errors 18.5x (BMR) and 3.2x (BSA) larger than what manual testing has discovered. Further, our study shows that the state-of-practice in medical app development needs further improvement, which requires a concerted effort from developers, app markets, and potentially regulatory entities. Our approach so far works for functions that are linearly-approximatable; we plan to expand the approach to a broader category of functions. We also plan to investigate a wider range of medical calculations and formulas.

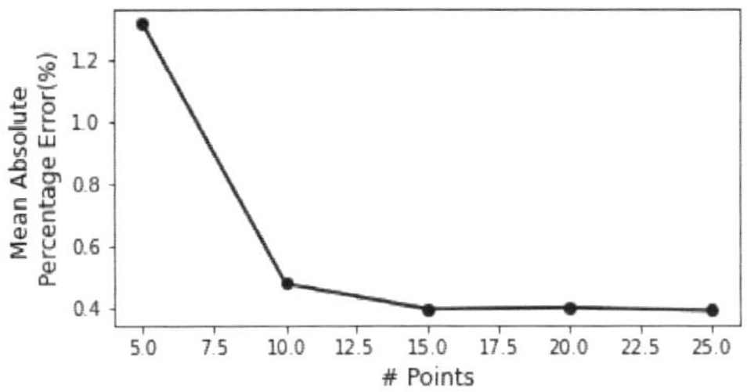

Figure 6. BMR: Impact of seed size on regression model accuracy for *com.microhealthllc.bmr_calculator*

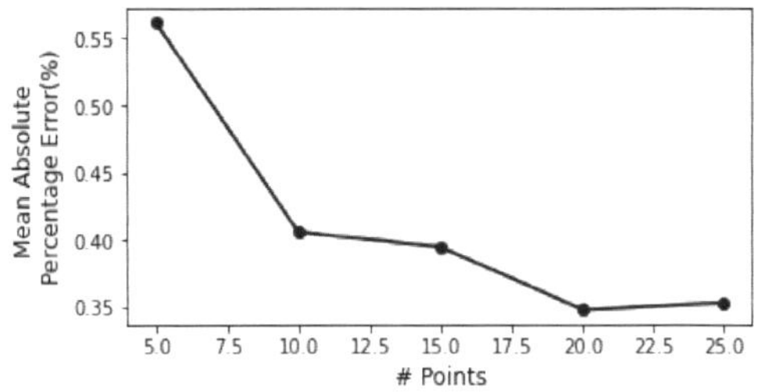

Figure 7. BSA: Impact of seed size on regression model accuracy for *com.medpresso.android.ui*

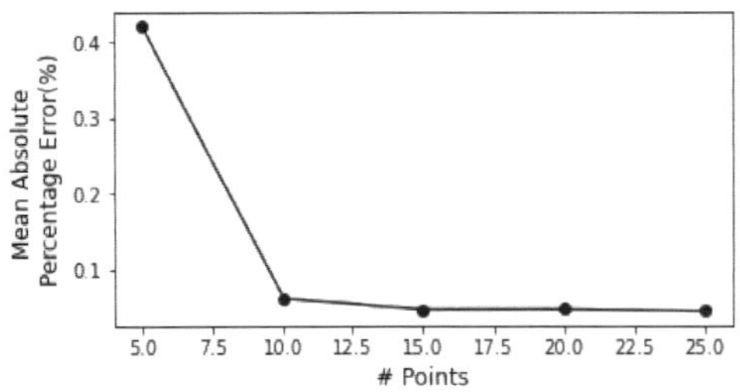

Figure 8. BMR: Seed size vs. model accuracy (all apps)

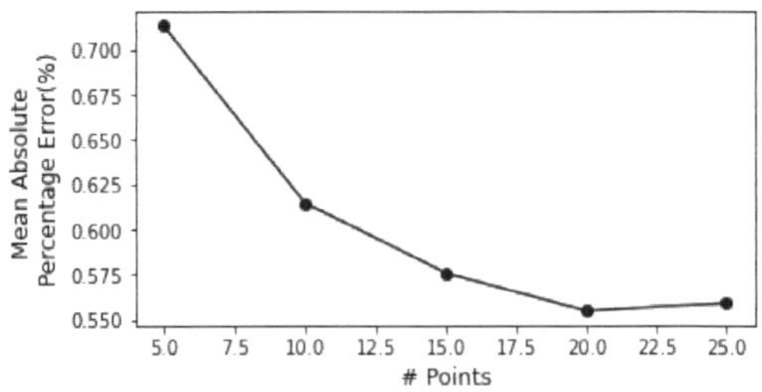

Figure 9. BSA: Seed size vs. model accuracy (all apps)

ACKNOWLEDGMENT

We thank the anonymous reviewers for their feedback. This material is based upon work supported by the National Science Foundation under Grant No. CCF-2106710.

REFERENCES

Almutawa, I., Neamtiu, I. (2024). Verification of body surface area-calculating Android apps. 16th International Conference on e-Health, 2024. To appear.

Bois, D. D. & Bois, E. F. D. (1916) A height-weight formula to estimate the surface area of man. Experimental biology and medicine 13 (4), 77–78.

Haffey, F. et al. (2013) A Comparison of the Reliability of Smartphone Apps for Opioid Conversion. *Drug safety*, *36*(2).

Harris, J. A. & Benedict, F. G. (1918) A Biometric Study of Human Basal Metabolism. *Proceedings of the National Academy of Sciences - PNAS.* 4 (12), 370–373.

Huckvale, K. et al. (2015) Smartphone apps for calculating insulin dose: A systematic assessment. *BMC medicine*. *13*(1).

Manini, T. M. (2010) Energy expenditure and aging. Ageing research reviews. 9 (1), 1–11.

Mifflin, M.D., St Jeor, S.T., Hill, L.A., Scott, B.J., Daugherty, S.A. and Koh, Y.O. (1990). A new predictive equation for resting energy expenditure in healthy individuals. The American Journal of Clinical Nutrition, 51(2), pp.241–247.

Moore, R. A., Waheed, A., & Burns, B. (2022). Rule of Nines. In StatPearls. StatPearls Publishing.

Mosteller, R. D. (1987) Simplified Calculation of Body-Surface Area. The New England journal of medicine. 317 (17)

Roza, A.M. and Shizgal, H.M. (1984). The Harris Benedict equation reevaluated: resting energy requirements and the body cell mass. *The American Journal of Clinical Nutrition*, *40*(1), pp. 168–182.

Samuel, R. et al. (2022) Could medical apps keep their promises? International Conference on eHealth. 173–181.

SciPy.org. (n.d.). scipy.stats.ttest_ind. Retrieved May 23, 2024, from https://docs.scipy.org/doc/scipy/reference/generated/scipy.stats.ttest_ind.html

World Health Organization (2021). Length/height-for-age. [online] www.who.int. Available at: https://www.who.int/tools/child-growth-standards/standards/length-height-for-age.

MEAL-TAKING ACTIVITY MONITORING IN THE ELDERLY BASED ON SENSOR DATA: COMPARISON OF UNSUPERVISED CLASSIFICATION METHODS

Abderrahim Derouiche[1], Damien Brulin[1], Eric Campo[1] and Antoine Piau[2]
[1]*LAAS-CNRS, University of Toulouse, CNRS, UPS, UT2J, Toulouse, France*
[2]*Center for Epidemiology and Population Health Research, CHU Toulouse, Toulouse, France*

ABSTRACT

In an era marked by a demographic change towards an older population, there is an urgent need to improve nutritional monitoring in view of the increase in frailty. This research aims to enhance the identification of meal-taking activities by combining K-Means, GMM, and DBSCAN techniques. Using the Davies-Bouldin Index (DBI) for the optimal meal-taking activity clustering, the results show that K-Means seems to be the best solution, thanks to its unrivalled efficiency in data demarcation, compared with the capabilities of GMM and DBSCAN. Although capable of identifying complex patterns and outliers, the latter methods are limited by their operational complexities and dependence on precise parameter configurations. In this paper, we have processed data from 4 houses equipped with sensors. The findings indicate that applying the K-Means method results in high performance, evidenced by a particularly low Davies-Bouldin Index (DBI), illustrating optimal cluster separation and cohesion. Calculating the average duration of each activity using the GMM algorithm allows distinguishing various categories of meal-taking activities. Alternatively, this can correspond to different times of the day fitting to each meal-taking activity. Using K-Means, GMM, and DBSCAN clustering algorithms, the study demonstrates an effective strategy for thoroughly understanding the data. This approach facilitates the comparison and selection of the most suitable method for optimal meal-taking activity clustering.

KEYWORDS

Clustering, Meal-Taking Activity, Elderly, Health Data, Davies Bouldin Index

1. INTRODUCTION

With the increase of elderly population in the world's comes the imperative to improve their life quality (Shlisky,2017), paying particular attention to the management of their daily nutritional needs (Holman,2020). Meticulous monitoring of the meal-taking habits of the elderly is essential to meet their nutritional needs and avoid the dangers associated with a poor nutrition. These dangers include sarcopenia, which compromises mobility; nutritional fragility, which increases susceptibility to infection and delays healing; and the aggravation of age-related chronic diseases such as cardiovascular disease, type 2 diabetes and certain forms of cancer (Volkert,2022). Digital technology is emerging as an effective method of monitoring meal-taking habits in real time (Suryadevara,2023). Aging is accompanied by a number of physiological changes that can affect nutritional status, such as changes in metabolism, reduced appetite, altered taste perception and difficulties in consuming food (Ghias,2023).

The use of technological innovations to monitor diet is a new approach to detect and predict risks (Burrows,2019). Thanks to technological advances, it is possible to observe eating habits as they happen, to identify nutritional problems quickly and to remedy them (Sempionatto,2021). Devices such as apps and smart objects play a key role in monitoring food and fluid intake, making it easier to adapt diets to the unique nutritional needs of the elderly. In this way, they ensure that their meal-taking habits correspond more closely to their nutritional needs (Armand, 2024).The use of digital technology in food management for the elderly offers a pioneering solution to the challenges posed by an aging population (Sakib,2023). This method not only raises the standard of living of the elderly through good nutrition, but also minimizes the risks of illness linked to poor eating habits and promotes a culture of healthy aging (Piccoli,2023). The integration of technological solutions in the nutritional care of the elderly marks a major step forward to a comprehensive and effective management of their nutritional health in the face of an aging society (Khalifa,2024).

In this context, the CART France initiative project (Derouiche, 2023) is emerging as a leader in the application of artificial intelligence to refine the identification of meal-taking habits in residential settings. Based on data collected as part of the CART project, this paper presents an innovative strategy that integrates K-Means (Arora,2016), Gaussian mixture model (GMM) (Cassimon,2023), and density-based spatial clustering of applications with noise (DBSCAN) (Ahmed,2016) algorithms for analyzing behavioral data related to residents' meal-taking habits. The Davies-Bouldin Index (DBI) was used to determine the optimal number of clusters to form (Made,2022). Once this has been done, the DBI is used again to determine the best clustering value and validate it by comparing the three clustering methods.

2. RELATED WORKS

In this section, we highlight three recent studies that use emerging technologies such as machine learning, deep learning, and wearable sensors to improve nutritional prediction, physical activity monitoring, and meal-taking identification. These research efforts demonstrate significant progress in accurately predicting nutritional intake from food images, improving activity and meal-taking recognition via wearable sensors, and understanding how older people adapt to maintain their nutrition despite age-related challenges.

B. Wang et al (Wang,2023) present a novel method for nutrition prediction from food images, which overcomes the challenges of direct regression by using a coarse-to-fine framework and structural linear smoothing loss. This approach significantly improves prediction accuracy over several benchmarks, illustrating its effectiveness in handling the wide range of nutrient values in foods and the complexity of food image data. The study highlights the advances of the method while recognizing the need for further research to address the variability of food intake environments and the potential problems associated with food waste management due to excessive control of nutrient intake.

G. Hussain et al (Hussain,2019) present a study on a wearable sensor system combining a smartwatch and necklace with a piezoelectric sensor for automated activity and food recognition in preventive healthcare. It achieves high accuracy in recognizing eight physical activities (e.g. going downstairs, eating, going upstairs, walking…) and six food categories, demonstrating the system's effectiveness. However, the study recognizes challenges such as the variability of food characteristics and the need for further improvement in sensor technology and algorithm optimization to increase recognition accuracy and adaptability in various real-life settings.

E. Setiawati et al. (Setiawati,2024) carried out a study using K-Means, K-Medoid, and DBSCAN algorithms to cluster obesity data and uncover patterns and relationships. The evaluation, based on the Davies-Bouldin Index (DBI), revealed that the K-Means algorithm with K2 clusters performed the best, achieving a DBI value of 0.604. The K-Medoid algorithm also performed well with a DBI value of 0.614, while the DBSCAN algorithm showed optimal performance with K3 clusters, achieving a DBI value of 1.040. The study concludes that the K-Means algorithm is the most effective for clustering obesity data. Nevertheless, limitations of the study include the exclusive use of the Davies-Bouldin Index for validation and the potential variability of results with different datasets.

These few illustrative examples show different approaches for accessing nutritional prediction and physical activity monitoring in the elderly. In the next section, we present the methodology used in our study.

3. MATERIAL AND METHOD

3.1 Data Collecting

In this study, the initial phase consists of collecting raw data from passive infrared motion sensors and contact sensors installed in the homes of elderly people living alone (average age of 82 with no particular pathology). These sensors generate binary data when activated, with the data collected over a 12-month period. The compiled database is in CSV format, and constitutes a large dataset. The next step is to develop a new algorithm designed to identify and monitor various types of nutritional activities within the home, focusing particularly on data from the kitchen or dining room, as shown in Figure 1. Data clustering is performed using three algorithms: K-Means, GMM, and DBSCAN.

Figures 2 and 3 show the process flow charts. The first flowchart details the steps involved in identifying and recording food-related activities from a CSV file, exploring the data to detect specific actions and calculating the duration of these activities. The second flowchart describes a general process for classifying data related to food activity using three different clustering algorithms (K-Means, GMM, and DBSCAN) to assess the accuracy of the results.

Figure 1. Contact and Motion Sensors in a Kitchen

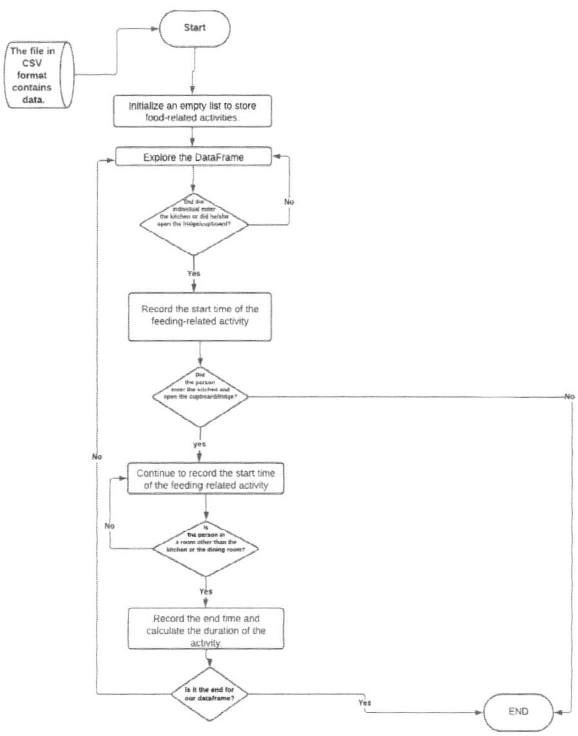

Figure 2. Nutrition Activity Tracking Flowchart

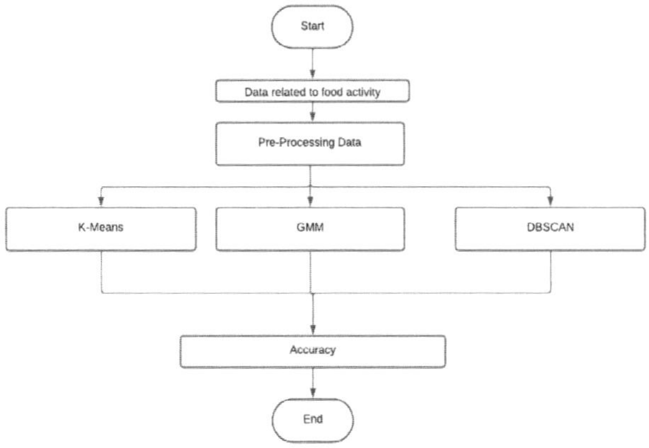

Figure 3. Process flow for Nutrition Activity Clustering Analysis

3.2 Clustering

Unsupervised clustering (Sabita,2021) appears to be an ideal method for discovering the intrinsic structures of food data, without the influence of predefined labels. It reveals unexpected classifications, such as the distinction between different types of food activities (meals and snacks), while also revealing hidden patterns. Furthermore, these methods objectively determine the optimal number of clusters based on mathematical criteria, thereby adjusting the complexity of the model to the relevance of the data. In our particular context, unsupervised clustering is proving very effective in establishing thresholds based on observed durations, making it easier to distinguish between meals and snacks.

3.2.1 K-Means

K-Means Clustering is a method used in data analysis and mining that operates in an unsupervised manner. It adopts a partitioning strategy to cluster data, aiming to group similar features within clusters while segregating dissimilar ones into separate clusters (Yuan,2019).

The main objective of K-Means is to enhance similarity within clusters while reducing it between distinct clusters. The crucial factor used to amplify data similarity within a cluster is the distance function, which evaluates the similarity of data by measuring the closest distance to the cluster center, as illustrated in equation 1.

$$D_e = \sqrt{(x_i - s_i)^2 (y_i - t_i)^2} \qquad (1)$$

3.2.2 GMM

Gaussian Mixture Model, commonly known as GMM, is a statistical framework represented by a mixture of densities. It is generally used to parametrically estimate the distribution of random variables, by conceptualizing them as a combination of several Gaussian components, also known as kernels. The aim is to determine the parameters of each Gaussian component, such as the mean, variance, and amplitude, optimizing them based on the maximum likelihood principle in order to obtain a distribution close to the desired one (El Attar,2012).

Gaussian Mixture Models (GMM) are limited by their sensitivity to initial conditions, the difficulty of choosing the correct number of Gaussian components, and the restrictive assumption that the data follow normal distributions (as illustrated in equation 2), which may not be valid for all data.

$$g(x, \Phi) = \sum_{k=1}^{g} \pi_k F(x, \theta_k) \qquad (2)$$

$g(x, \Phi)$: This is the overall GMM probability density function for a data point x, given the model parameters.
g: The number of components (Gaussian distributions) in the mixture.

$\sum_{k=1}^{g}$: This is the summation sign, which indicates that the next term must be added to all the g components of the mixture.
π_k:The mixing coefficient for the Kth component. It represents the weight of the Kth Gaussian in the mixture and satisfies $0 \leq \pi_k \leq 1$ and $\sum_{k=1}^{g} \pi_k = 1$.This ensures that the total probability distribution sums to 1.
$F(x, \theta_k)$: The probability density function of the kth Gaussian component. This function is parameterized by θ_k, which typically includes the mean and covariance of the kth Gaussian.

3.2.3 DBSCAN

DBSCAN stands out in scenarios where K-means and GMM show limitations, thanks to its ability to detect clusters of various shapes and skillfully manage outliers and noise. Unlike the need to predefine cluster numbers as in K-means and GMM, DBSCAN uses the density of the data to discover this parameter autonomously, making it uniquely suitable with datasets with complex and non-linear structures. Additionally, its ability to handle fluctuations in data density gives it a significant advantage in processing effectively with data of varying densities, a challenge often faced by conventional methods focused on identifying clusters of a specific geometric shape.

$$N_{Eps} = \{q \in D / dist(p, q) < Eps\} \quad (3)$$

N_{Eps}: This represents the Epsilon-neighborhood of a point p. It is the set of the points that fall within a radius Epsilon from point p.
q: A point within the dataset D.
D: The complete dataset in which clustering is being performed.
dist(p,q): The distance function between points p and q.
Eps: The Epsilon parameter, which is the radius of the neighborhood around point p.

4. RESULTS AND ANALYSIS

4.1 Comparison of Clustering Algorithms

The Davies-Bouldin scores, computed using the K-Means approach as shown in Figure 4, reveal optimal quantities of clusters that vary from household to household, highlighting the diversity of behavior patterns related to food consumption. Household 2385 shows optimal clustering at 10 clusters, while household 2445 is optimal with 8 clusters. For household 2373, the data clearly give an optimal clustering with 10 clusters. For household 2356, the indicators suggest maximum clustering performance at 6 clusters as well.

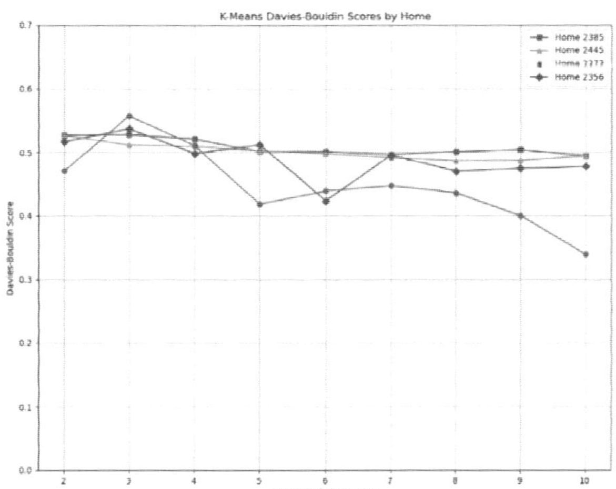

Figure 4. K-Means Davies Bouldin scores for 4 houses

The GMM clustering analysis and Davies-Bouldin scores presented in Figure 5 suggest that increasing the number of clusters often leads to more pronounced separations, particularly noticeable at K=10 for households 2385 and 2373, where the most distinct demarcations occur. Household 2445 finds its optimal clustering definition for K=9, while household 2356 is optimal with K=7.

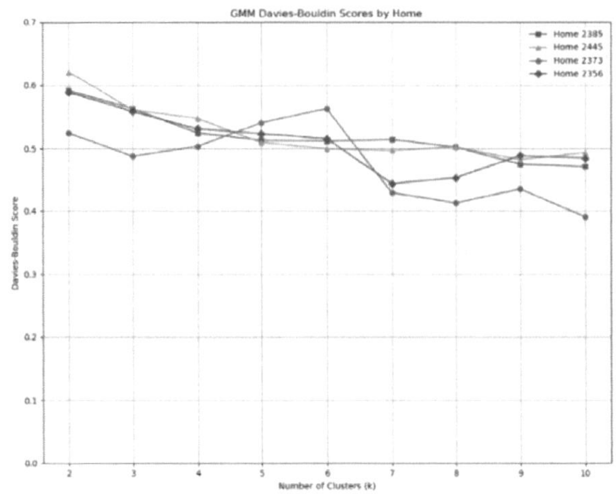

Figure 5. GMM Davies Bouldin scores for 4 houses

Davies-Bouldin scores derived from DBSCAN clustering, as shown in Figure 6, give optimal Eps values for discerning food-related activity clusters within households. Households 2385 and 2356 show better cluster separation for Eps values of 10 and 9 respectively. Household 2445 show better cluster separation for Eps value 4 and household 2373 show better cluster separation for EPS value 3.

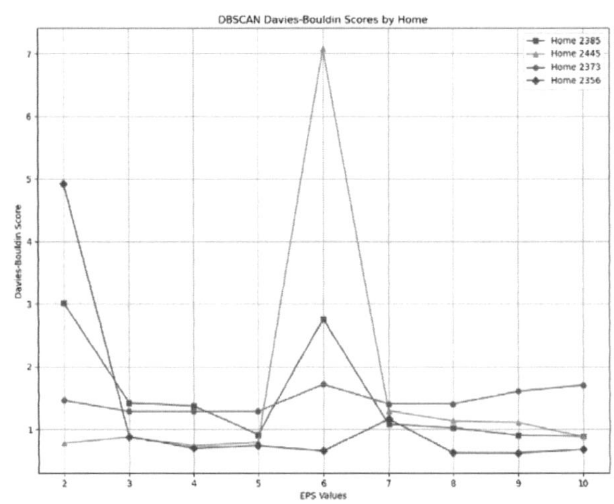

Figure 6. DBSCAN Davies Bouldin scores for 4 houses.

These results indicate the diversity of meal-taking activity patterns and help determine the ideal number of activity types for each household.

4.2 Discussion

When comparing clustering methods to identify meal-taking activities, GMM and K-Means perform better as the number of clusters increases. K-Means, in particular, stands out for lower DBI scores when the number of clusters is higher, suggesting better differentiation between activities such as meals and snacks. The effectiveness of DBSCAN varies considerably with Eps values, requiring fine-tuning of parameters for optimal clustering. This comparative study shows that K-Means appears to be the most consistent method for distinguishing the types of meal-taking activities in different households (Figure 4). Meal-taking patterns, due to their intricate and at times multimodal essence, showcase a broad spectrum of diversity. Leveraging the K-means algorithm to ascertain the ideal number of clusters, the GMM algorithm is subsequently used. This algorithm adeptly harnesses a collection of Gaussian curves, offering the agility needed to encompass the assorted variability. This methodology not only facilitates a detailed depiction of the multifaceted nature of meal-taking behaviors but also enables the computation of the mean duration associated with each habit-related activity. This approach allows us to estimate the means and variances of the Gaussians as shown in Figure 7. Calculating the mean duration of individual activities through the GMM algorithm facilitates the differentiation of diverse categories of meal-taking behaviors (4 categories of meal-taking behaviors shown in Figure 7). Alternatively, this could pertain to distinct times of the day associated with each meal-taking activity.

Although the analysis can identify clusters and trends, the practical application of these results to improve elderly nutrition requires additional interventions and monitoring steps beyond data analysis.

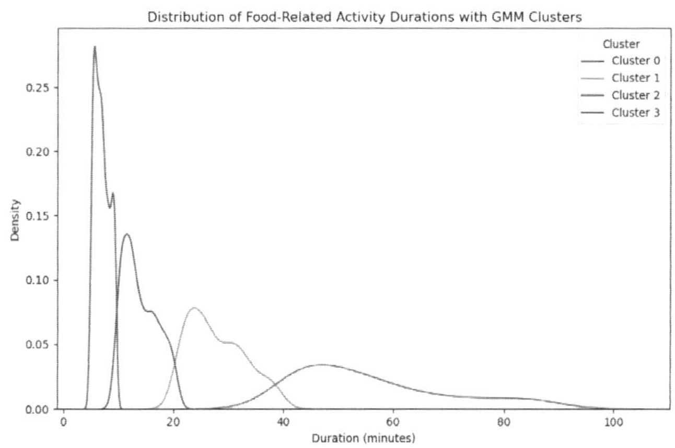

Figure 7. Distribution of Nutrition-Related Activity Durations with GMM

5. CONCLUSION

This research focuses on determining the optimal number of clusters in the context of monitoring meal-taking activity. Employing three different clustering algorithms (K-Means, GMM, DBSCAN) is an effective strategy that enables a thorough understanding of the data and facilitates the comparison and selection of the most suitable method. It concludes that the K-Means algorithm seems to be the best choice for clustering due to clear data separation and consistent performance across different households. GMM's flexibility is advantageous for complex eating patterns, while DBSCAN's outlier detection is crucial for identifying atypical meal-taking activities. However, K-Means may falter with complex cluster shapes and outliers; GMM faces high computational resources; and DBSCAN's performance is highly dependent on precise parameter tuning. Future alert systems for meal-taking activity changes could benefit from an overall approach combining these methods, complemented by time series and predictive analysis for comprehensive, forward-looking meal-taking activity monitoring.

REFERENCES

Ahmed, K. N., & Razak, T. A. (2016). An overview of various improvements of DBSCAN algorithm in clustering spatial databases. International Journal of Advanced Research in Computer and Communication Engineering, 5(2), 360-363.

Ali El Attar. Estimation robuste des modèles de mélange sur des données distribuées. Apprentissage. Université de Nantes, 2012. Français.

Arora, P., & Varshney, S. (2016). Analysis of k-means and k-medoids algorithm for big data. Procedia Computer Science, 78, 507-512.

Burrows, T. L., & Rollo, M. E. (2019). Advancement in dietary assessment and self-monitoring using technology. Nutrients, 11(7), 1648.

Cassimon, D., Fadare, O., & Mavrotas, G. (2023). The impact of food aid and governance on food and nutrition security in Sub-Saharan Africa. Sustainability, 15(2), 1417.

Derouiche, A., Brulin, D., Campo, E., & Piau, A. (2023, June). Home Assessment of Physical and Physiological Parameters: Collaborative Research on Aging Using Technology. In JetSan 2023.

Fardet, A., & Rock, E. (2014). Toward a new philosophy of preventive nutrition: from a reductionist to a holistic paradigm to improve nutritional recommendations. Advances in nutrition, 5(4), 430-446.

Fortin, M., et al. (2020). The impact of artificial intelligence on health care and health research: A foresight scanning. Journal of Medical Internet Research, 22(11), e18244.

Ghias, K., Jiang, Y., & Gupta, A. (2023). The impact of treatment-induced dysgeusia on the nutritional status of cancer patients. Clinical Nutrition Open Science, 50, 57-76.

Govindasamy, K., & Velmurugan, T. (2018). Analysis of student academic performance using clustering techniques. International Journal of Pure and Applied Mathematics, 119(15), 309-323

Holman, B. W., Fowler, S. M., & Hopkins, D. L. (2020). Red meat (beef and sheep) products for an ageing population: A review. International journal of food science & technology, 55(3), 919-934.

Hussain, G., Maheshwari, M. K., Memon, M. L., Jabbar, M. S., & Javed, K. (2019). A CNN based automated activity and food recognition using wearable sensor for preventive healthcare. Electronics, 8(12), 1425.

Khalifa, M., & Albadawy, M. (2024). Artificial intelligence for diabetes: Enhancing prevention, diagnosis, and effective management. Computer Methods and Programs in Biomedicine Update, 100141.

Made, I., Bimantara, S., & Supriana, W. (2022). Case based reasoning (CBR) for obesity level estimation using K-Means indexing method, 11(4), 145-156.

Piccoli, G. B., Cederholm, T., Avesani, C. M., Bakker, S. J., Bellizzi, V., Cuerda, C., & Barazzoni, R. (2023). Nutritional status and the risk of malnutrition in older adults with chronic kidney disease–implications for low protein intake and nutritional care: a critical review endorsed by ERN-ERA and ESPEN. Clinical Nutrition, 42(4), 443-457.

Sabita, R. (2021). Everything you need to know about machine learning. Analytics Vidhya.

Sakib, S. N. (2023). Salutogenic marketing in the elderly. Advances in medical diagnosis, treatment, and care (AMDTC) book series, 117-143.

Sempionatto, J. R., Montiel, V. R. V., Vargas, E., Teymourian, H., & Wang, J. (2021). Wearable and mobile sensors for personalized nutrition. ACS sensors, 6(5), 1745-1760.

Setiawati, E., Fernanda, U. D., Agesti, S., Iqbal, M., & Herjho, M. O. A. (2024). Implementation of K-Means, K-Medoid and DBSCAN Algorithms In Obesity Data Clustering. IJATIS: Indonesian Journal of Applied Technology and Innovation Science, 1(1), 23-29.

Shlisky, J., Bloom, D. E., Beaudreault, A. R., Tucker, K. L., Keller, H. H., Freund-Levi, Y. & Meydani, S. N. (2017). Nutritional considerations for healthy aging and reduction in age-related chronic disease. Advances in nutrition, 8(1), 17-26.

Suryadevara, C. K. (2023). Revolutionizing dietary monitoring: a comprehensive analysis of the innovative mobile app for tracking dietary composition. International Journal of Innovations in Engineering Research and Technology, 10(88), 44-50.

Theodore Armand, T. P., Kim, H. C., & Kim, J. I. (2024). Digital Anti-Aging Healthcare: An Overview of the Applications of Digital Technologies in Diet Management. Journal of Personalized Medicine, 14(3), 254.

Volkert, D., Beck, A. M., Cederholm, T., Cruz-Jentoft, A., Hooper, L., Kiesswetter, E., & Bischoff, S. C. (2022). ESPEN practical guideline: Clinical nutrition and hydration in geriatrics. Clinical Nutrition, 41(4), 958-989.

Wang, B., Bu, T., Hu, Z., Yang, L., Zhao, Y., & Li, X. (2023). Coarse-to-Fine Nutrition Prediction. IEEE Transactions on Multimedia, 26, 3651-3662.

Yuan, C., & Yang, H. (2019). Research on K-value selection method of K-means clustering algorithm. J, 2(2), 226-235.

INFORMATION SYSTEM FOR CALCULATING THE HIGHEST ALLOWED WHOLESALE PRICE OF MEDICINAL PRODUCT IN CROATIA

Maja Fatiga[1,2] and Sanja Grčić Plečko[1,3]
[1]*Agency for Medicinal Products and Medical Devices, Zagreb, Croatia*
[2]*Faculty of Organization and Informatics, Varaždin, Croatia*
[3]*Faculty of Information Studies in Novo Mesto, Novo Mesto, Slovenia*

ABSTRACT

In the Republic of Croatia, the method of calculating the highest wholesale price of a medicinal product is prescribed and defined by law, and falls under the jurisdiction of the Agency for Medicinal Products and Medical Devices of Croatia. The calculation process is complex, involving a reference method of price calculation where, during the calculation of the price of a particular medicinal product, it is necessary to compare the prices of the same or similar medicinal product from reference countries, based on predefined criteria. As support to the process of calculating the highest wholesale price of a medicinal product, an information system was developed. The development, functionalities, and impact of that information system are described in this article. It is important to emphasize that the data model for medicinal products registry in the application is based on the ISO IDMP standard and relevant SPOR RMS lists were introduced.

KEYWORDS

Medicinal Product Price, Information System, Reference Price Method/System

1. INTRODUCTION

In the Republic of Croatia, it is prescribed by the provisions of the Act on Amendments to the Medicinal Products Act (Official Gazette No. 100/18) (Official Gazette, 2018) that before a packaging of prescription medicinal product is placed on the market, it is mandatory to determine the highest allowed wholesale price (further on mention also as highest price) of that medicinal product. This process falls under the jurisdiction of the Agency for Medicinal Products and Medical Devices of Croatia (HALMED), which is the National Competent Authority (NCA) responsible for regulating medicinal products and medical devices in Croatia. (HALMED, 2024a)

The Agency assumed jurisdiction in 2019 from the Croatian Health Insurance Fund (Hrvatski zavod za zdravstveno osiguranje, HZZO). After taking over jurisdiction and adjusting business processes within HALMED to incorporate activities of the acquired authority, considering the complexity of the process of calculating the highest allowed price of medicinal product, it became evident that developing a new information system would be necessary to support this process.

The development of the information system commenced with the publication of a public procurement notice in mid-2021 and was completed by the end of 2022. The system, known within HALMED by the acronym ICL from its Croatian name „*Izračun cijene lijeka*" (Medicinal Product Price Calculation), adopts the so-called reference pricing method used in several European Union countries such as Germany, Sweden, the Netherlands, Spain, Denmark, Poland, Italy, and Slovenia, as well as in global countries like New Zealand, Australia, or British Columbia in Canada (Bojnec and Kajdiž, 2012). The reference pricing method means that prices from different countries for the same medicinal product are used as a reference price in the price calculation.

This paper briefly describes the business process covered by the newly developed information system, outlines the phases and challenges of the system's development, and provides detailed descriptions of the modules and functionalities comprising the system. It highlights the benefits derived from the development of the information system as well as ongoing efforts to improve system functionality.

2. INFORMATION SYSTEM FOR CALCULATING THE HIGHEST ALLOWED WHOLESALE PRICE OF MEDICINAL PRODUCT IN CROATIA

2.1 Business Process of Calculating the Highest Allowed Wholesale Price of Medicinal Product In Croatia

Rules for determining the highest allowed wholesale price of a medicinal product in the Republic of Croatia are defined in the Ordinance on criteria for determining the highest allowed wholesale price of a medicinal product and exceptionally higher than the highest allowed wholesale price of a medicinal product and annual calculation of the price of a medicinal product (Official Gazette 33/2019) (Official Gazette, 2019) and the Ordinance on amendments to the Ordinance on the criteria for determining the highest permitted wholesale price of a medicinal product, the exceptional increase in the maximum permitted wholesale price of a medicinal product and the annually calculated price of a medicinal product (Official Gazette 72/2023) (Official Gazette, 2023).

The rules are also available on HALMED's web site (HALMED, 2024d), (HALMED, 2024c), (HALMED, 2024b) where the whole business process of calculating the highest allowed price of medicinal product in Croatia is explained.

The highest price of a medicinal product is the highest price that a prescription medicinal product can achieve in wholesale trade and is determined separately for each strength, pharmaceutical form, and packaging of the medicinal product. Medicinal product packaging refers to a medicinal product with the same pharmaceutical and dose form and strength, but different pack size.

The determination of the highest price of a medicinal product is carried out in two ways: a) based on the request of the marketing authorization holder for the packaging of medicinal product intended to be marketed for the first time or of wholesaler with the authorization for parallel importation/entry of the medicinal product and b) in the process of the annual calculation of medicinal product prices.

The request for determining the highest price of a medicinal product must be submitted for the packaging of the medicinal product intended for the first time to be placed on the market in the Republic of Croatia. As an exception to this rule, the marketing authorization holder for placing the medicinal product on the market or wholesale trade with authorization for parallel importation/entry of the medicinal product is not obliged to submit a request in case the highest price of the medicinal product determined for a medicinal product with the same active substance, strength, pharmaceutical form, and packaging is used. In that case, the marketing authorization holder for placing the medicinal product on the market or wholesale trade with authorization for parallel importation/entry of the medicinal product is obliged to notify the Agency of the use of the highest allowed wholesale price of the medicinal product before placing the respective packaging of the medicinal product on the market. Based on such notification, the Agency issues a confirmation of determining the highest allowed wholesale price of the medicinal product.

The Agency conducts the annual calculation procedure of the highest allowed wholesale prices of medicinal products for all packaging of prescription medicinal products that are in circulation in the Republic of Croatia on the day of the start of the annual calculation or have a determined highest price of the medicinal product. The start date of the annual calculation of the highest allowed wholesale prices of medicinal products is the first day of February. The Agency will inform marketing authorization holders for placing the medicinal product on the market or wholesale trade with authorization for parallel importation/entry of the medicinal product about the calculated amount of the highest allowed wholesale price of the medicinal product within the deadline defined by the Ordinance. If it considers that the price of the medicinal product was not calculated in accordance with the provisions of the Ordinance, the marketing authorization holder for placing the medicinal product on the market or wholesale trade with authorization for parallel importation/trade of the medicinal product may, within the deadline defined by the Ordinance, request the adjustment of the calculated price of the medicinal product. The annual calculation procedure of the highest allowed wholesale prices of medicinal products ends with the publication of prices on the Agency's website.

The highest price of a medicinal product is determined by the following priority of selection:

1) calculation based on the average reference price of the medicinal product, and if there is no reference product, then

2) based on the valid wholesale price of that product, and if there is no reference product or valid highest price of the medicinal product, then

3) based on the wholesale price of that product in at least three European Union/European Economic Area countries that are not on the list of five reference countries listed in Ordinance, and if there is no reference product, valid highest allowed wholesale price of the medicinal product, or wholesale price of the medicinal product available for at least three EU/EEA member states, then

4) based on a proposal for the price of the medicinal product based on the manufacturer's price.

The reference wholesale price of a medicinal product is the price of the medicinal product determined based on data on the prices of reference products in the reference EU member states (respecting the specified order of countries as a priority in consideration: Italy, Slovenia, Czech Republic, Spain, and France).

To calculate the average reference wholesale price of the medicinal product, it is necessary to determine the average of three reference wholesale prices of the medicinal product, from three different reference countries. Exceptionally, in case there are no three determined reference wholesale prices of the medicinal product, it is possible to determine the average reference wholesale price of the medicinal product as the average of two reference wholesale prices of the medicinal product (from two reference countries), or if there are no two determined reference wholesale prices of the medicinal product, the average reference wholesale price of the medicinal product is considered the reference wholesale price of the medicinal product determined for one of the reference countries.

Any medical product from a reference country with the same active substance, strength, pharmaceutical form, and packaging as the product for which the price is determined is considered a reference product.

If a product of the same size packaging as the product for which the price is determined does not exist in the comparative country, the calculation includes the packaging of the product closest in number of unit doses to the packaging for which the price is determined. In case there are several packaging that are simultaneously closest in number of unit doses to the packaging for which the price is determined, all such packaging are included in the calculation.

If products of the same type of packaging with respect to the active substance, strength, pharmaceutical form, and packaging size are in circulation in the Republic of Croatia, after determining the highest allowed wholesale price of the medicinal product for one of these packaging, the price is also applied to other identical packaging.

2.2 Methods of the Development

2.2.1 Phases and Development of The Information System

The development of a complex system like ICL required extensive preparations and analysis, as well as defining the general direction in which the development would proceed. Before the development, numerus workshops with business users, external contractor and HALMED IT department were held. The goal of this workshops were for external contractor to get familiar with the business process of calculating highest allowed price of medicinal product, to analyse the International Organization for Standardization (ISO) for the identification of medicinal products (IDMP) standard (European Medicines Agency, 2024a) which was predetermined as a standard that the data model must follow, to define the data model, to make a mock-up screens of user interface and to make a plan for initial import of data. During the development phases, workshops were also held to define in more detail the functionalities that were being developed during that phase. For each of the development phase, functional specification was prepared by an external contract and approved by an internal business users. At the end of the phase, developed functionalities were tested to ensure that all the functionalities were correctly implemented.

The development approach was phased, and four phases were defined for the gradual development of the ICL system. The phases were as follows: 1) developing the application's user interface, the codebooks to be used, and defining user permission groups within the application, 2) developing the data model of the application and describing the functionalities of the medicinal product registry and reference medicinal products module, 3) developing the module supporting the annual calculation process of the highest allowed price, and 4) describing and developing functionalities for other types of procedures for calculating the highest allowed price of medicinal product. (**Error! Reference source not found.**).

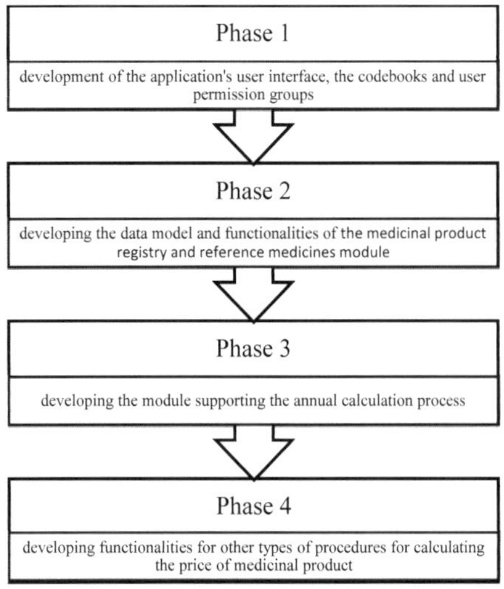

Figure 1. Development phases of the ICL application

Phase 1 was focused on developing the user interface and functionalities to enable system administration and record changes in the application. Permission groups were defined for which it was known at that time that they would need to be developed depending on the restrictions and roles of users in the application. Other permission groups that were found necessary through further development of the application were added later. It was predefined that the majority of the codebook to be used would be imported from the SPOR service (European Medicines Agency, 2024b), with a few specific codebooks for the ICL application that will be added.

In Phase 2, the data model for recording medicinal product data was defined. It was predetermined that the data model must follow ISO IDMP standards. Two modules were developed: the first module developed was a module with a (1) list of medicinal products retrieved from another HALMED system where nationally authorised medicinal products are recorded, and (2) a list of medicinal products authorised under the centralised authorisation procedure (CAP). This two type of medicinal product are part of module called Medicinal Product Registry in ICL. The second module developed in this phase was a module for recording referent medicinal products from referent countries used for medicinal product price calculation.

Phase 3 involved the development of a module for the annual calculation of medicinal product prices and all accompanying functionalities related to this annual price calculation.

Phase 4 encompassed the development of functionalities that follow other procedures and methods of medicinal product price calculation that are not part of the annual calculation.

2.2.2 The System Functionalities, Modules and Use

The ICL system consists of 5 modules: the first module is called "Medicinal Product Registry", the second is "Reference Medicinal Products", the third is "Annual Calculation", the fourth is "Procedures", and the fifth is the "Administration" module. The first four modules are specific modules covering the business process of medicinal products pricing calculation, while the fifth is for user administration and assigning viewing and modifying rights in the application, system logs and data change logs, and allowing the administration of codebooks if such action is possible for a specific codebook. The ICL system is connected via services with automated filing system, document management system and case management tracking system from which nationally authorised medicinal product data is imported to ICL. (Figure 2).

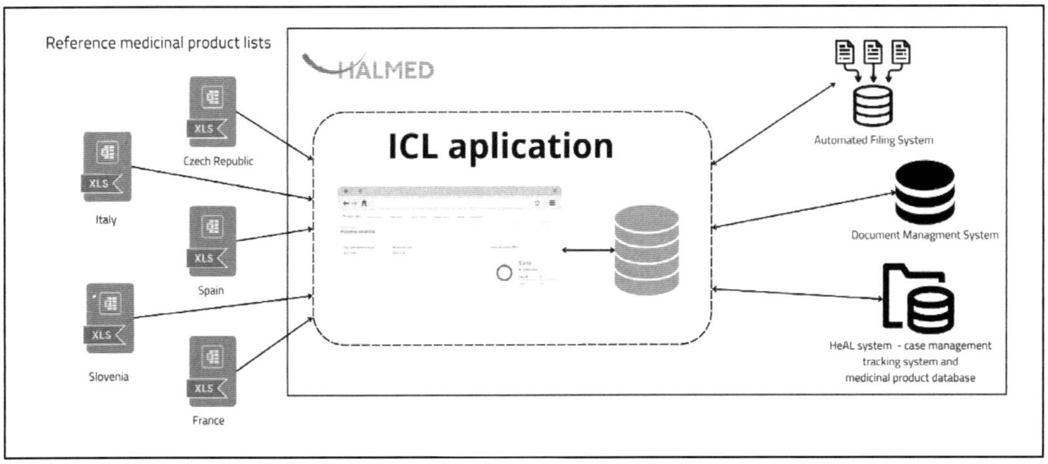

Figure 2. ICL's structure scheme

The first module enables the viewing and administration of medicinal product registry. This registry comprise a list of all prescription nationally authorised medicinal products and products approved through the centralised authorisation procedure available in Croatia. The module displays data in two lists: the list of medicinal products and the list of medicinal product packaging. Since the price is calculated for each packaging, the list of packaging contains most of the information related to the medicinal product price. Data on nationally authorised medicinal products are transferred from another HALMED system used in the medicinal product marketing authorisation processes. The packaging data recorded there is synchronized daily into the ICL system, and ICL users can view all changes made to that data through logs. The data coming from that other system cannot be edited and changed in the ICL system because it is data that has been validated and approved through the medicinal product authorisation process. This dataset has been expanded with specific data needed for price calculation, such as the number of containers per packaging, the number of unit forms per container, or the number of unit forms per packaging. Medicinal products authorised through the centralized authorisation procedure are manually entered into the ICL system when a request for the price calculation of such a packaging occurs, as HALMED does not authorise CAP products, and there are currently no records of such medicinal products in any other HALMED system. All data on CAP medicinal products can be edited and changed in the ICL system.

The Reference Medicinal Products module allows the import of medicinal products lists with respective countries' wholesale prices, obtained from resources defined in Ordinance. All pricelists for five reference countries provide HALMED with a list of reference products in the form of an Excel file, which is then uploaded into the ICL system. The structure of these Excel files differs for each reference country. It was agreed that only a subset of attributes common to all would be displayed in the joint list of reference products in ICL, and all imported data is saved in ICL database and can be viewed for each reference medicinal product by using an option in the interface. Additional specific attributes of reference medicinal product necessary for calculation have been added to ICL for more accurate matching of the medicinal product with reference products. These additional attributes of reference products are manually entered when a new reference products appears. This module also keeps track of price changes, allowing the display of historical price changes for each reference packaging.

The Annual Calculation module is used for the business process of annual calculation of medicinal product prices, which occurs once a year and is mandatory for HALMED as prescribed and defined by law. The calculation starts every year on February 1st, and it is defined which packaging is subject to annual price calculation. The module consists of a list of all annual calculations performed since the creation of the ICL application, plus the year before the creation of ICL application. For that year, the existed data were imported and used as a basis for the first annual calculation conducted through the ICL application. Each annual calculation displays a list of packaging included in the current year's annual calculation. Users can manually add and change the list of packaging depending on inclusion criteria up to the start date of the annual calculation, after which only removing a packaging from the annual calculation is possible. Upon creating a new annual calculation, the application checks, according to predefined rules, which packaging should be included in the annual calculation and compares those medicinal products with reference products. Based on the found reference products, the price of the packaging is calculated. Medicinal products for which no reference products are found are further checked by responsible ICL application users, and if there are indeed no reference products, the price is calculated based on other parameters prescribed by law. The module also provides administrative support for the annual calculation process, allowing the creation of notifications sent to marketing authorization holders about the calculated annual price, as well as direct filing from the ICL system to the filling system and saving the documents to document management system.

The Procedures module supports all other procedures of price calculation, such as Requests for Price Calculation in case a packaging is placed on the market for the first time, the Confirmation procedure for taking over the price of a medicinal product that is identical to the medicinal product taking over the price according to defined criteria, or the procedure for requesting prices higher than the highest allowed price. Depending on the type of procedure, some procedures use the same functionalities of matching with reference products and calculating prices, while others record only the necessary data. All procedures allow the creation of appropriate documents and filing through the service.

The Administration module allows the addition of system users and user rights management. It also enables the viewing of system logs and data change logs. The module also allows the administration of Organizations, i.e., entities that are Marketing Authorization Holders or Local Representatives of foreign Marketing Authorization Holders. In this module, it is possible to view the codebooks used and to add and modify existing codebooks if such an option is allowed for a specific codebook.

3. DISCUSSION

The ICL application was developed with the aim of supporting the business processes of calculating the highest allowed price of medicinal product wholesale and assisting users who carry out that process. The necessity of this application is evident from the number of records contained within it. For example, in April 2024, the number of packing of medicinal product included in annual calculation 2024 was 5,349 and the number of reference medicinal product was 235,187. (Table 1Table 1. Type and quantity of records in ICL application (April, 2024.)

The calculation process within the annual calculation takes 40 days, after which it is necessary to inform the marketing authorization holders about the calculated price, which means comparing 5,349 packaging with reference products while respecting the sequential priority of reference countries during this period. This means that sometimes it is sufficient to compare with the first three reference countries, but if reference products are not found in those countries, products from the fourth or even fifth referent country are compared.

Table 1. Type and quantity of records in ICL application (April, 2024.)

Type of records in ICL	Number of records
1. Packaging of medicinal product included in 2024 annual calculation	5 349
2. Reference Medicinal products	235 187
2.1 Reference Medicinal products from Italy	156 014
2.2 Reference Medicinal products from Slovenia	5 392
2.3 Reference Medicinal products from Czech Republic	11 094
2.4 Reference Medicinal products from Spain	33 679
2.5 Reference Medicinal products from France	29 008

This is precisely what makes the application special. It was designed with the purpose of automating part of the comparison and matching process between medicinal product and reference products based on prescribed criteria such as pharmaceutical form, active substance, and strength.

The process of comparison and matching is possible only if the terms that are compared are the same. So the first step before matching is to pair the terms from different reference list with national term. Since there are 5 reference countries, the lists of reference products in Excel files come in different structures, languages, with varying amounts of data describing the product and none of those lists use standardised terms from SPOR RMS lists. Initially, by importing data from previous annual calculations from an auxiliary application used for calculation, ICL application attempted to capture as many links as possible between such paired terms in the ICL application. The users of the application played a significant role in verifying all paired terms. Such paired terms are stored in a separate code book used for the application to "learn" how to independently link terms. For example, for eye ointment, that has standardised term (ID 100000073772) on SPOR RMS list “Pharmaceutical Dose Form” that is used in HALMED’s Medicinal Product Registry, on Italian list term "unguento ocular" is used, on Spanish list term "pomada oftálmica" is used and on Slovenian list term "mazilo za oči". All those terms are linked to the Croatian term "mast za oko", and every time such a pharmaceutical form appears in a reference product, the application knows which Croatian term it refers to. These connections, established not only for pharmaceutical forms but also for substance names, ATC codes, and container types, constitute the backbone that enables automatic pairing of medicinal products in the ICL application. Users play a crucial role in linking these terms, as the application alerts them if a new reference products appears with a new term. In such cases, the user links it to the corresponding Croatian term, i.e. standardised term from SPOR RMS list, creating a connection that will be used in the application from that moment on.

When a medicinal product packaging is intended to be placed on the market, a calculation of the highest allowed price for that medicinal product is requested. During this calculation, reference products that match according to defined criteria are thoroughly checked. The application offers possible reference products based on the matching method described above, and the responsible user verifies and confirms the selection of the reference products. This aspect of using selected reference products for calculating the price is automated by the application.

The described method of linking reference terms to standardised SPOR RMS terms used on Croatian Medicinal Product Registry is the reason why approximately 93% of packaging included in the annual calculation have already a calculated price when creating a new annual calculation.

Since the application uses SPOR dictionaries, all Croatian terms are unambiguously linked to codes from these dictionaries. Although terms from reference price lists are not aligned with SPOR dictionaries, in the SPOR dictionary where all terms are in English, some terms have translations into other languages.

4. CONCLUSION

The development of the ICL application has significantly streamlined the process of calculating the highest allowed medicinal product price, as well as the process of matching with medicinal products from reference countries due to easier searchability of reference products based on any criterion. The advantage of the application is its ability to remember paired terms from various reference lists of different countries with Croatian terms, thus offering possible reference products as candidates during matching. Because of services that connect ICL with file management system and document file system, all the administrative part of calculating the medicinal product price process can be done within ICL application. The ICL application helps to enhance transparency and allows for easy tracking of every step of price calculation, as well as the history of price changes for a medicinal product.

With such historical data, trends in medicinal product price changes can be easily monitored and the impact of medicinal product prices on market availability of medicinal product can be observed and investigated as it is described in similar researches in (Anggriani, 2018), (Mardetko and Kos, 2018) and (Kajdiž and Bojnec, 2012). For instance, studies in the pharmaceutical and healthcare sectors are conducted to better understand the impacts of medicinal product prices on healthcare accessibility, the financial aspects of treatment, and to develop policies ensuring fair access to medications. (Schneider and Vogler, 2019).

On the other hand, the effectiveness of the ICL application depends greatly on data integrity. Responsible employees who use the application are responsible for data cleanliness. The more accurate and correctly matched the data is, the accurate the application will find all possible candidates on the lists from reference countries, and the matching process will also be more reliable.

Considering the short history of application usage, users are still in the phase where they see opportunities for improving the application and adding new functionalities to facilitate their work tasks through daily use. The ICL application is regularly upgraded, with the goal of eventually enabling not only internal use but also external use by developing functionalities that will allow submitting requests for calculating the highest allowed medicinal product price directly through extranet portal connected to the ICL application.

REFERENCES

Anggriani, Y. (2018) 'The Role of Medicine Pricing Policy for Improving the Affordability of Medicines', *Jurnal Ilmu Kefarmasian Indonesia*, 16(2), p. 172. doi:10.35814/jifi.v16i2.550.

Bojnec, Š., & Kajdiž, R. (2012). *Price Regulation for Medicinal Products in Slovenia.* In Knowledge and Learning: Global Empowerment; Proceedings of the Management, Knowledge and Learning International Conference 2012 (pp. 185-192). International School for Social and Business Studies, Celje, Slovenia.

European Medicines Agency (2024a) *Data on medicines (ISO IDMP standards): Overview.* Available at: https://www.ema.europa.eu/en/human-regulatory-overview/research-development/data-medicines-iso-idmp-standards-overview (Accessed: 8 April 2024).

European Medicines Agency (2024b) *European Medicine Agency: SPOR.* Available at: https://spor.ema.europa.eu/sporwi/ (Accessed: 8 April 2024).

HALMED (2024a) *Agency for Medicinal Products and Medical Devices of Croatia. About HALMED.* Available at: https://www.halmed.hr/en/O-HALMED-u/ (Accessed: 8 April 2024).

HALMED (2024b) *Agency for Medicinal Products and Medical Devices of Croatia. Annual calculation of medicinal product prices.* Available at: https://www.halmed.hr/en/Promet-proizvodnja-i-inspekcija/Najvisa-dozvoljena-cijena-lijeka-na-veliko/Godisnji-izracun-cijene-lijeka/ (Accessed: 8 April 2024).

HALMED (2024c) *Agency for Medicinal Products and Medical Devices of Croatia. Highest permitted wholesale price of a medicinal product and exceptional increase of the price.* Available at: https://www.halmed.hr/en/Promet-proizvodnja-i-inspekcija/Najvisa-dozvoljena-cijena-lijeka-na-veliko/Najvisa-dozvoljena-cijena-lijeka-na-veliko-i-iznimno-povecanje-cijene/ (Accessed: 8 April 2024).

HALMED (2024d) *Agency for Medicinal Products and Medical Devices of Croatia. Medicinal product pricing system.* Available at: https://www.halmed.hr/en/Promet-proizvodnja-i-inspekcija/Najvisa-dozvoljena-cijena-lijeka-na-veliko/Sustav-odredivanja-cijena-lijekova/ (Accessed: 8 April 2024).

Kajdiž, R. and Bojnec, Š. (2012) 'Ali regulacija in kritje cen zdravil vplivata na javne izdatke za zdravila? (Do the price regulation and reimbursement affect public expenditures for medicinal products?)', *Zdravniski Vestnik*, 81(9), pp. 618–625.

Mardetko, N. and Kos, M. (2018) 'Influence of generic reference pricing on medicine cost in Slovenia: A retrospective study', *Croatian Medical Journal*, 59(2), pp. 79–89. doi:10.3325/CMJ.2018.59.79.

Official Gazette (2018) *Act on Amendments to the Medicinal Products Act (Official Gazette No. 100/18).* Available at: https://narodne-novine.nn.hr/clanci/sluzbeni/2018_11_100_1931.html (Accessed: 8 April 2024).

Official Gazette (2019) *Ordinance on criteria for determining the highest allowed wholesale price of a medicinal product and exceptionally higher than the highest allowed wholesale price of a medicinal product and annual calculation of the price of a medicinal product.* Available at: https://narodne-novine.nn.hr/clanci/sluzbeni/2019_04_33_705.html (Accessed: 8 April 2024).

Official Gazette (2023) *Ordinance on amendments to the regulation on criteria for determining the highest allowed wholesale price of a medicinal product and exceptionally higher than the highest allowed wholesale price of a medicinal product and annual calculation of the price.* Available at: https://narodne-novine.nn.hr/clanci/sluzbeni/2023_07_72_1191.html (Accessed: 8 April 2024).

Schneider, P. and Vogler, S. (2019) 'Studies of Medicine Price Developments', *Medicine Price Surveys, Analyses and Comparisons: Evidence and Methodology Guidance*, pp. 165–178. doi:10.1016/B978-0-12-813166-4.00009-7.

ENABLERS OF STAKEHOLDER INVOLVEMENT IN THE DIGITALISATION OF THE UK HEALTHCARE SECTOR: AN ACTIVITY THEORY PERSPECTIVE

Boroto Hwabamungu and Paul Shepherd
University of Bath
Claverton Down, Bath BA2 7YA, United Kingdom

ABSTRACT

This study is part of a larger research project exploring stakeholder involvement impact on the adoption of digital technologies. The literature indicates that there is a need for further research on the role, involvement and management of stakeholder in the implementation of digital technologies in the healthcare sector. Aiming to address part of this research gap, this study investigated factors that enable stakeholder involvement in the digitalization of the UK healthcare sector using activity theory as the theoretical framework. The study followed a qualitative interpretive approach in which semi-structured interviews were used to collect data. Ten interviews were conducted and analysed using thematic analysis, both manually and through the qualitative software analysis tool NVivo. The findings indicate that there are six clusters of factors that enable stakeholder involvement in the digitalisation of the UK healthcare sector: (1) digitalisation ownership and culture, (2) systemic approach to digitalisation complexity, (3) leadership and management digital competencies and innovativeness, (4) communication and involvement mechanisms and tools, (5) stakeholder trust in digitalisation process and leadership and, (6) stakeholder perceived digitalisation benefits. These factors' influence is enacted through three layers: firstly, the digitalisation process leadership and management layer, secondly, the communication layer and, thirdly, the behaviour change layer. Further research is being undertaken to develop best practice guidelines for stakeholder involvement for successful digitalization.

KEYWORDS

Stakeholder, Digital Health, Involvement Enabler, Digitalisation, Healthcare Sector

1. INTRODUCTION

The disruptive nature of digital technologies and the benefits of the adoption of these innovative technologies across industries and sectors are well documented (Verhoef et al., 2021). Different organisations have consequently invested in the acquisition of new digital technologies. This has entailed embarking on a digitalisation journey, taking advantage of new technological innovations such as the internet of things, artificial intelligence, and big data (Verhoef et al., 2021). However the adoption of these technologies within organisations, a critical component of digitalisation (Stofberg et al., 2021), and the achievement of digital technology adoption objectives remain major challenges across industries in general. This is particularly the case for the healthcare sector globally and in the UK. The ability of digital technologies to transform healthcare services (Gjellebaek et al., 2020) is a key driver for digitalisation in the sector (Raimo et al., 2023).

While progress has been made in other industries, the health care sector (Kraus et al., 2021) is still lagging behind in terms of adoption (Stoumpos et al., 2023). The peculiarities of the healthcare sector are a major barrier to digitalisation (Hermes et al., 2020). Yet a gap in the literature exists on progress in the adoption of digital technologies in the healthcare sector (Raimo et al., 2023) and there is a need for further research to understand the socio-technical factors that affect its digitalisation (Habran et al., 2018). This should be understood beyond the technological component to include processes, activities and actions aimed at involving all stakeholders to improve digital technologies implementation (Wyatt et al., 2020).

Digitalisation in the healthcare sector is undertaken in a multi-stakeholder and multidisciplinary environment. The involvement of multiple stakeholders is essential for optimum digitalisation outcome and better health care provision (Li and Carayon, 2021). Developing an understanding of the intricacies of digitalisation and digital transformation in the healthcare sector necessitates research on the role of different stakeholders and the management thereof (Stoumpos et al., 2023). Moreover there is a need for in-depth research focusing on the micro, meso and macro levels in the digitalisation of the healthcare sector (Velthoven and Cordon, 2019). This study therefore aimed to contribute to the development of this understanding by exploring the enablers of stakeholder involvement in the digitalisation of the UK healthcare sector.

2. BACKGROUND

Organisations across sectors are facing digital transformation, digitalisation and digital technology adoption pressures driven by the ubiquitous nature of digital technologies. They include robotics and process control automation, industrial internet of things, additive manufacturing e.g., 3D printing, augmented and virtual reality, simulation, data and systems integration, big data analytics, industrial security, cognitive computing and artificial intelligence, mobility and wearables and, cloud based platforms (Turner et al., 2021). This has revolutionised businesses leading to the emergence of what is generally referred to as the fourth industrial revolution (4IR) (Osunsanmi et al., 2020, Woodhead et al., 2018). The trend has spread across all sectors including the healthcare sector with the establishment of the Healthcare 4.0 notion.

Huge strides are being made to accelerate digitalisation in the healthcare sector (Habran et al., 2018). In the context of the United Kingdom (UK) National Health Insurance (NHS), digitalisation has been identified as a transformative necessity to improve efficiencies throughout the health system and to empower healthcare practitioners and patients (Wyatt et al., 2020). While there are established advantages of adopting these technologies, the healthcare sector still faces great digitalisation challenges (Gardas, 2022). In the case of digitalisation in hospitals, (Burmann et al., 2023) identified "subject-related, object-related and context-related" factors that influenced the adoption of digital technologies including skills, education, affinity for digitalisation, future workplace, participation and perceived benefit. There exist different factors at micro, meso and macro levels that influence the adoption of innovative digital technologies in the UK healthcare sector (Asthana et al., 2019).

Digitalisation in the healthcare sector, a multi-stakeholder undertaking (Kokshagina, 2021), is affected by a tri-dimensional complexity: the healthcare system complexity, the complexity of digitalisation in the sector and the complexity of human and cultural factors (Stoumpos et al., 2023). Multiple stakeholder influence on digitalisation (Brunetti et al., 2020) exacerbates this complexity. A stakeholder is defined by Freeman (1984) as "any group or individual who can affect or is affected by the achievement of the organisation's objectives". The literature highlights the relevance of stakeholder involvement in digitalisation in general and the healthcare sector in particular. Velthoven and Cordon (2019) for example argue that stakeholder involvement contributes to the alignment of the different stakeholders' digitalisation needs and expectations and the adoption of digital technologies in the healthcare.

Digitalisation in the healthcare sector is prone to multiple challenges (Mabillard and Mattijs, 2021) such as the complexity and breadth of the stakeholder landscape and their involvement (Nilsen et al., 2020). Yet stakeholder involvement is critical in the healthcare sector's socio-technical ecosystem for successful adoption of digital technologies and sustainable healthcare (Pereno and Erikson, 2020). To alleviate these challenges, this study identified enablers of stakeholder involvement in the digitalisation of the UK healthcare sector.

3. METHODOLOGY

This research has followed an interpretive case study approach (Yin, 2009) in which semi-structured interviews (Myers, 2009) were used to collect data. A total of ten interviews were conducted with participants recruited across different levels and structures of the UK healthcare system. Participants included two digitalisation leaders/managers, one former nurses, two clinicians, two digital health consultants, two digitalisation academics and one pharmaceutical industry consultant. Seven interviews were conducted remotely using MS Teams and three were conducted in-person. The interviews were recorded and then transcribed. Activity theory (Mursu et al., 2007, Gormley et al., 2020) with its socio-technical exploratory capabilities (Crawford and Hasan, 2006), was used as the theoretical framework to guide the development of the interview questions and the analysis of the collected data. The data was analysed using thematic analysis (Castleberry and Nolen, 2018), both manually and using qualitative software analysis tool NVivo in an iterative process. Interview transcripts were read, and excerpts marched to activity theory concepts. This preliminary analysis resulted in the identification of twenty-four initial themes. These themes were then coded and grouped based on their similarities. In this process some concepts' relevance could be associated to more than one subgroup. They were coded as FESI (Factor Enabling Stakeholder Involvement) followed by two digits separated by a dot. The first digit indicates the level of analysis and the second the numerical order of each theme. Table 1 provides a summary of the thematic analysis outcome.

4. FINDINGS

The iterative thematic analysis resulted in the identification of six major enablers of stakeholder involvement in the digitalisation of the UK healthcare sector. The identified enablers were then integrated into a conceptual model as illustrated in Figure 1.

Table 1. Summary of iterative thematic analysis findings

INITIAL THEMES	THEMES GROUPING	FINAL THEMES
Organisational innovativeness and leadership innovative mindset **[FESI 1.1]** Digital champion culture with its positive and negative **[FESI 1.2]** Patients and clinician trust in digital technologies **[FESI 1.3]** Leadership commitment to successful digitalisation **[FESI 1.4]** Digitalisation project hierarchical initiation **[FESI 1.5]** Digitalisation team lead connection relationships **[FESI 1.6]** Digitalisation project initiation communication approach **[FESI 1.7]** Management's clinical and digitalisation expertise **[FESI 1.8]** Perceived health risks **[FESI 1.9]** Project duration **[FESI 1.10]** Clinician trust in leadership and management approach to digitalisation digital technologies **[FESI 1.11]** Digitalisation approach **[FESI 1.12]** Teams size **[FESI 1.13]** Project size and digitalisation complexity management **[FESI 1.14]** Unit specific digitalisation **[FESI 1.15]** Leadership innovativeness and approach **[FESI 1.16]** Stakeholder perceived t impact on daily tasks **[FESI 1.17]** Clinician initiated digitalisation initiative **[FESI 1.18]** Management digitalisation competencies **[FESI 1.19]** Stakeholder trust in digitalisation leader **[FESI 1.20]** Multi-stakeholder involvement activities facilitation and facilities location **[FESI**	Digital champion culture with its positive and negative **[FESI 1.2]** Digitalisation project hierarchical initiation **[FESI 1.5]** Unit specific digitalisation **[FESI 1.15]** Clinician initiated digitalisation initiative **[FESI 1.18]**	Digitalisation ownership and culture **[FESI 2.1]**
	Project duration **[FESI 1.10]** Digitalisation approach **[FESI 1.12]** Project size and digitalisation complexity management **[FESI 1.14]** Leadership innovativeness and approach **[FESI 1.16]**	Systemic approach to digitalisation complexity **[FESI 2.2]**
	Organisational innovativeness and leadership innovative mindset **[FESI 1.1]** Leadership commitment to successful digitalisation **[FESI 1.4]** Digitalisation team lead connection relationships **[FESI 1.6]** Management's clinical and digitalisation expertise **[FESI 1.8]** Management's clinical and digitalisation expertise **[FESI 1.8]** Management digitalisation competencies **[FESI 1.19]** Leadership innovativeness and approach **[FESI 1.16]**	Leadership and management digital competencies and innovativeness **[FESI 2.3]**
	Digitalisation project initiation communication approach **[FESI 1.7]** Teams size **[FESI 1.13** Multi-stakeholder involvement activities facilitation and facilities location **[FESI 1.21]** Levels of communication and target messages to different stakeholder groups **[FESI 1.22]**	Communication and involvement mechanisms and tools **[FESI 2.4]**
	Patients and clinician trust in digital technologies **[FESI 1.3]** Clinician trust in leadership and management approach to digitalisation digital technologies	Stakeholder trust in digitalisation process and leadership **[FESI**

1.21] Levels of communication and target messages to different stakeholder groups **[FESI 1.22]** Stakeholder perception of digitalisation initiator's credentials **[FESI 1.23]** Perceived digitalisation benefits to clinicians and patients **[FESI 1.24]**	**[FESI 1.11]** Stakeholder trust in digitalisation leader **[FESI 1.20]** Stakeholder perception of digitalisation initiator's credentials **[FESI 1.23]**	**2.5]**
	Perceived health risks **[FESI 1.9]** Stakeholder perceived t impact on daily tasks **[FESI 1.17]** Perceived digitalisation benefits to clinicians and patients **[FESI 1.24]** Perceived digitalisation benefits to clinicians and patients **[FESI 1.24]**	Stakeholder perceived digitalisation benefits **[FESI 2.6]**

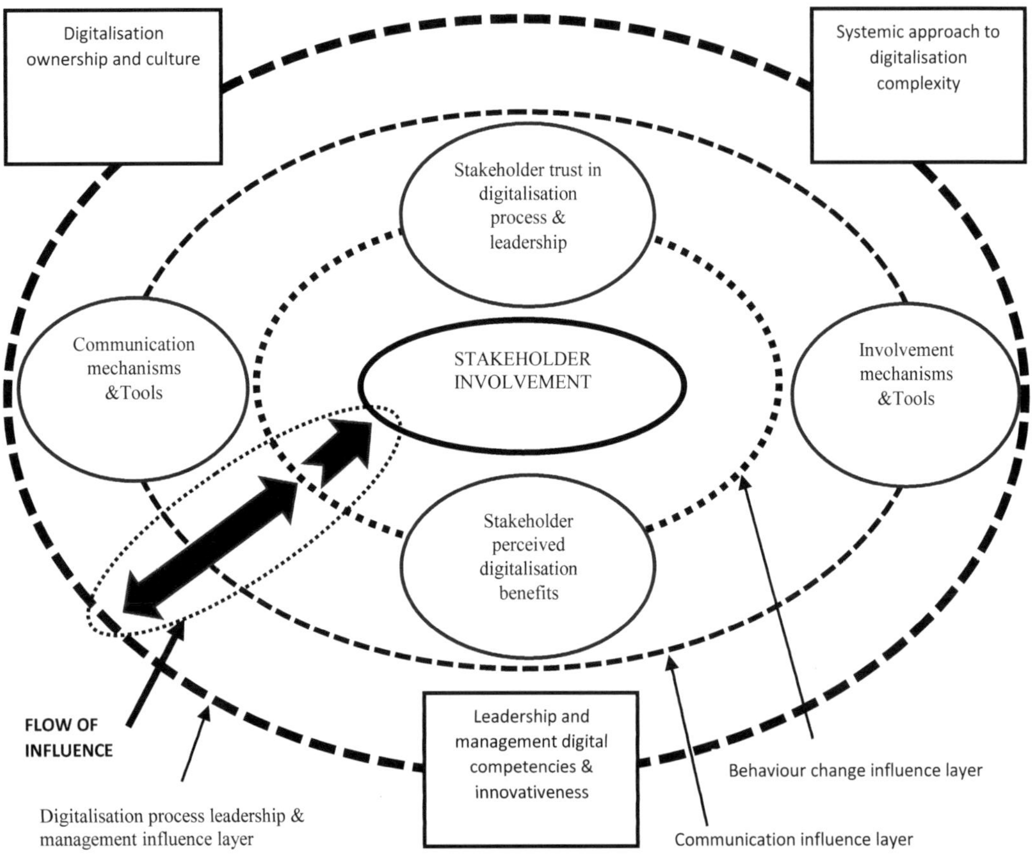

Figure 1. Model of enablers of stakeholder involvement in the digitalisation of the healthcare sector

There are six major enablers of stakeholder involvement in the digitalisation of the UK healthcare sector: (1) digitalisation ownership and culture, (2) systemic approach to digitalisation complexity, (3) leadership and management digital competencies and innovativeness, (4) communication and involvement mechanisms and tools, (5) stakeholder trust in digitalisation process and leadership and, (6) stakeholder perceived digitalisation benefits. These factors manifest themselves at three layers of influence: firstly, the digitalisation process leadership and management layer, secondly, the communication layer and, thirdly, the behaviour change layer. The flow of influence between the enabling factors and/or the involvement of stakeholder can be multi-directional or mono-directional within a digitalisation activity system.

4.1 Digitalisation Ownership and Culture

Digitalisation ownership and culture play a critical role in enabling stakeholder involvement in the digitalisation of the healthcare sector. It emerged that depending on who in the digitalisation process initiated a project, they are likely to engage in stakeholder involvement activities during the process. The breadth of involved stakeholders' landscape will be dictated by the digitalisation project initiator and/or owner's involvement prerogatives which are directly linked to the organisational culture and the other identified involvement enablers. The digitalisation culture intrinsically influences stakeholder involvement practices and digitalisation sub-activities and strongly affects the attainment of digitalisation objectives as observed between clinician-initiated and top-management initiated digitalisation projects. One participant highlighted the following observation:

> *"Your culture is important... culture in organisation, culture in ecosystem... When having one senior leader or two or more demonstrate commitment, the chances of success are high".* ***(Participant E)***

4.2 Systemic Approach to Digitalisation Complexity

Having a systemic approach to address the complexity of digitalisation facilitate the involvement of stakeholder. Such an approach allows for the alignment of management and leadership adopted model of digitalisation and the breath of related processes and sub-activities such as stakeholder involvement. New digital technologies implementation models are required in the current changing digitalisastion context. In this regard, participants made the following observations:

> *"They work with all stakeholders, they call it a whole system approach, working with the whole system to maximise the outcome rather than working in silos... Looking at how you can do it across all levels"* ***(Participant I)***

> *"There is a need for a mindset change, a new mindset different from the old school approach... It's all about people and behaviour and engaging people... The technical dimension can be easily solved but it's the behaviour change and it takes time"* ***(Participant F)***

> *"The approach helps bringing people across the organisation on board: IT, procurement, governance people, research and ethics bodies... Shift from compliance to commitment is needed".* ***(Participant J)***

4.3 Leadership and Management Competencies and Innovativeness

Leaders' and managers' understanding of information technology maturity, the exponential pace of digital technologies innovation and the implications on the need for change in mindset and approach to digitalisation is crucial for stakeholder involvement. A digitally competent manager is likely to approach digitalisation using a participatory approach that encourage the involvement of the different stakeholders and not limiting digitalisation decisions on few stakeholder groups such as suppliers of digital technologies. Two digitalisation leaders described the implications of the innovativeness and approach on stakeholder involvement as follow:

> *"If you want to transform an industry, it's no longer about one organisation performance, it's all about the ecosystem and subsystem... The most important things are shares vision and shared values... there have to be a shared vision that all stakeholder bye into".* ***(Participant I)***

> *"I think that the management team weren't people that worked in hospital, and I think that was a real problem and they didn't seem to have taken the time to actually look at how hospitals worked... So, I don't think you have to have a hospital background... as long as you make sure you know how the place works".* ***(Participant D)***

4.4 Communication and Involvement Mechanisms and Tools

Communication and involvement mechanisms and tools are an additional involvement enabler. In this regard, the availability and choice of stakeholder involvement facilities has been critical where location at hospital or designated community facility has enabled involvement. Stakeholder involvement mechanisms and tools such as stakeholder mapping and the relevance of stakeholder involvement can be critical as highlighted by one participant:

> *"Yeah, if you do your stakeholder map properly and you gauge all your stakeholders, I won't say you can't go wrong, but it really does reduce the risk. But some organisations just don't recognise the importance, they don't even recognise some people of stakeholders.* ***(Participant C)***

> *"So actually if they had spoken to the housekeepers from the start, that would have made it much easier. So it's not always the senior management that the important people, although they might be the people that are paying for it".* ***(Participant A)***

It is therefore important to have adequate communication and involvement mechanisms and tools in place for good stakeholder involvement in the digitalisation of the healthcare sector.

4.5 Stakeholder Trust in the Digitalisation Process and Leadership

Trust in the digitalisation process and the leadership plays a critical role in involving stakeholders in the digitalisation process. Developing stakeholder trust in the digitalisation process and leadership has proven to contribute to good relationships between different stakeholders, leading to good perception of digital technologies benefits and their adoption. Participants made the following observations:

> *"I can talk to people from parts of the system and understand what they are saying... It gives credibility when they are talking to a person they know and who knows what he is talking about... It works, call it leadership or something else but my approach of building strong relationship with people make the solution work".* ***(Participant J)***

> *"Normally these things go through who you know, your connections, who you trust and the respect you have... other people will give you promises but it becomes a matter of faith only... It's about the trust and the recommendations from other people from you network"* ***(Participant H)***

4.6 Stakeholder Perceived Digitalisation Benefits

Stakeholders' perception of digitalisation benefits is another enabler of stakeholder involvement. The perception can be created through multiple engagements with different stakeholder groups in the form of digitalisation awareness workshop. This can also be the result of the digitalisation team's own pre-identification of potential stakeholders for whom digitalisation would be of interest. One participant made the following observation in this regard:

> *"We try to reach people from communities and groups who are likely to benefit so that we get perspectives from different people... It's important to get a range of views".* ***(Participant E)***

5. DISCUSSION

Stakeholder involvement is essential to the achievement of digitalisation objectives in general and in the healthcare sector in particular. Identifying sector specific involvement enablers contributes to the development of adequate industry specific guidelines and informed stakeholder involvement practices. In the context of the healthcare sector, Pereno and Erikson (2020) argues for a systemic approach to healthcare decision making and to address the complexity of stakeholder involvement. In this study, a systemic approach to digitalisation complexity has been identified as one of the major stakeholder involvement enablers. An adequate approach to digitalisation is key to the adoption of innovative digital health technologies (Kraus et al., 2021).Digitalisation ownership and culture has also been identified as another involvement enabler. Habran et al. (2018) argues that digitalisation projects initiated by healthcare practitioners were relevant to healthcare service provision and followed an approach that encouraged stakeholder involvement. In this regard, Mishra and Sharma (2022) noted that patients adopted digital technologies differently and that their involvement was likely to contribute to better adoption outcomes.

In the continuously evolving digital technologies ecosystem, successful digitalisation is dependent on adequate digital leadership attributes and the acquisition of new competencies in the healthcare sector (Laukka et al., 2022). This is aligned with this study findings regarding the enabling role of two identified factors: leadership and management digital competencies and innovativeness, and communication and involvement mechanisms and tools. Digital competencies and communication skills are critical to stakeholder involvement and successful digitalisation in the changing digital health ecosystem (Konttila et al., 2019). Moreover, the transformative implications of management decision-making on the healthcare system in general, and on digital transformation endeavours in particular, are crucial to the attainment of digitalisation objectives (Gjellebaek et al., 2020). Similarly, an organisation's innovativeness and keenness to take digital technology adoption risks is dependent on its organisational culture and leadership (von Oorschot, 2021). Early innovators (Newman et al., 2021) and innovation champions (Ernstsen et al., 2021) are therefore critical.

This study's findings also highlighted elements of trust, trust enactment and trust implications on stakeholder involvement behavioural change and attitudes towards digitalisation process and adoption of digital technologies. In its complexity and multi-dimensionality, trust transcends systemic levels to enable behavioural change and the enactment of positive or negative actions based on the types of relationships within a system (Blöbaum, 2016). Personal and professional relationships of digitalisation leaders have been essential in catalysing stakeholder involvement leading to meaningful engagement, improved perceptions on digital technology benefits, and enabling the building of trust in digital technologies and the digitalisation process.

An understanding of the micro, meso, and macro levels factors, their interdependencies and the implications thereof on enabling adequate communication and meaningful engagements between the different stakeholders at different hierarchical and organisational levels for optimal digitalisation processes is therefore necessary as argued by Kokshagina (2021). This study's proposed model provides exploratory capabilities in this regard and can contribute to better digital technology adoption outcome in the healthcare sector.

6. CONCLUSION

The findings of this study have highlighted the existence of systemic, organisational and individual factors that enable the involvement of stakeholders in the digitalisation of the healthcare of the UK. This study categorised these enablers in six major factors: (1) digitalisation ownership and culture, (2) systemic approach to digitalisation complexity, (3) leadership and management digital competencies and innovativeness, (4) communication and involvement mechanisms and tools, (5) stakeholder trust in digitalisation process and leadership and, (6) stakeholder perceived digitalisation benefits. The proposed model provides a broad perspective of the enablers of stakeholder involvement in the digitalisation of the healthcare of the UK.

This study suggests that the involvement of stakeholders in the digitalisation of the healthcare sector can be enhanced when digitalisation leaders and managers take cognisance of the enablers of involvement using this study's proposed model. The model can be used as an assessment tool and guiding framework to design interventions at micro, meso, and macro level aimed at facilitating stakeholder involvement in practice and to mitigate non-involvement risks on digitalisation. This study contributes to the development of a better understanding of stakeholder involvement complexity by providing insight into the enablers of stakeholder involvement in digitalisation. While the proposed model can potentially be applied to other sectors and industries with similar stakeholder involvement risks in a complex digitalisation environment, its generalisability still needs to be tested and is the object of the authors' ongoing investigation.

ACKNOWLEDGEMENT

The work reported in this paper was undertaken as part of the Made Smarter Innovation: Centre for People-Led Digitalisation, at the University of Bath. The project is funded by the Engineering and Physical Sciences Research Council (EPSRC) Grant EP/V062042/1.

REFERENCES

Asthana, S., Jones, R., & Sheaff, R. (2019). Why does the NHS struggle to adopt eHealth innovations? A review of macro, meso and micro factors. BMC Health sevices research, 19, 984. https://doi.org/10.1186/s12913-019-4790-x

Blöbaum, B. (2016). Trust and communication in a digitalized world: Models and concept of trust research. Springer.

Brunetti, F., Matt, D. T., Bonfanti, A., Longhi, A. D., Pedrini, G., & Orzes, G. (2020). Digital transformation challenges: strateges emercing from a multi-stakeholder approach. The TQM journal, 32(4), 697-724.

Burmann, A., Schepers, S., & Meister, S. (2023). Acceptance facotrs of digitalization in hospital: a mixed-methods study. Health and technology, 13, 843-859.

Castleberry, A., & Nolen, A. (2018). Thematic analysis of qualitative research data: is it as easy as it sounds? Currents in pharmacy teaching and learning, 10(6), 807-815.

Crawford, K., & Hasan, H. M. (2006). Demonstrations of the activity theory framework for research in information systems. Australian journal of information systems, 13(2), 49-68.

Ernstsen, S. N., Whyte, J., Thuesen, C., & Maier, A. (2021). How innovation champions frame the future: three vision for digital transformaiton of construction. Journal of construction engineering and management, 147(1).

Freeman, R. E. (1984). Strategic management: a stakeholder approach. Pitman.

Gardas, B. B. (2022). Organisational hinderances to healthcare 4.0 adoption: a multi-criteria decision analysis framework. Journal of multi-criteria decision analysis, 29(1-2), 186-195. https://doi.org/10.1002/mcda.1766

Gjellebaek, C., Svensson, A., Bjorkquist, C., Fladeby, N., & Grunden, K. (2020). Management challenges for future digitalization of healthcare services. Futures, 124, 102636.

Gormley, G. J., Kajamaa, A., Conn, R. L., & O'Hare, S. (2020). Making the invisible visible: a place for utilising activity theory within in situ simulation to drive healthcare organizational development? Advances in simulation, 5(29). https://doi.org/10.1186/s41077-020-00148-8

Habran, E., Saulpic, O., & Zarlowski, P. (2018). Digitalisation in healthcare: an analysis of projects proposed by practitionners. British journal of healthcare management, 24(3), 150-155.

Hermes, S., Riasanow, R., Clemons, E. K., Bohm, M., & Krcmar, H. (2020). The digital transformation of the healthcare industry: exploring the rise of emerging platform ecosystems and their influence on the role of patients. Business research, 13, 1033-1069. https://doi.org/10.1007/s40685-020-00125-x

Kokshagina, O. (2021). Managing shifts to value-based healthcare and value digitalization as a multi-level dynamic capability development process. Technological forecasting & social change, 172, 121072.

Konttila, J., Siira, H., Kyngas, H., Lahtinen, M., Elo, S., Kaariainen, M., Kaakinen, P., Oikarinen, A., Yamakawa, M., Fukui, S., Utsumi, M., Higami, Y., Higuchi, A., & Mikkonen, K. (2019). Healthcare professionals' competencies in digitalisation: a systematic review. Journal of clinical nursing, 28(5-6), 745-761.

Kraus, S., Schiavone, F., Pluzhnikova, A., & Invernizzi, A. C. (2021). Digitalisation in healthcare: analysing the current state-of-research. Journal of Business Research, 123, 557-567.

Laukka, E., Polkki, T., & Kanste, O. (2022). Leadership in the context of digital health services: a concept analysis. Journal or nursing management, 30, 2763-2780.

Li, J., & Carayon, P. (2021). Health care 4.0: a vision for smart and connected health care. IISE Transactions on healthcare systems engineering, 11(3), 171 - 180. https://doi.org/10.1080/24725579.2021.1884627

Mabillard, V., & Mattijs, J. (2021). Dgitalistion of healthcare policies and services: tensions and challenges. Solavay Brussels school of economics and management, Universite libre de Bruxelles.

Mishra, V., & Sharma, M. G. (2022). Digital transformation evaluation of telehealth using convergence, maturity and adoption. Health policy and technology, 11, 100684.

Mursu, A., Luukkonen, I., Toivanen, M., & Korpela, M. (2007). Activity theory in information systems research and practice: theoretical underpinning for an information systems development model. Information research 12, Paper 311.

Myers, M. D. (2009). Qualitative resarch in business management. Sage publications.

Newman, C., Edwards, D., Martek, I., Lai, J., Thwala, W. D., & Rillie, I. (2021). Industry 4.0 deployment in the construction industry: a bibliometric literature review and UK-based case study. Smart and sustainable built environment, 10(4), 557-580. https://doi.org/10.1108/SASBE-02-2020-0016

Nilsen, E. R., Stendal, K., & Gullslett, M. K. (2020). Implementation of eHealth technology in community health care: the complexity of stakeholder involvement. BMC Health sevices research, 20, 395.

Oorschot, S. v. (2021). Digitalisation and temporary agencies: impact on the business model and internal organisation. Research in hospitality management, 11(3), 241-248. https://doi.org/10.1080/22243534.2021.2006926

Osunsanmi, T. O., Aigbavboa, C. O., Oke, A. E., & Liphadzi, M. (2020). Appraisal of stakeholders' willingness to adopt construction 4.0 technologies for construction projects. Built environment project and asset management, 10(4), 547 - 565.

Pereno, A., & Erikson, D. (2020). A multi-stakeholder perpective on sustainable healthcare: from 2030 onwards. Futures, 122, 102605.

Raimo, N., Turi, I. D., Albergo, F., & Vitolla, F. (2023). The drivers of the digital transformation in the healthcare industry: an empirical analysis iin Italian hospitals. Technovation, 121, 102558.

Stofberg, L., Strasheim, A., & Koekemoer, E. (2021). Digitalisation in the workplace: the role of technology on employee engagement and creativity teams. In N. Ferreira, I. L. Potgieter, & M. Coetzee (Eds.), Agile coping in the digital workplace: emerging issues for research and practice. Springer.

Stoumpos, A. I., Kitsios, F., & Talias, M. A. (2023). Digital transformation in healthcare: technology acceptance and its applications. International journal of environmental research and public health, 20, 3407.

Turner, C. J., Oyekan, J., Stergioulas, L., & Griffin, D. (2021). Utilizing industry 4.0 on the construction site: challenges and opportunities. IEEE transations on industrial informatics, 17(2), 746-756.

Velthoven, M. H. V., & Cordon, C. (2019). Sustainable adoption of digital health innovations: perspectives from a stakeholder workshop. Journal of medical internet research, 21(3). https://doi.org/10.2196/11922

Verhoef, P. C., Broekhuizen, T., Bart, Y., Bhattacharya, A., Dong, J. Q., Fabian, N., & Haenlein, M. (2021). Digital transformation: a multidisciplinary reflection and research agenda. Journal of Business Research, 122, 889-901.

Woodhead, R., Stephenson, P., & Morey, D. (2018). Digital construction: from point solutions to IoT. Automation in construction, 93, 35-46. https://doi.org/10.1016/j.autcon.2018.05.004

Wyatt, D., Lampon, S., & McKevitt, C. (2020). Delivering healthcare's 'triple aim': electronic healteh records and healthresearch participant in the UK National Health Service. Sociology of health & illness, 42(6), 1312-1327.

Yin, R. K. (2009). Case study research: design and methods. Sage publications.

USING DESIGN METHODOLOGIES TO EXPLORE A PRE-SURGICAL SYSTEM

Marko Cigljarev[1], Yefri Ventura[1], Christopher Pandolfi[1], Xavier Massé[1], Ana Rita Morais[1] and Angela Jerath[2]

[1]*George Brown College School of Design, 3 Lower Jarvis St, Toronto, ON M5A 3Y5, Canada*

[2]*Sunnybrook Health Sciences Centre, 2075 Bayview Ave, Toronto, ON, M4N 3M5, Canada*

ABSTRACT

The integration of advancing healthcare technologies within complex legacy systems contributes to conditions for suboptimal design. Specifically focusing on the pre-surgical phase, this paper provides insights into the pivotal role user-centred design can play in enhancing the healthcare sector. Using design methodologies like ecosystem mapping, participatory workshops, and information visualization we can better comprehend healthcare systems, generate solution ideation, and effectively communicate findings. By leveraging these tools, the aim is to offer recommendations on improving healthcare systems, emphasizing the significance of multidisciplinary collaboration and user-centered design.

KEYWORDS

User-Centred Design, Ecosystem Map, Participatory Workshop, Information Visualization, Pre-Surgical

1. INTRODUCTION

Patient safety remains a major focal point in healthcare research (Acevedo & Kuo 2021; Wacker 2020; Buchert & Bulter 2020). There has been an estimated 40% increase in adverse events due to medical errors in the past 30 years (Nauman et al. 2021). Studies show that approximately 50% of adverse events are attributed to preventable medical errors, a statistic that has remained unchanged for ten years (Schwendimann et al. 2018; de Vries et al. 2008). Over the past two decades, it has been found that a major cause of medical error is attributable to poorly designed healthcare systems which contribute to an environment conducive to human error (Kohn, Corrigan & Donaldson 2000). Consequently, there is a strong focus on research that improves healthcare systems to ensure patient safety and a need to incorporate user-centred design processes to understand the patient experience and guide effective design interventions. In this study, surgical procedures, specifically procedures associated with pre-operative care, are explored through systems design as well as user-centred design approaches.

Surgical outcomes are largely affected by the patient's physical and psychological conditions (Levett & Grimmett 2019). The pre-surgical phase, which begins when the patient agrees to surgical treatment up until surgery, presents a crucial opportunity to enhance the patient's condition, thereby minimizing surgical risks and complications. This phase entails coordination among various hospital departments and necessitates thorough communication and planning among specialized clinics, pharmacies, laboratories, healthcare providers, the patient, and their caregiver(s). This complexity is further highlighted by the administrative stress put on the healthcare system with the increased demand for surgical care.

This paper introduces an innovative approach that combines systems analysis and user journey mapping techniques, engages participants through workshops, and utilizes information visualization to explore the pre-surgical processes. As part of a three-year project, we set out to understand pre-operative care, develop design solutions, and communicate the obtained knowledge; based on user feedback and ideation. The findings present an in-depth understanding of the complexity of socio-technical systems within healthcare and inform the practicality and feasibility of user-centred design processes.

2. USER-CENTRED DESIGN

The exponential growth in technology has enabled parallel technological advances in medicine. This constant addition of new devices, software, and processes to legacy systems is a contributing factor to poor system design (Adams et al. 2017). User experience researchers and designers use various methods to help identify design opportunities and innovations of products and services.

These design research techniques can be seen in co-designing care programs (Valaitis et al. 2019), health technology applications (LeRouge et al. 2013), and in Ku and Lupton's publication (2020) for the emerging field of healthcare design. Mapping exercises are used to better understand and improve existing systems. In nursing school, concept mapping has been used to enhance critical thinking skills and meaningful learning (Wilson et al. 2015). In healthcare research, exercises involving both professionals and laypeople have been used to guide brainstorming sessions and share ideas surrounding healthcare concepts (Leyns et al. 2018). Using maps to visually represent things is an effective way to provide context, facilitate discussion, (Drysdale et al. 2019) and generate insight when assessing the healthcare system.

Regardless of context, it is evident that mapping relationships and concepts within complex systems is a valuable way to gain a comprehensive understanding and make improvements. This investigation will explore three methods to show the effectiveness of design thinking in assessing surgical systems: ecosystem mapping, participatory design workshops, and qualitative data analysis through information visualization.

3. METHODOLOGY

In the initial phase of our study, we thoroughly reviewed the information and systems architecture to gain insight into the intricate network of interactions within the healthcare system. This involved understanding how people and technologies interact, both within and outside the hospital. These reviews provided a foundational understanding of the pre-surgical system, nested within broader surgical and national healthcare frameworks. By comprehensively examining the socio-technical context, we uncovered key touchpoints and interactions, laying the groundwork for subsequent methodologies aimed at enhancing the pre-surgical care experience (Figure 1).

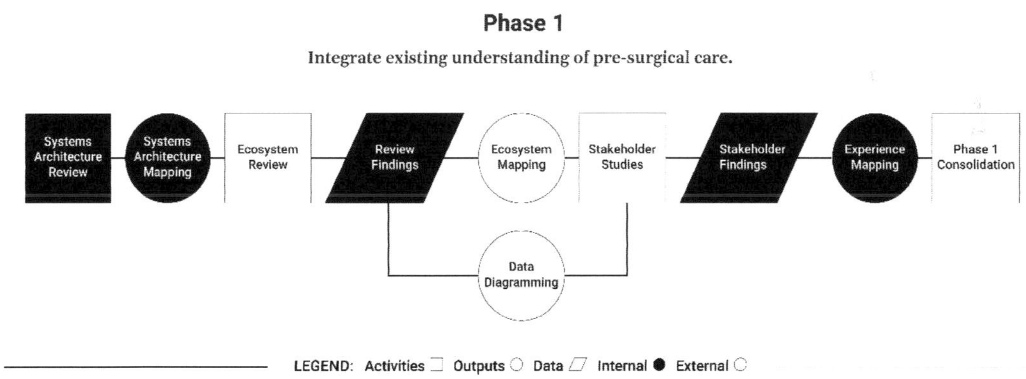

Figure 1. Illustrates the multifaceted research framework utilised for the pre-surgical system's initial scope. It involved activities, outputs and data for internal team use or external sharing with stakeholders.

3.1 Ecosystem Giga Mapping

Ecosystem giga maps simplify the display of information, concepts, and their interrelationships, enhancing the ability to grasp and manage super-complexity. Through mapping, the team gains insight into both individual components and the holistic system, fostering a comprehensive understanding. Interactions between users, tools, infrastructure, tasks, and data reveal both high-level exchanges and small details. Viewing an ecosystem giga map at the macro level unveils the system's complexity, including hubs, patterns (repetitions, proximity, symmetry, and intersections), shortcuts, dead ends, and other intricate paths.

On an ecosystem giga map, displaying users, paths, products, services, and data flows enables designers to highlight the busiest intersection points. These points occur where user paths, interfaces, inputs, system outputs, and data flows overlap. Like pillars supporting the load of a building, intersection points may also reveal the stress inflicted on one specific part of the system. The ecosystem giga map provides a visual way to extract insights to guide brainstorming, creative thinking, and, ultimately, design solutions.

3.2 Participatory Design Workshops

Participatory design workshops are collaborative sessions where individuals gather at key moments during the design process, using specific tools to ensure quality results that are feasible and meet their needs. They serve as a platform for education, research, or problem-solving across various disciplines. Participants are encouraged to tackle problems using design thinking methodologies and collaborative design practice, gaining new perspectives that can generate innovative outputs.

The design workshops engaged participants from various roles such as patients, familial support, nurses, anesthesiologists, and administration; ensuring a diverse range of perspectives and expertise. This fostered collaboration and ensured that resulting interventions addressed the needs and concerns of all stakeholders involved. These workshops harnessed the power of individuals to generate ideas, collect rich qualitative data, and provide a deeper understanding of the pre-surgical process.

3.3 Qualitative Data Analysis and Information Visualization

Qualitative data analysis is a method used to examine and interpret non-numerical data. It involves systematically organizing, and interpreting textual, visual, or auditory data collected through methods such as interviews, observations, or surveys. Qualitative data analysis was used as an aid for understanding complex human experiences, behaviors, and perspectives in the pre-surgical process. It offers depth and context to research findings, providing nuanced insights in ways that quantitative methods alone may overlook.

Information visualization is a powerful tool for representing complex concepts as simplified visuals. By presenting qualitative data in visual formats such as charts, tables, graphs, maps, and diagrams, we can quickly grasp the meaning behind the data and make informed decisions. These visualizations are a tool for comparison that enabled us to efficiently analyze and discover areas within the pre-surgical care system that either required the most attention or could be highly impacted by our design solutions.

A key factor of systems design thinking is time and scale. The above methods were ideal for exploring the healthcare system because they allowed the team to gain insight into the experiences of users from the perspective of timelines and places. They offered both a high-level picture as well as the detailed aspects of the process. Furthermore, these tools were able to highlight connections and relationships between people, data, and systems. Each method influenced the next and contributed to the goal of understanding how the process could be made safer.

4. FINDINGS

With a comprehensive understanding of the presurgical process our focus shifted towards developing interventions to address key frustrations within the pre-surgical care system. This phase involved exploring design solutions through workshops and ideation sessions. These methodologies allowed us to delve deeper into the findings, informing our design ideation phase (Figure 2). Supported by insights from user feedback, we explored targeted solutions and high-level systemic approaches involving multiple integrated products and services.

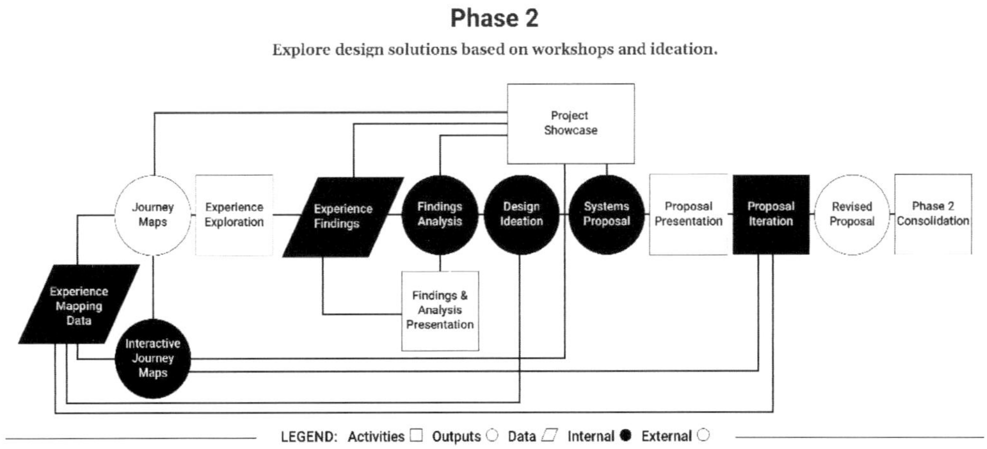

Figure 2. Displays the multidisciplinary approach implemented during Phase 2 of the research timeline. During this phase we engaged in more external collaboration, providing a holistic perspective at identifying potential interventions

4.1 Ecosystem Giga Map

The individual user maps were combined to form one giga map. Although each connection is considered generally unreadable, the purpose is to gain a holistic view of the surgical process and allow the design team to extract insights (Figure 3). Connections in the giga map show the relationships and dependence of elements within the surgical system to each other. The final ecosystem giga map was used to detect patterns, define areas where more substantial knowledge is needed, and uncover potential points of intervention. The combined map illustrates the enormous amount of data exchange needed to have a successful operating system. The complexity of the giga map visually displays the complexity of the surgical system. Some of this can be attributed to redundant tasks such as repetition of patient education and getting the patient's medical history.

One of the most important observations made was that there are minimal connections at the interface level compared to the rest of the giga map. This could indicate one of two options: (1) the design team's lack of understanding of how these interfaces relate to the rest of the system, or (2) shows the complete dependence the surgical system has on a single software application alone. This application is a database accessed by clinicians as a viewer, searcher, and for query of information. As mentioned earlier, the copious number of lines leading to this tool could reveal stress inflicted to this part of the system. This then prompts the design team to explore if the intersection point should be bypassed with a parallel journey leading to a new product and service, or if shortcuts should be designed to ease the user experience leading up to an existing product and service.

These insights would not have been possible without using ecosystem mapping to provide a visual representation of a complex system. They have made the process easier to understand so that the design team could uncover existing gaps and identify opportunities for improvement to affect change within the surgical process.

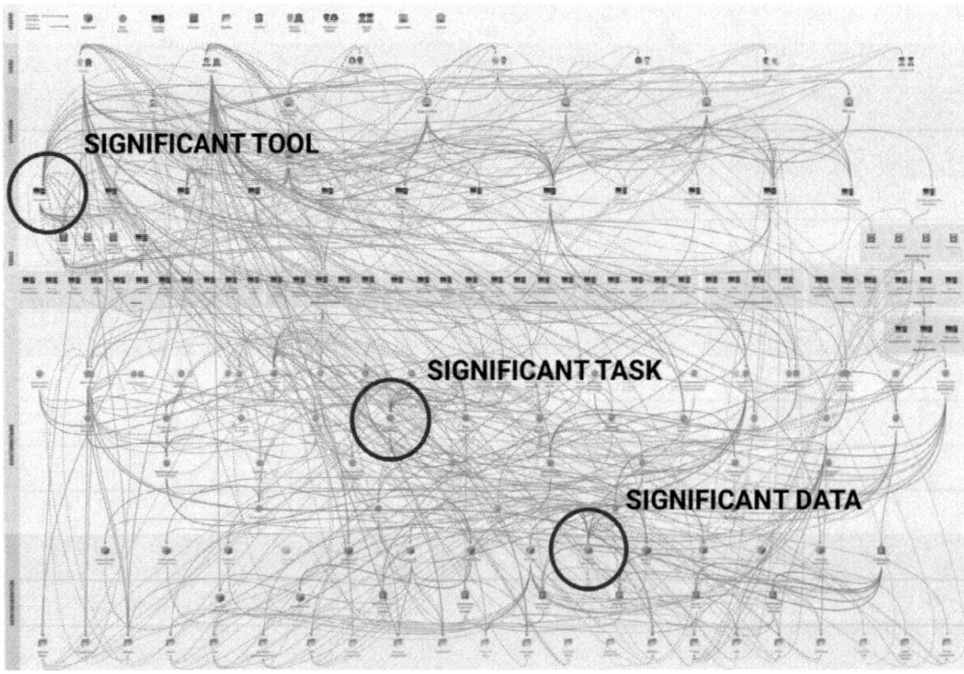

Figure 3. Final ecosystem giga map showing areas with a high volume of connections and significance

4.2 Participatory Design Workshops

The project conducted two design workshops, engaging participants in the process. The first involved 17 participants from across five roles including patients, familial support, nurses, anesthesiologists, and administration. Each participant was given a workshop kit with instructions and resources to record their own journey maps. Journey mapping provides a detailed visualization of a specific user's tasks, tools, thoughts, and emotions all set on a timeline. This is a highly effective tool that helps designers build for the end user and not for themselves.

Completing this exercise provided a detailed account of the pre-surgical experiences of real-world users and outlined the effects of large healthcare systems on the individual scale. By comparing the different individual experiences, we gained crucial insight into specific challenges and sources of frustration.

For the second workshop, 12 participants, 11 of which were involved in the first session, engaged with the design team for solution ideation. The team utilised the digital white board tool Miro as a space for collaboration with participants. Each participant was taken through an examination of five simplified journey maps, one map per role, and together with the design team they brainstormed potential interventions (Figure 4). One key factor was the deliberate omittance of frustrations on the maps. This was to eliminate any bias and allow ideation based purely on the pre-surgical process. Instead, participants were asked to think about the user journeys with some prompt questions under the themes of time, tools, and empathy. This encouraged deep thinking through the lens of pace, relationships, and being human.

The team recorded the thoughts, comments, and ideas of each participant and was left with a plethora of potential interventions. To organize the information from these brainstorming sessions, the team used a tool called affinity mapping. By combining the inputs from the different individuals based on commonalities, groups of similar ideas were identified and new relationships between ideas began to form. The advantage of participatory design is the ability to leverage the collective's strength to recognize the value of each idea.

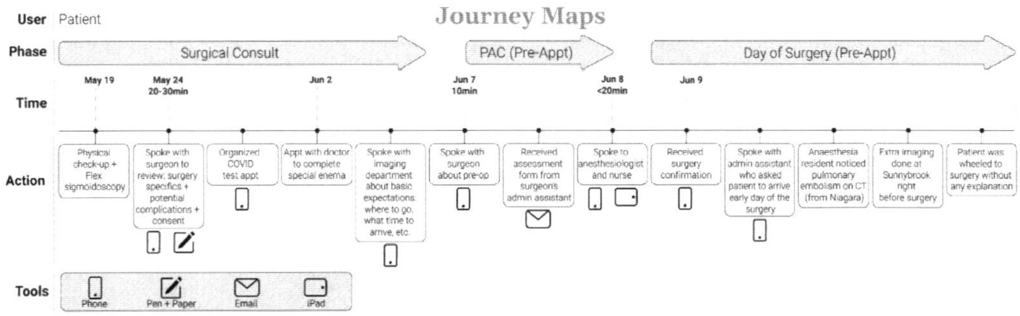

Figure 4. Simplified patient journey map used for workshop two. This map chronologically guides participants through the activities of a specific role, employing time, phases, and tools as key identifiers throughout the journey

4.3 Qualitative Data Analysis and Information Visualization

Initially, our analysis of participant experience data aimed to understand touchpoints and interactions, culminating in the development of detailed interactive journey maps. This tool was a means for direct comparison (Figure 5). These maps allowed for overlaying individual journeys and categorization by various factors. Furthermore, they provided a comprehensive overview of the pre-surgical care experience, enabling us to craft scenarios that identify moments of resistance. Leveraging user experience research techniques, we explored precedents from various large system models, ranging from blockchain to airport navigation, to intervene and address frustrations within the pre-surgical care system.

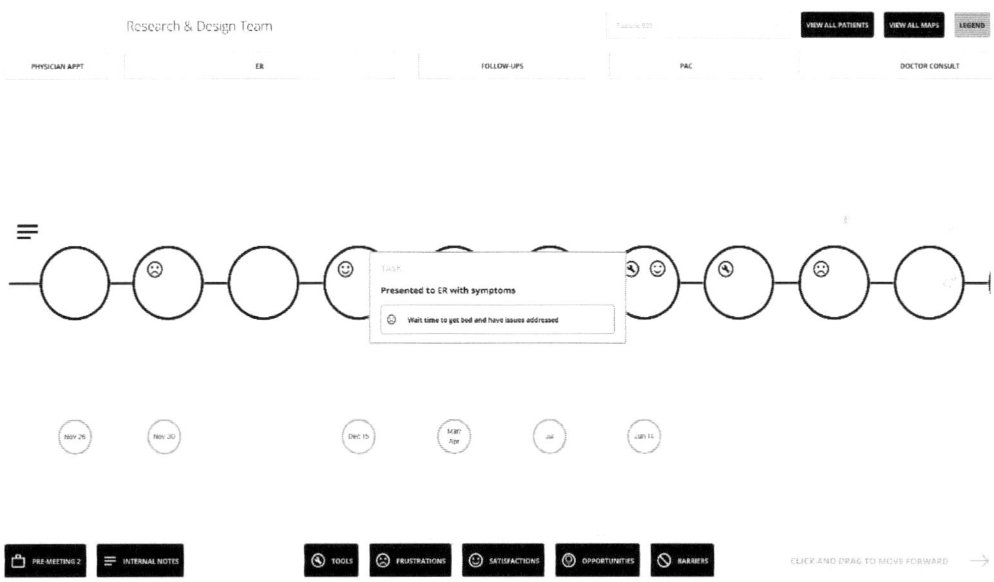

Figure 5. Screenshot of an interactive journey map. By clicking on a specific moment within a journey, internal users could access frustrations, opportunities, barriers, and more, as well as compare key aspects between individuals

Following this, we meticulously analyzed the findings from workshop one and two to synthesize them into solution and frustration matrices. These matrices categorized similar solutions and frustrations by their affinities and organized them by location and time, helping to pinpoint the exact moments within the pre-surgical care process that required attention. The solutions matrix identified high volume intervention points within the system, while the frustration matrix quantified the alleviation potential of each intervention.

This data organization process, leading to the creation of the matrix, facilitated a comprehensive understanding of participant insights, enabling us to visually represent the distribution of solutions and frustrations across different phases and locations of the pre-surgical care journey. Additionally, the matrix served as a foundation for further analysis, informing the development of targeted interventions aimed at enhancing the pre-surgical care experience. Furthermore, the matrix was complemented by a rose chart, which overlaid all the quantified data, providing an elegant tool for comparison (Figure 6). The rose chart allowed us to break down the varying roles expressing certain frustrations or solutions, along with the number of participants who expressed the same sentiments, in a simple way.

By presenting data in visually appealing formats such as interactive journey maps, matrices, and rose charts, we were able to offer clear and concise representations of complex data, aiding stakeholders in understanding and interpreting key insights. This clarity fostered shared understanding, facilitating meaningful discussions and alignment towards common goals. Ultimately, the visualization of qualitative data played a pivotal role in driving informed decision-making and refining how we might enhance the pre-surgical care experience.

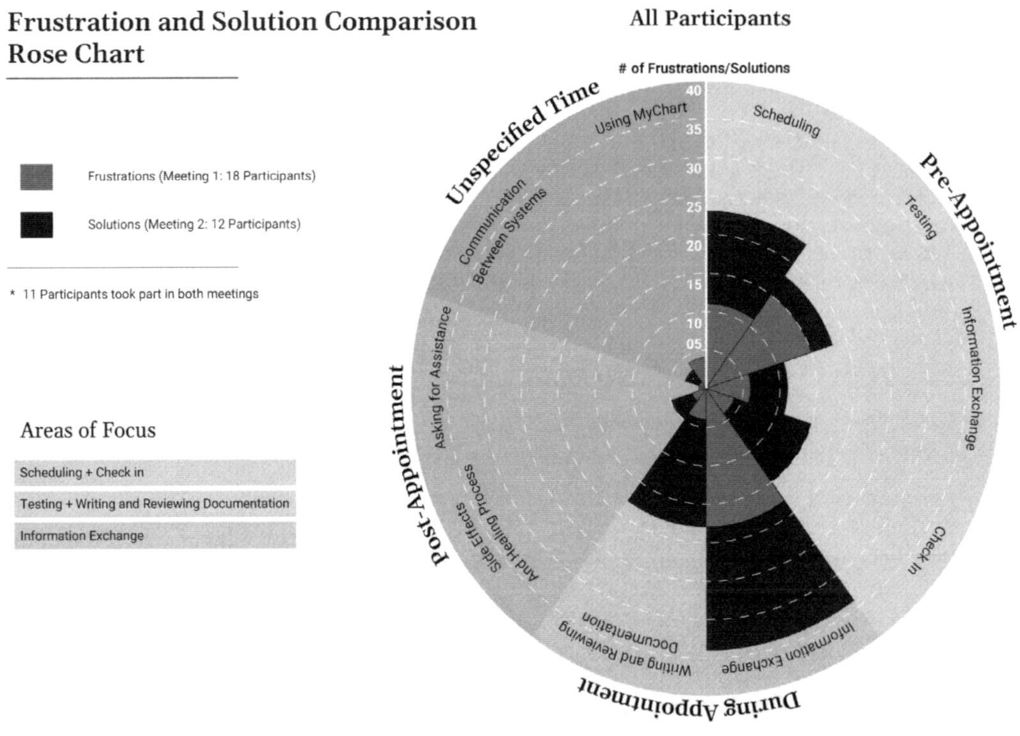

Figure 6. A rose chart overlaying solutions and frustrations. Each group is categorized by their affinities and phases, the size and volume clearly identify areas of focus

5. REFLECTION

With a comprehensive understanding of potential interventions and a detailed data analysis, a new direction was determined for the project. Informed by our design research, we are developing a guide on how to innovate the healthcare sector and leverage technology to address the current issues within healthcare services and systems. This guide will feature a comprehensive list of devices and processes designed to tackle the pain points identified in our research. As a high-level overview meant to be published in the Healthtech space, the goal is to showcase and demonstrate potential developments that could shape the future of healthcare. This is the current phase of the project and is set to conclude August 2024.

To begin the evaluation of ideas from phase two, the team built an interactive survey that would be completed by staff at the participating hospital. The survey consisted of scenarios that relate to the frustrations identified in the first phase of the project and provides 4 options of potential solutions. Participants are asked to rank the solutions 1 through 4, with number 1 being the most effective and 4 being the least effective at addressing the above challenges. After each question, participants can write any comments they have about their selection. Once the survey is complete, our team will be able to refine the interventions and begin assembling the guide that will highlight the most effective and efficient ideas found in our research.

Using design thinking we set out to develop a highly effective targeted solution that could aid in the pre-surgical care process but ended up finding the complexity of the healthcare system required for multiple integrated systems and products. Design methodologies proved an effective way of finding areas of opportunity and interventions within the healthcare system. Entering the final phase of the project a new direction was established. Rather than developing a single solution it was determined that we will highlight many potential solutions worth developing that could elevate human experience and in the process showcase how design thinking can enhance healthcare systems. These design tools allow for a holistic understanding of the problem space and consequently foster innovation rather than limit it to an isolated aspect.

6. CONCLUSION

This paper contributes towards a novel approach to using a combination of established systems design and user-centred design methodologies in exploring a complex socio-technical system in healthcare such as the pre-surgical system. This approach and the following methods can be applied for future use-cases when identifying opportunities and informing design guidelines for systems improvement.

These recommendations include: 1) involving multidisciplinary perspectives ranging from engineering, health sciences and education, health care delivery improvement, and health care technologies 2) methodologies with quantitative and qualitative approaches 3) integration of sociotechnical systems with a focus on human-technology interactions 4) the design of person-centered systems taking into account human needs and recognition of the impact of social disparities and inequities on health. This last point requires a methodology such as user-centred design which places humans (patients, caregivers, and health care workers) at the center of healthcare delivery. Human factors are a valuable discipline in healthcare because it focuses on optimizing work systems and the interaction between social and technical elements within these systems. Combining both human factors' systems engineering with user-centred design practice ensures a more interdisciplinary methodology by connecting socio-technical systems, human-technology interactions, and designing from a person-centred lens.

ACKNOWLEDGEMENT

This project has spanned four years and involved the contributions of many individuals. We extend our heartfelt gratitude to all the healthcare professionals, patients, and their families for their invaluable participation and support. Our team has been truly multidisciplinary, and this work would not have been possible without the expertise and dedication of Nastaran Dadashi, Eva Aboagye, Aleksandra Kmieciak, Sabrina Curutan, Eirene Keh, Karlo Matt Rian Ong, Leah Zhou, and Julian Mutis Vera. To everyone involved, we express our deepest appreciation.

REFERENCES

Acevedo, E. and Kuo, L. E., 2021. The Economics of Patient Surgical Safety. *Surgical Clinics of North America,* Vol. 101, No. 1, pp 135–48.

Adams, K. et al, 2017. An Analysis of Patient Safety Incident Reports Associated with Electronic Health Record Interoperability. *Applied Clinical Informatics,* Vol. 8, No. 2, pp 593–602.

Buchert, A. R. and Butler, G. A., 2020 Pathways and Guidelines: An Approach to Operationalizing Patient Safety and Quality Improvement. *Patient Safety and Quality Improvement in Healthcare*, 245–54.

Drysdale, K. et al, 2019. Mapping Experiences of Serodiscordance: Using Visual Methodologies to Construct Relationality in Families Living with or Affected by Stigmatized Infectious Disease. *Qualitative Health Research,* Vol. 30, No. 5, pp 793–808.

Kohn, L.T. et al, 2000. *To Err Is Human: Building a Safer Health System.* National Academies Press. Washington (DC), USA.

Bon, K. and Lupton, E., 2020. *Health Design Thinking: Creating Products and Services for Better Health*. Cooper Hewitt, New York City, USA.

LeRouge, C. et al, 2013. User Profiles and Personas in the Design and Development of Consumer Health Technologies. *International Journal of Medical Informatics,* Vol. 82, No. 11.

Levett, D. Z. H. and Grimmett, C., 2019. Psychological factors, prehabilitation and surgical outcomes: evidence and future directions. *Anaesthesia,* Vol. 74, Suppl.1, pp 36-42.

Leyns, C. C. et al, 2018. Using Concept Mapping to Identify Policy Options and Interventions towards People-Centred Health Care Services: a Multi Stakeholders Perspective. *International Journal for Equity in Health*, Vol. 17, No. 1.

Nauman, J. et al, 2020. Global Incidence and Mortality Trends due to Adverse Effects of Medical Treatment, 1990-2017: A Systematic Analysis from the Global Burden of Diseases, Injuries and Risk Factors Study. Cureus, 12(3), e7265.

Schwendimann, R. et al, 2018. The occurrence, types, consequences and preventability of in-hospital adverse events - a scoping review. *BMC health services research*, Vol. 18, No. 521.

Valaitis, R. et al, 2019. Health TAPESTRY: Co-Designing Interprofessional Primary Care Programs for Older Adults Using the Persona-Scenario Method. *BMC Family Practice,* Vol. 20, No. 1.

de Vries, E. N. et al, 2008. The Incidence and Nature of in-Hospital Adverse Events: A Systematic Review. *Quality and Safety in Health Care,* Vol. 17, No. 3, pp 216–23.

Wacker, J., 2020. Measuring and Monitoring Perioperative Patient Safety: A Basic Approach for Clinicians. *Current Opinion in Anaesthesiology,* Vol. 33, No. 6, pp 815–22.

Wilson, J. et al, 2015. Concept Mapping. *Qualitative Health Research,* Vol. 26, No. 8, pp 1151–61.

Short Papers

IMPLEMENTING BUSINESS INTELLIGENCE IN HIGHER EDUCATION: A ROADMAP FOR DASHBOARD DEVELOPMENT

Romeu Sequeira[1], Paulo Alves[2], Arsénio Reis[1,3] and Frederico Branco[1,3]
[1]*School of Science and Technology, University of Trás-os-Montes and Alto Douro*
Quinta dos Prados, 5000-801 Vila Real, Portugal
[2]*Research Centre in Digitalization and Intelligent Robotics (CeDRI), Instituto Politécnico de Bragança*
5300-253 Bragança, Portugal
[3]*Institute for Systems and Computer Engineering, Technology and Science (INESC TEC)*
4200-465 Porto, Portugal

ABSTRACT

This study presents a roadmap for the adoption of Business Intelligence (BI) systems in Higher Education Institutions (HEIs), focusing on the development of dashboards tailored to the specific needs of this educational sector. It includes a methodology for designing and implementing dashboards that address the unique complexities and challenges faced by HEIs. The strategic objective is to improve decision-making and resource management in HEIs by providing solutions that are flexible and adaptable to the dynamic changes in the educational environment. Using interviews with key stakeholders in a HEI as a method of gathering requirements and information, this study highlights the basis for the design and development of dashboards and emphasises the importance of a thorough understanding of the needs of HEIs. A case study of a Portuguese HEI illustrates the significant impact of BI systems, highlighting both the quantifiable benefits and the challenges of customisation to meet the specific needs of HEIs. It also explores the future possibilities of these technologies, foreseeing a transformation in the operational and strategic management and decision-making processes of HEIs. This work contributes to the academic literature and provides a practical guide for institutions looking to integrate BI technologies to optimise their performance and ensure a competitive position in the ever-changing education sector.

KEYWORDS

Business Intelligence, Higher Education, Data Engineering, Dashboard Development, Decision-Making

1. INTRODUCTION

In a contemporary scenario defined by the rapid evolution of Information Technologies (IT), Higher Education Institutions (HEIs) are faced with the need to adapt to a context increasingly dominated by data analysis and management. Bianchi and Sousa (2016), together with Khouja et al. (2018), underline the essentiality of IT governance in HEIs, emphasising the importance of establishing effective communication between IT departments, business entities and stakeholders. Within this framework, the governance of Business Intelligence (BI) systems stands out for incorporating crucial elements such as strategic alignment, funding, project management and ensuring data quality. Furthermore, the adoption of best practices, supported by clear guidelines and evaluation methods, is seen as fundamental to maximising the return on investment in BI projects, ensuring more effective institutional management and improved academic performance (Combita Niño et al., 2020). In this context, the deployment of BI solutions in HEIs is recognised as a key strategy for empowering data-informed decisions, thus raising the quality of management and academic performance.

The central aim of this article is to present a detailed roadmap for implementing BI systems in HEIs, with a particular focus on the development of dashboards. This innovative approach aims to overcome the technical challenges associated with installing BI systems in HEIs, as well as deepen the understanding and practical applicability of these technologies in this sector. Thus, this work aims to provide a practical guide to help HEIs effectively implement BI systems, in particular by proposing dashboards that take into account the particularities and specific requirements of the higher education sector.

In this context, the research is based on the Design Science Research (DSR) methodology, which is recognised for its ability to integrate theory and practice through the development and evaluation of innovative artefacts. This methodology is well suited to the needs of HEIs, enabling not only the development of solutions that effectively respond to the challenges identified but also their adaptation to the specific needs of the higher education context. This ensures that the implementation of BI systems is relevant and effective. An important milestone in this study was the validation of the proposed roadmap for the implementation of BI systems, using a Portuguese HEI as a case study. This step made it possible to test the applicability of the roadmap in a real environment, and at the same time to adapt and improve it based on concrete feedback.

2. DASHBOARD PROPOSAL

Dashboards are essential tools for visualizing complex data, enabling users to understand relevant information quickly and effectively (Hart, 2021). In the context of HEIs, the adaptation of these dashboards to specific contexts plays a crucial role in supporting informed decision-making processes. Adapting dashboards to the needs of each HEI can enhance the efficient use of data for continuous improvement, as demonstrated by Guerra et al. (2020), Kunjan et al. (2019), Leitner and Ebner (2017), and Vázquez-Ingelmo et al. (2018).

Developing context-specific dashboards in HEIs involves a customization process that takes into account the different types of data and services offered by each institution. Well-designed dashboards contribute to efficient management, informed decision-making, and continuous improvement in various areas of an HEI, such as course management, student performance monitoring, and administrative efficiency analysis (De Laet et al., 2020; Guerra et al., 2020; McCoy & Rosenbaum, 2019).

This work highlights the importance of tailoring dashboards to the specificities of each area within the HEI to promote more informed and effective management across the institution. Through a series of interviews with various key stakeholders in the HEI, the process of developing dashboards in our study included the following stages:

1) Identifying contexts - Reviewing literature and research to identify the specific contexts in which dashboards are needed in higher education (Guerra et al., 2020; Vázquez-Ingelmo et al., 2021);

2) Understand user needs - Conduct interviews, surveys or focus groups with stakeholders in each context to understand their specific needs, goals and challenges. This will help gather requirements and information on which to base the design and development of dashboards (Guerra et al., 2020; McCoy & Rosenbaum, 2019);

3) Define dashboard objectives - Based on the identified contexts and user needs, define the objectives of each dashboard. Determine what information and knowledge the dashboards should provide to support decision-making and improve outcomes in each specific context (Vázquez-Ingelmo et al., 2021);

4) Selecting appropriate visualisations - Select appropriate visualisations and data representations that effectively communicate the desired information and insights. Consider the types of data available, the audience and the objectives of each dashboard (Vázquez-Ingelmo et al., 2020);

5) Design the layout and navigation of the dashboard - Design the layout and navigation of each dashboard to ensure ease of use and intuitive access to relevant information. Consider the organisation of data, the layout of views, and the inclusion of filters or interactive elements (Vázquez-Ingelmo et al., 2020);

6) Develop data integration and visualisation - Develop the necessary data integration processes to collect and consolidate data from different sources relevant to the context. Use appropriate data visualisation tools and techniques to present data in a meaningful and actionable way (Vázquez-Ingelmo et al., 2021);

7) Test and iterate - Conduct usability testing and gather user feedback to assess the effectiveness and usability of each dashboard. Iterate the design and functionality based on user feedback to ensure the dashboards meet the specific needs of each context (McCoy & Rosenbaum, 2019; Vázquez-Ingelmo et al., 2021);

8) Implementation and maintenance - Implement the dashboards in an appropriate technology infrastructure that ensures the security and accessibility of the data. Establish a maintenance plan to regularly update and improve the dashboards based on evolving needs and user feedback (Vázquez-Ingelmo et al., 2020).

2.1 Integration of Low-Fidelity User Experience Approach with Dashboard Personalisation

Tang et al. (2022) highlight the importance of using personalised dashboards in decision-making in higher education, and it is important to empower decision-makers to personalise their dashboards. Thus, developing dashboards for each of the contexts of information use requires a careful and personalised approach to meet the specific needs and requirements of each area. Therefore, it is important to personalise dashboards according to the specificities of each area to promote more informed and effective management across the HEI.

This leads to the low-fidelity UX approach used in our work, which we propose for the use of low-fidelity prototypes or models in the context of user experience (UX) design and evaluation. Low-fidelity refers to prototypes that can be created quickly and inexpensively, often using simple sketches such as wireframes or basic mockups, without detailed visual design or interactive functionality (Gardey et al., 2020). A low-fidelity UX approach aims to gather initial feedback and information from users and stakeholders in the design process. It allows designers to focus on the overall structure, layout and functionality of the user interface, rather than on visual details or complex interactions (Gardey et al., 2020).

By using low-fidelity prototypes, designers can iterate quickly and make changes based on user feedback, ensuring that the final product meets users' needs and expectations. This approach is particularly useful in the early stages of design when different design concepts have been explored and users' initial impressions have been gathered (Gardey et al., 2020).

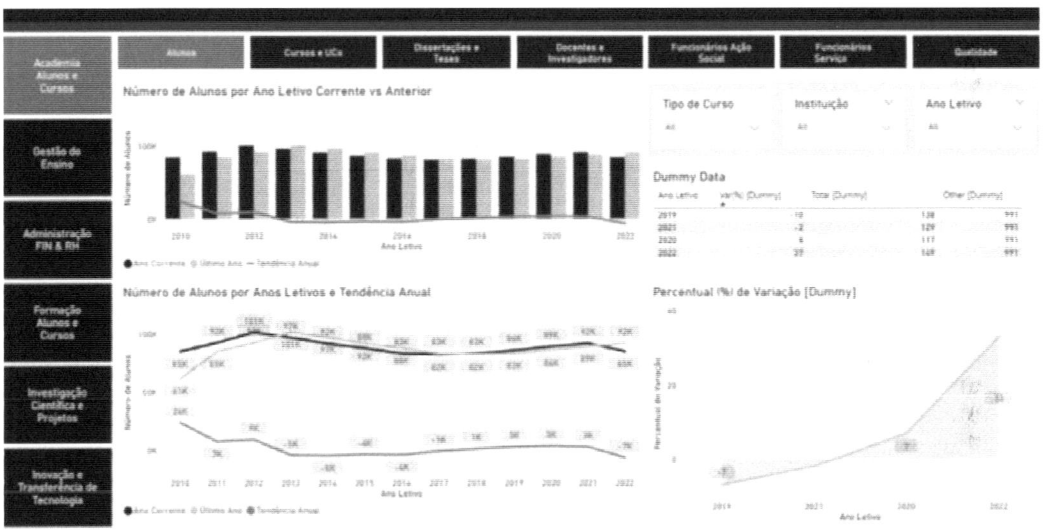

Figure 1. Dashboard mockup

Figure 1 shows a dashboard mockup of the HEI BI system, developed using the low-fidelity UX approach mentioned above.

Overall, the low-fidelity UX approach is an effective way of involving user groups in the identification of needs, as well as in the design and validation of dashboards, helping to ensure that proposed solutions meet user expectations and are functional and effective for each HEI context (Berni & Borgianni, 2021; McCoy et al., 2020; Mehjabeen et al., 2021; Narh-Bana et al., 2021; Rajkumar & Kogilavani, 2021; Regazzoni et al., 2021).

3. CONCLUSION

The roadmap proposed in this study focuses on the development of personalised dashboards to improve decision-making in HEIs. This methodology serves as a practical guide to the development of customised dashboards, which are essential to the successful implementation of data strategies. These dashboards are designed to meet the unique needs and specific requirements of the higher education sector.

This study also illustrates the transformative impact of BI systems in higher education, specifically as evidenced by the implementation of a customised BI solution at a HEI in Portugal, with a particular focus on the development of dashboards. The benefits identified are significant, including optimisation of the decision-making process, more efficient management of educational resources and a significant increase in the ability to adapt strategically to the constant dynamics within the higher education sector. However, implementation faces challenges and limitations, mainly due to the extensive effort required to customise BI solutions to the specifics of each HEI. This highlights the need for rigorous planning in the design and implementation of these systems, where the development of dashboards plays an important role.

In addition, the applications derived from this study propose a redefinition of how HEIs can use data to foster a culture of academic excellence and innovation, suggesting a transformation in data-driven higher education management. This work not only contributes to the existing body of knowledge but also provides a crucial framework for future research into the effective integration of BI technologies in higher education, paving the way for a new era in higher education management in which data-driven decision-making becomes a central pillar of academic excellence and innovation.

ACKNOWLEDGMENT

The authors acknowledge the work facilities and equipment provided by CeDRI (UIDB/05757/2020 and UIDP/05757/2020) to the project team.

REFERENCES

Berni, A., & Borgianni, Y. (2021). Making Order in User Experience Research to Support Its Application in Design and Beyond. Applied Sciences, 11(15), 6981. https://doi.org/10.3390/app11156981

Bianchi, I. S., & Sousa, R. D. (2016). IT Governance Mechanisms in Higher Education. Procedia Computer Science, 100, 941–946. https://doi.org/10.1016/j.procs.2016.09.253

Combita Niño, H. A., Cómbita Niño, J. P., & Morales Ortega, R. (2020). Business intelligence governance framework in a university: Universidad de la Costa case study. International Journal of Information Management, 50, 405–412. https://doi.org/10.1016/j.ijinfomgt.2018.11.012

De Laet, T., Millecamp, M., Ortiz-Rojas, M., Jimenez, A., Maya, R., & Verbert, K. (2020). Adoption and impact of a learning analytics dashboard supporting the advisor—student dialogue in a higher education institute in Latin America. British Journal of Educational Technology, 51(4), 1002–1018. https://doi.org/10.1111/bjet.12962

Gardey, J. C., Garrido, A., Firmenich, S., Grigera, J., & Rossi, G. (2020). UX-Painter: An Approach to Explore Interaction Fixes in the Browser. Proceedings of the ACM on Human-Computer Interaction, 4(EICS), 1–21. https://doi.org/10.1145/3397877

Guerra, J., Ortiz-Rojas, M., Zúñiga-Prieto, M. A., Scheihing, E., Jiménez, A., Broos, T., De Laet, T., & Verbert, K. (2020). Adaptation and evaluation of a learning analytics dashboard to improve academic support at three Latin American universities. British Journal of Educational Technology, 51(4), 973–1001. https://doi.org/10.1111/bjet.12950

Hart, D. (2021). Conceptualizing the systemic evaluation of dashboards in quality enhancement processes in higher education. Systems Research and Behavioral Science, 38(6), 817–832. https://doi.org/10.1002/sres.2736

Khouja, M., Rodriguez, I. B., Ben Halima, Y., & Moalla, S. (2018). IT Governance in Higher Education Institutions: A Systematic Literature Review. International Journal of Human Capital and Information Technology Professionals, 9(2), 52–67. https://doi.org/10.4018/IJHCITP.2018040104

Kunjan, K., Doebbeling, B., & Toscos, T. (2019). Dashboards to Support Operational Decision Making in Health Centers: A Case for Role-Specific Design. International Journal of Human-Computer Interaction, 35(9), 742–750. https://doi.org/10.1080/10447318.2018.1488418

Leitner, P., & Ebner, M. (2017). Development of a Dashboard for Learning Analytics in Higher Education. In P. Zaphiris & A. Ioannou (Eds.), Learning and Collaboration Technologies. Technology in Education, 10296, 293–301. https://doi.org/10.1007/978-3-319-58515-4_23

Lemon, C., Huckvale, K., Carswell, K., & Torous, J. (2020). A Narrative Review of Methods for Applying User Experience in the Design and Assessment of Mental Health Smartphone Interventions. International Journal of Technology Assessment in Health Care, 36(1), 64–70. https://doi.org/10.1017/S0266462319003507

McCoy, C., & Rosenbaum, H. (2019). Uncovering unintended and shadow practices of users of decision support system dashboards in higher education institutions: Uncovering unintended and shadow practices of users of decision support system dashboards in higher education institutions. Journal of the Association for Information Science and Technology, 70(4), 370–384. https://doi.org/10.1002/asi.24131

Mehjabeen, S., Matin, M., Gupta, R. D., Sutradhar, I., Mazumder, Y., Kim, M., Sharmin, S., Islam, J., & Sarker, M. (2021). Fidelity of kangaroo mother care services in the public health facilities in Bangladesh: A cross-sectional mixed-method study. Implementation Science Communications, 2(1), 115. https://doi.org/10.1186/s43058-021-00215-9

Narh-Bana, S. A., Kawonga, M., Chirwa, E. D., Ibisomi, L., Bonsu, F., & Chirwa, T. F. (2021). Fidelity of implementation of TB screening guidelines by health providers at selected HIV clinics in Ghana. PLOS ONE, 16(9), e0257486. https://doi.org/10.1371/journal.pone.0257486

Rajkumar, P., & Kogilavani, S.V. (2021). Opinion mining for user experience evaluation model using kernel-naive Bayes classification algorithm. Journal of Engineering Research, 9. https://doi.org/10.36909/jer.11319

Regazzoni, F., Pagani, S., Cosenza, A., Lombardi, A., & Quarteroni, A. (2021). A physics-informed multi-fidelity approach for the estimation of differential equations parameters in low-data or large-noise regimes. Rendiconti Lincei - Matematica e Applicazioni, 32(3), 437–470. https://doi.org/10.4171/RLM/943

Tang, Y. W., Toh, S., & Leow, A. S. (2022). Dashboards for Decision Making in Higher Education. ASCILITE Publications, 321–329. https://doi.org/10.14742/apubs.2019.279

Vázquez-Ingelmo, A., García-Peñalvo, F. J., & Therón, R. (2018). Domain engineering for generating dashboards to analyze employment and employability in the academic context. Proceedings of the Sixth International Conference on Technological Ecosystems for Enhancing Multiculturality, 896–901. https://doi.org/10.1145/3284179.3284329

Vázquez-Ingelmo, A., García-Peñalvo, F. J., & Therón, R. (2021). Towards a Technological Ecosystem to Provide Information Dashboards as a Service: A Dynamic Proposal for Supplying Dashboards Adapted to Specific Scenarios. Applied Sciences, 11(7), 3249. https://doi.org/10.3390/app11073249

Vázquez-Ingelmo, A., García-Peñalvo, F. J., Therón, R., & Conde, M. Á. (2020). Representing Data Visualization Goals and Tasks through Meta-Modeling to Tailor Information Dashboards. Applied Sciences, 10(7), 2306. https://doi.org/10.3390/app10072306

LSTM NETWORK MAPPING MODEL FOR TEMPERATURE INDUCED-CRACK WIDTH VARIATION OF CONCRETE CABLE-STAYED BRIDGE

Xujia Liu, Youliang Ding, Chunfeng Wan and Xiaonan Zhang
Key Laboratory of Concrete and Pre-stressed Concrete Structures of the Ministry of Education, Southeast University, Nanjing 210096, China
College of Civil Engineering, Southeast University, Nanjing 210096, China

ABSTRACT

The big data (BD) technology is indispensable to realize the structure health monitoring (SHM) of the bridge. Through wielding the BD in SHM, the structural state quickly assessment of the structure is achieved. In this paper, the temperature-induced-crack width variation of a reinforced concrete cable-stayed bridge is discussed, and the mapping model between temperature and crack width is established based on the long short-term memory network. The results show that the model has a good mapping effect. The model can provide a reference for bridge operation, structural safety and durability assessment.

KEYWORDS

Big Data, Structure Health Monitoring, Crack Width, Long Short-Term Memory Network, Mean Square Error

1. INTRODUCTION

The structure health monitoring (SHM) (Y. Bao, et al., 2019) for quick assessment of the bridge state can be realized by using big data (BD). As an important part of urban traffic, the safety and durability of bridges are directly related to the safety of urban traffic. Through the bridge SHM, the parameters of the bridge can be monitored in real-time, the potential safety risks can be found and solved in time (D. Dan, Y. Ying, L. Ge, 2021), and the safety of urban traffic can be improved.

The crack is the natural enemy of the safety and durability of the reinforced concrete cable-stayed bridge (RCC-SB). The spatial shape of cracks is also the key object of monitoring and detection. The temperature field variation of the primary factors inducing the spatial shape of cracks for the RCC-SB. The spatial shape of the crack is mapped by temperature, which is of reference significance for the bridge health assessment.

In this paper, the crack width variation is taken as the object, and the mapping between temperature and crack is realized by using a long short-term memory (LSTM) (*Sherstinsky A., 2020*) network model by the data of six temperature sensors. Firstly, the LSTM network model architecture is introduced, and the crack and temperature data are processed by the sliding window average method and Baworth low-pass filter, and then the model is trained and verified.

2. ARCHITECTURE OF LSTM NETWORK MODE

LSTM is a classical deep-learning algorithm architecture. The LSTM structure consists of four parts, including a memory cell, input gate, output gate and forget gate as shown in Figure 1. The LSTM network model is shown in Figure 2. The f_t, i_t, $\tilde{c}_t$ *and* o_t in Figure 1 are defined as

$$f_t = \sigma(W_f \cdot [h_{t-1}, x_t] + b_f) \quad (1)$$

$$i_t = \sigma(W_i \cdot [h_{t-1}, x_t] + b_i) \quad (2)$$

$$\tilde{c}_t = \tanh(W_c \cdot [h_{t-1}, x_t] + b_c) \quad (3)$$

$$o_t = \sigma(W_o \cdot [h_{t-1}, x_t] + b_o) \quad (4)$$

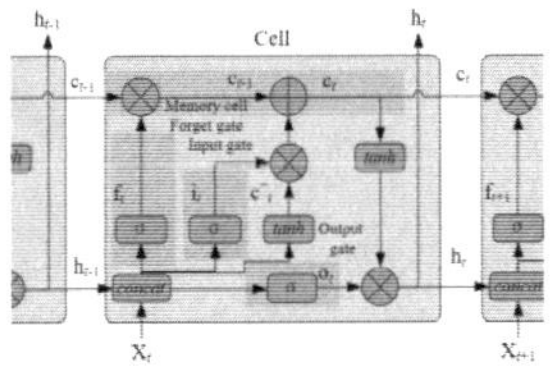

Figure 1. LSTM Cell

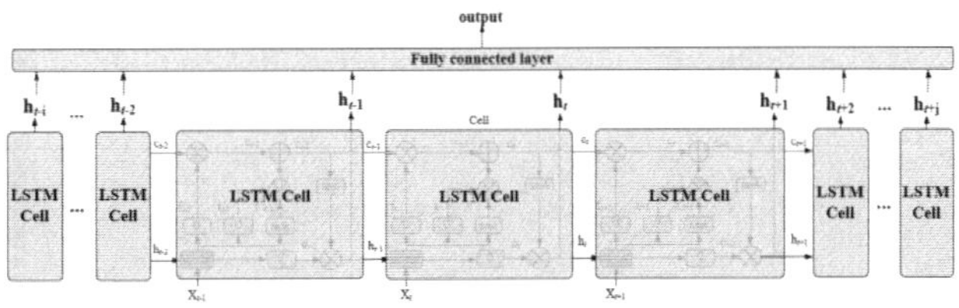

Figure 2. LSTM network model

3. TEMPERATURE INDUCED-CRACK WIDTH VARIATION

3.1 Case

The data was derived from the SHM of an RC cable-stayed bridge. The structural type of the bridge, the girder cross-section of the bridge, and the location of the sensor arrangement are shown in Figure 3. There are eight temperature sensors T1~T8 and a crack width sensor C1 on the inner surface of the cross-section bottom plate. The sampling frequency of sensors is 1 / 300 Hz. The bridge began service in Dec. 2005 and is a full floating RC cable-stayed bridge with a length of 152 m + 370 m + 152 m. The main girder of the Huai'an Bridge is a double box girder section, and the bottom plate is arranged with prestressed tendons. The girder is provided with a concrete diaphragm every 2.5 m ~ 6 m. The pavement of Huai'an Bridge was lightened in 2017, and the overall weight of its girder was reduced by 1500 tons. The temperature sensors T3, T5, T6 and T8 are located on the surface outside the girder, and others are arranged on the surface inside the girder. The crack width sensor is erected on the surface inside the girder.

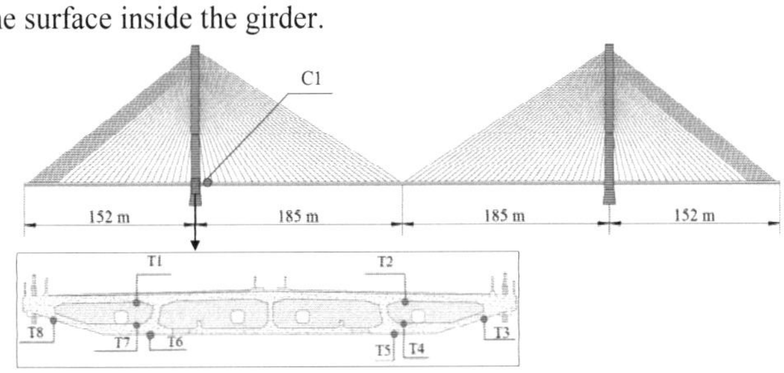

Figure 3. Bridge elevation view and Sensor arrangement

3.2 Temperature and Crack Width

The temperature and crack width data on the time dimension are shown in Figure 4 and Figure 5. The time-history curves of temperature and crack width shown in Figure 4 and Figure 5 pass through the slip window mean filter and the Barworth low-pass filter. It can be discovered that the temperature (T3, T5, T6 and T8) of the surface outside the girder is more sensitive than that of the surface inside. The variation trend of temperature and crack width is opposite, and the crack width decreases with temperature.

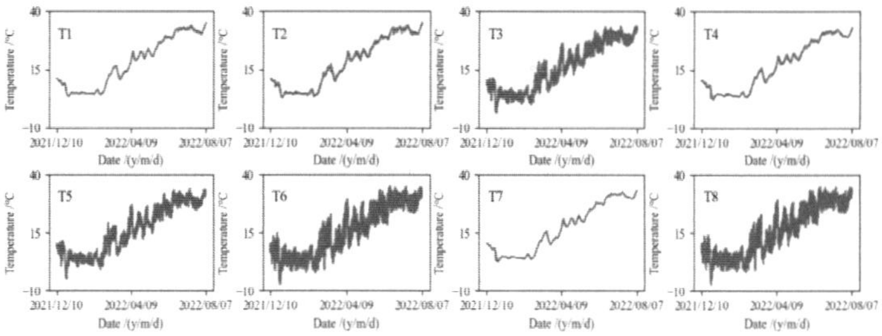

Figure 4. Temperature variation from Dec.10, 2021 to Aug.07, 2022

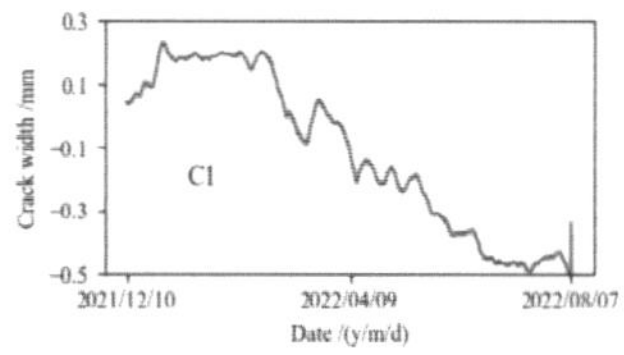

Figure 5. Crack width variation from Dec.10, 2021 to Aug.07, 2022

3.3 Regression Analysis

The time series long used in the model is one day and the batch size is 512, and the learning rate is 0.0001. The mean square error and true and prediction values of the train and test model are shown in Figure 6 and Figure 7, respectively. We select the model of the minimum mean square error, and draw the time-series curves of true and prediction of the train and validation set as shown in Figure 7.

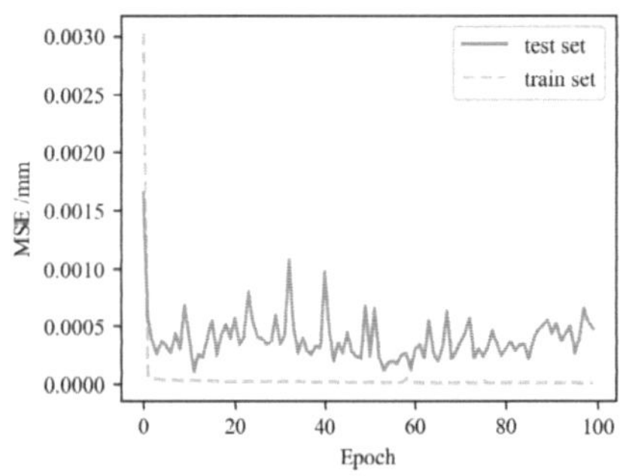

Figure 6. Mean square error

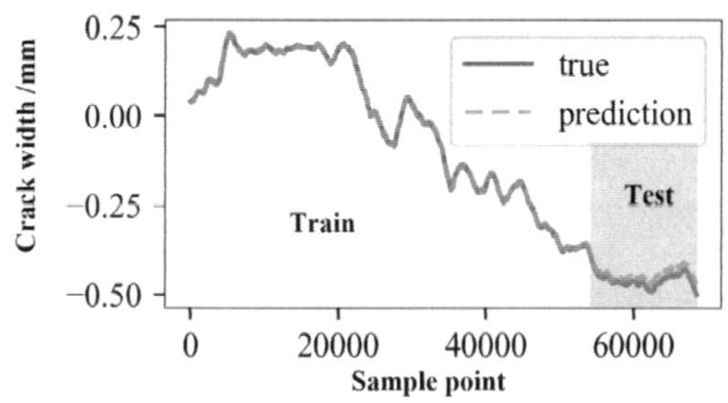

Figure 7. True and prediction values of the train and validation model

4. CONCLUSION

From the time-history curves of the temperature and crack width variation, it can be discovered that the crack width decreases with the increase of the temperature for the RC cable-stayed bridge.

The application of the LSTM network model to the mapping of temperature-induced crack width variation of the RC cable-stayed bridge has a good effect. When the training epoch of the model is 11, the mean square error of the test set reaches the minimum value of 1.09e-4, meanwhile, the mean square error of the training set is 3.05e-5.

ACKNOWLEDGEMENT

This research has been supported by the National Natural Science Foundation of China (Grant No. 52378288).

REFERENCES

Dan, D., Ying, Y., Ge, L. (2021). Digital twin system of bridges group based on machine vision fusion monitoring of bridge traffic load. IEEE Trans. Intell. Transp. Syst.

Sherstinsky, A. (2020). Fundamentals of Recurrent Neural Network (RNN) and Long Short-Term Memory (LSTM) Network. Physica D, 404, 132306.

Bao, Y., Chen, Z., Wei, S. et al. (2019). The state of the art of data science and engineering in structural health monitoring. Engineering, 5, 234–242.

DIGITAL MAPPING OF VEHICLE LOADS TO STRUCTURAL STRAINS BASED ON M-H SAMPLING AND BILSTM NETWORK

Xiaonan Zhang[1], Youliang Ding[1,2] and Chunfeng Wan[1,2]
[1]*School of Civil Engineering, Southeast University, Nanjing, 210096, China*
[2]*Key Laboratory of Concrete and Prestressed Concrete Structures of Ministry of Education, Southeast University, Nanjing 210096, China*

ABSTRACT

The weigh-in-motion (WIM) system and the structural health monitoring (SHM) system have been used as two separate modules playing different roles in structural operation and providing different information for maintenance. With the advent of the big data (BD) era, various artificial intelligence networks and methods can integrate the WIM system with the SHM system based on the correspondence between the two systems. This study explores the digital mapping of vehicle loads to bridge strains using BD from WIM and SHM. The correspondence of vehicle load to strain is realized by Metropolis-Hastings (M-H) sampling, and the complex nonlinear model of vehicle load-strain is established by Bi-directional long short-term memory (BiLSTM) network. The proposed method is verified with the continuous monitoring data of a concrete box girder bridge. The results show that the vehicle load-strain mapping model can be built accurately, which can provide a reference for vehicle restriction and bridge assessment.

KEYWORDS

Big Data, Structure Health Monitoring, Weigh-In-Motion, Bi-Directional Long Short-Term Memory Network, Digital Mapping

1. INTRODUCTION

With the continuous development of sensor technology, the weigh-in-motion (WIM) system and the structural health monitoring (SHM) system are widely used in structural condition monitoring and safety performance assessment (R Wu, Jahanshahi, 2020). In previous research and applications, the WIM system and the SHM system are usually used as two separate modules to perform different functions in structural services.

A bridge weigh-in-motion (BWIM) system focusing on traffic flow can effectively collect information such as vehicle weight and speed (Y Yu. et al, 2016). Among the many factors that affect the safety condition and remaining life of in-service bridges, vehicle load is the main live load component that leads to bridge damage (W He. et al, 2019). SHM system is now widely used in bridge condition monitoring. During the bridge service, the SHM system can collect the response of the bridge under various factors, especially those caused by vehicles. Some studies combine these two systems for bridge condition assessment (R Hou. et al, 2020), but most of them have relied on cameras and have not explored the digital mapping relationship between the two in depth.

The advent of the big data (BD) era and the continuous development of machine learning allows artificial intelligence networks to be used to monitor and evaluate the response of bridges under traffic loads (X Zhang. et al, 2024). Deep learning performs well in finding complex relationships and capturing features between data (S Kim. et al, 2009). Therefore, the relationship between vehicle loads and load effects can be well explored by neural networks.

Due to the lack of research on integrating WIM and SHM systems for vehicle load prediction, this study aims to investigate the digital mapping between vehicle loads and structural strains. The main contributions of this study can be summarized as follows: (a) The problem of difficult correspondence between vehicle loads and vehicle-induced strains is solved by Metropolis-Hastings (M-H) sampling. (b) The nonlinear mapping model between vehicle load and strain is built by Bi-directional long short-term memory (BiLSTM).

2. METHODOLOGY

2.1 M-H Sampling

The Metropolis-Hastings (M-H) sampling is used to solve the problem that the mapping model requires length-consistent vehicle loads and strains.

The M-H sampling is a sub-method of Markov chain Monte Carlo (MCMC)[37]. MCMC sets an acceptance rate $\alpha(i,j)$ for the random state transition matrix P and probability distribution $\pi(x)$ to satisfy

$$\pi(i)P(i,j)\alpha(i,j)=\pi(j)P(j,i)\alpha(j,i)\,, \tag{1}$$

where i and j represent different states. M-H sampling sets the $\alpha(i,j)$ as

$$\alpha(i,j)=\min\{\frac{\pi(j)P(j,i)}{\pi(i)P(i,j)},1\}\,. \tag{2}$$

Compared with MCMC, the M-H sampling increases the $\alpha(i,j)$ and improves the sampling efficiency.

2.2 BiLSTM

Compared with the recurrent neural network (RNN), three gates are added in the hidden layer h of long short-term memory (LSTM), namely forget gate, input gate, and output gate, while a new cell state is added as well. The structure of LSTM is shown in Figure 1(a).

The LSTM network can be calculated by the following equations:

$$\text{Forget gate}\quad \boldsymbol{f}_t=\sigma(\boldsymbol{W}_f\cdot[\boldsymbol{h}_{t-1},\boldsymbol{x}_t]+\boldsymbol{b}_f)\,, \tag{3}$$

$$\text{cell state}\quad \tilde{\boldsymbol{C}}_t=\tanh(\boldsymbol{W}_s\cdot[\boldsymbol{h}_{t-1},\boldsymbol{x}_t]+\boldsymbol{b}_s)\,, \tag{4}$$

$$\text{Updated cell states}\quad \boldsymbol{C}_t=\boldsymbol{f}_t*\boldsymbol{C}_{t-1}+\boldsymbol{i}_t*\tilde{\boldsymbol{C}}_t\,, \tag{5}$$

$$\text{Output gate}\quad \boldsymbol{o}_t=\sigma(\boldsymbol{W}_o[\boldsymbol{h}_{t-1},\boldsymbol{x}_t]+\boldsymbol{b}_o)\,, \tag{6}$$

$$\text{Output}\quad \boldsymbol{h}_t=\boldsymbol{o}_t*\tanh(\boldsymbol{C}_t)\,, \tag{7}$$

where $\boldsymbol{x}_t$ represents the input at time t, $\boldsymbol{h}_{t-1}$ denotes the hidden state value at time t-1, $\boldsymbol{W}$ and $\boldsymbol{b}$ denote the weight matrix and bias vector, respectively; σ denotes the Sigmoid function and tanh denotes the tangent hyperbolic function; the symbol ‘ * ’ represents the Hadamard product.

BiLSTM network is proposed that can combine the information of the input sequence in both forward and backward directions based on the LSTM network. The basic network architecture of BiLSTM used in this study is shown in Figure 1(b), which consists of five layers: a sequence input layer, a BiLSTM hidden layer, a fully connected layer, a softmax layer, and a regression layer.

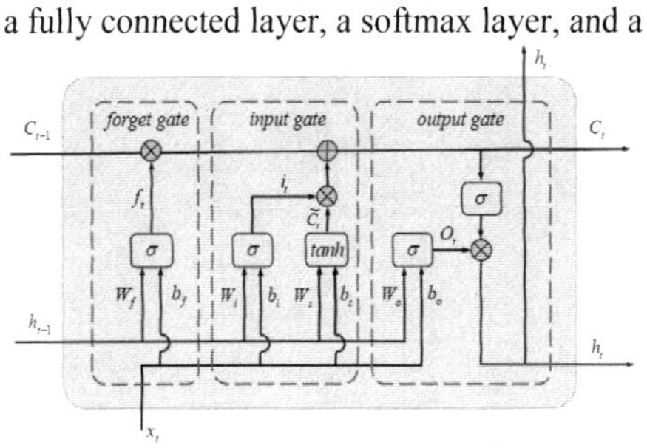

(a)

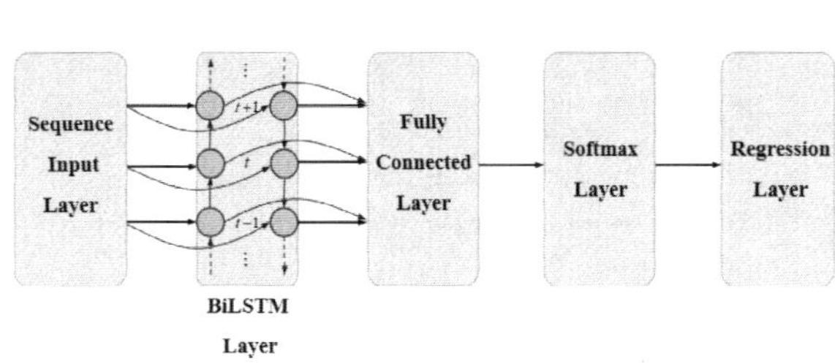

(b)

Figure 1. The structure of LSTM network and BiLSTM network: (a) internal unit structure of LSTM network, (b) structure of BiLSTM network

3. EXPERIMENTAL VERIFICATION

3.1 Engineering Background

Continuous monitoring data from a 25 m continuous prestressed concrete box girder bridge are used to validate the proposed method. The bridge with twin decks is equipped with a health monitoring system in the middle span. The sensor locations are shown in Figure 2(a). The WIM system is used to record the traffic flow information across the bridge, as shown in Figure 2(b). The sampling frequency of strain sensors is 20 Hz.

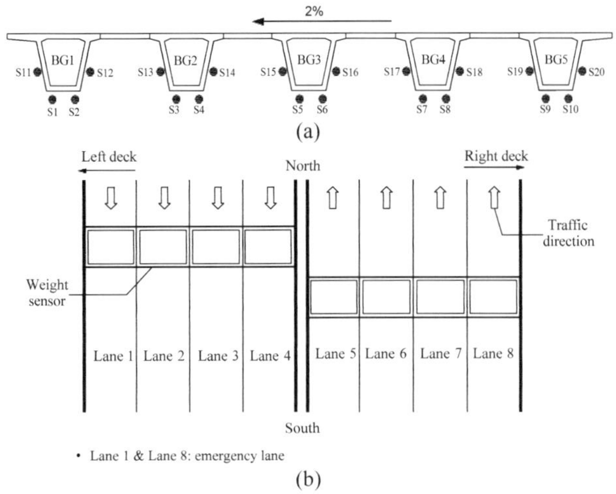

Figure 2. Sensors and WIM arrangement of the concrete box girder bridge: (a) strain sensors, (b)WIM and lane division

3.2 Fitting and Sampling of Vehicle Loads and Vehicle-Induced Strains

The monitored strains of S7 are decomposed into signals of different frequency bands using the discrete wavelet transform. By eliminating the low-frequency temperature-induced strains, the high-frequency vehicle-induced strains are extracted. The maximum value of each vehicle-induced strain is extracted to investigate the mapping relationship. One year's vehicle-induced strains and vehicle loads are statistically analyzed using gaussian mixture model (GMM). As shown in Figure 3, the distributions of both vehicle-induced strains and vehicle weights have three peaks with similar distribution characteristics and can be well-fitted with the 3-order GMM.

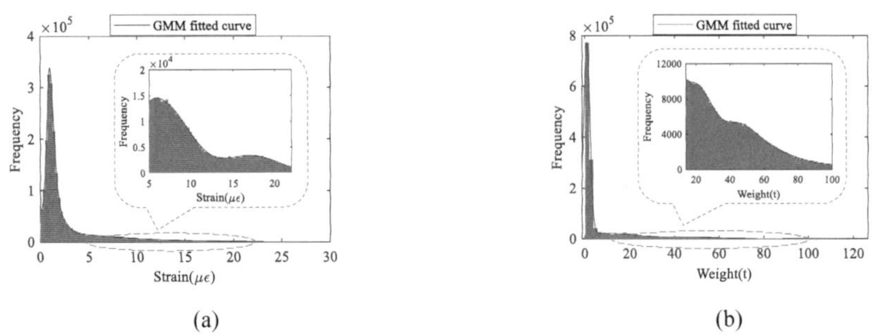

Figure 3. Fitting of vehicle-induced strains and vehicle weights: (a) vehicle-induced strains, (b) vehicle weights

The sampling results that obey the original distribution by M-H sampling are shown in Figure 4. It can be seen that the samples obtained are almost identical to the original data.

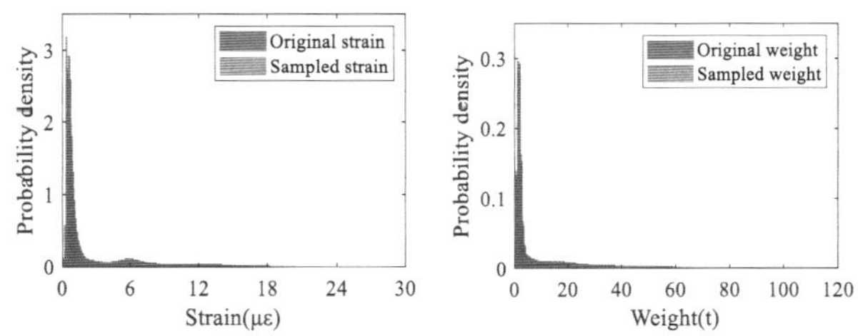

(a) (b)

Figure 4. The results of M-H sampling: (a) vehicle-induced strains, (b) vehicle loads

3.3 The Modeling of Vehicle Load-Strain Nonlinear Correlation

After obtaining samples of vehicle loads and vehicle-induced strains with consistent length, the samples are sequentially corresponded and a five-layer BiLSTM network is used to capture the relationship between the vehicle loads and vehicle-induced strains. The solution optimizer used is Adam and the initial learning rate is 0.01. A certain number of vehicle-induced strains and vehicle loads is required as the training set, validation set and test set of the regression model.

Figure 5(a) shows the predicted vehicle loads obtained by the trained BiLSTM regression model compared with the measured vehicle loads. It can be seen from Figure 5(a) that the predicted vehicle loads is very close to the measured vehicle loads in terms of overall trends and details. The fitting curve of the predicted vehicle loads and the measured vehicle loads is given in Figure 5(b). It can be seen that the correlation between the two is extremely high because of the almost linear fitting curve. The model has a prediction accuracy of 99.1%. Therefore, the trained BiLSTM regression model can accurately predict the corresponding vehicle loads based on the vehicle-induced strains.

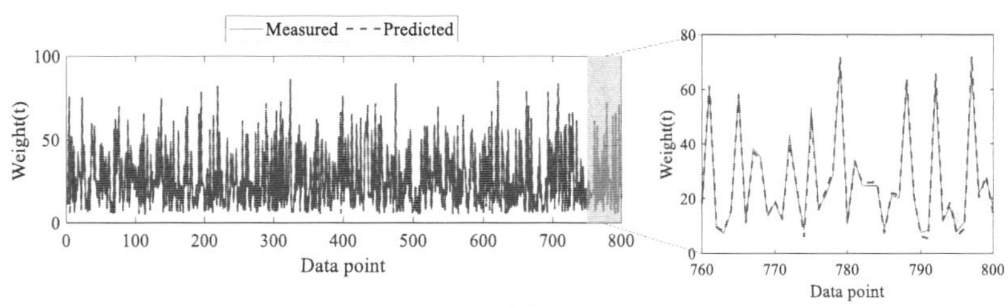

(a)

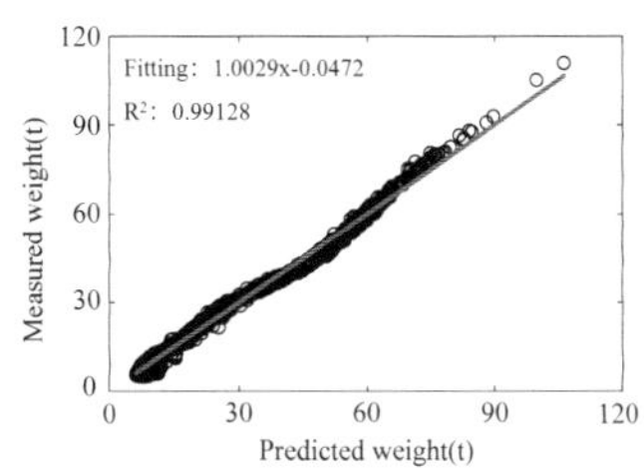

(b)

Figure 5. Prediction results of vehicle loads:(a) typical regression results for measured and predicted vehicle loads, (b) fitted curve of measured and predicted vehicle loads

4. CONCLUSION

In this paper, a digital mapping model of vehicle loads and vehicle-induced strains is explored through M-H sampling and the BiLSTM network. The digital mapping model integrates the WIM system with the SHM system, after which the data can be better utilized for guiding the management and maintenance of the bridge. A real concrete bridge is used as an engineering case to verify the feasibility of the method. The conclusions drawn from this study are summarized as follows:

1. The distributions of vehicle-induced strains and vehicle loads have the same statistical characteristics, which can be fitted by the 3-order GMM. Samples of vehicle loads and vehicle-induced strains with the same length can be obtained by M-H sampling.

2. There is a complex nonlinear relationship between vehicle loads and vehicle-induced strains, which can be well modeled by BiLSTM network. The established vehicle load-strain model can obtain the corresponding vehicle load using the vehicle-induced strain.

In this paper, only the effect of vehicle weight is considered when investigating the digital mapping relationship between vehicle loads and structural strains. Although vehicle weight is the most important factor affecting the response, the rest factors such as vehicle type, vehicle speed, and road roughness also affect this mapping relationship. The effects of multiple factors can be considered in subsequent studies.

ACKNOWLEDGEMENT

This research was supported by the National Natural Science Foundation of China (Grants. 52378288).

REFERENCES

He W. et al, (2019). Virtual Axle Method for Bridge Weigh-in-Motion Systems Requiring No Axle Detector. *Journal of Bridge Engineering*. Vol. 24, No. 9.

Hou R. et al, (2020). Cyber-physical system architecture for automating the mapping of truck loads to bridge behavior using computer vision in connected highway corridors. *Transportation Research Part C-Emerging Technologies*. Vol. 111, pp 547-571.

Kim S. et al, (2009). Vehicle Signal Analysis Using Artificial Neural Networks for a Bridge Weigh-in-Motion System. *Sensors*. Vol. 9, No. 10, pp 7943-7956.

Wu RT, Jahanshahi MR, (2020). Data fusion approaches for structural health monitoring and system identification: Past, present, and future. *Structural Health Monitoring-an International Journal*. Vol. 19, No. 2, pp 552-586.

Yu Y. et al, (2016). State-of-the-art review on bridge weigh-in-motion technology. *Advances in Structural Engineering*. Vol. 19, No. 9, pp 1514-1530.

Zhang XN. et al, (2024). Mixed Skewness Probability Modeling and Extreme Value Predicting for Physical System Input-Output Based on Full Bayesian Generalized Maximum-Likelihood Estimation. *IEEE Transactions on Instrumentation and Measurement*. Vol. 73.

CYBER-PHYSICAL CLOUD TESTBED ON RASPBERRY PI MICROCOMPUTERS

Adolf Kamuzora[1], Zakarya Farou[1], Tamás Jursonovics[2], Peter Pogrzeba[2] and Imre Lendák[1]
[1]*Department of Data Science and Engineering, Institute of Industry - Academia Innovation, ELTE Eötvös Loránd University, Pázmány Péter sétány 1/C, H-1117, Budapest, Hungary*
[2]*Deutsche Telekom, 53227 Bonn, Landgrabenweg 151, Germany*

ABSTRACT

The objective of this research project was to create a cyber-physical cloud testbed based on a small cluster of Raspberry Pi's (RPi) microcomputers running Docker Swarm. RPi's are low-cost microcomputers capable of running Linux and/or Android. Using Docker allowed us to rapidly run light-weight containers with different applications and Swarm mode was used to dynamically assign cluster nodes and balance loads. The key output of this project is a cyber-physical testbed which can be seamlessly scaled to tens of RPis and managed with limited administrative effort, while also facilitating resource utilization monitoring, analytics, deployment of various orchestrations and load balancing. This cyber-physical environment consisting of a cluster of RPis will allow us to conduct experiments and compare the energy efficiency of different content delivery network (CDN) configurations and algorithms. With additional hardware resources we will be able to run more sophisticated simulations closely reflecting real-life cloud deployments in the context of CDNs.

KEYWORDS

Raspberry Pi, Preboot Execution Environment, Netboot, Orchestration, Docker Swarm

1. INTRODUCTION

With rapid advancements in technology, now you can go to Amazon, or any cloud providers and spin virtual machines in seconds. However, a cloud is a network of computers difficult to visualize or demonstrate. With Raspberry Pi's, we can build a low cost, versatile clustered-computing environment capable of being deployed in different configurations. In this case, to have a cloud testbed that can be demonstrated and experimented with in various use cases (Mendki, P., 2018; Ahmed, A. and Pierre, G., 2018; Bellavista, P. and Zanni, A., 2017; Hajji, W. and Tso, F.P., 2016).

In this project, we set up a cyber-physical demonstration environment of Raspberry Pis capable of booting from the network using the Preboot eXecution Environment (PXE). We also deployed Docker Swarm for RPi cluster management. With this setup, the configuration and orchestration of these low-cost devices becomes more convenient, simplifying cluster provisioning, maintenance, deployment of various services and collection of metrics. Our contribution in this work is the deployment of an agnostic demo-rig set-up of constrained nodes and orchestration that can be conveniently scaled and managed forming a cloud testbed for various use cases.

The rest of this report is organized as follows: Section 2 presents the background of high-level concepts, Section 3 presents a description of the materials and methods for the rig setup, with illustration and implementation provided in Section 4. Finally, Section 5 concludes the paper and introduces future research work.

2. BACKGROUND

2.1 Preboot eXecution Environment (PXE)

The Preboot eXecution Environment (PXE) is most commonly used within data centers to automatically deploy root images to new servers for installation or in some cases, run disk-less servers/nodes whereby the boot image is loaded into memory rather than installed to local storage (Johnston, S.J. and Cox, S.J., 2017). This is important as it enables us to swiftly boot RPi devices without the need for OS installation and configuration. Once plugged to the network, the device boots promptly.

PXE framework is implemented on the client side within a Network Interface Card BIOS extension. This firmware includes a minimal UDP/IP stack, a Preboot client module using DHCP and a TFTP client module which together form the API used by network boot programs (NBP). This allows boot programs to interact and work with services offered by a PXE server. For a device to PXE boot, relevant instructions are needed where a DHCP server is involved to store and serve this information (Stewart, C., 2021). This process is illustrated in Figure 1.

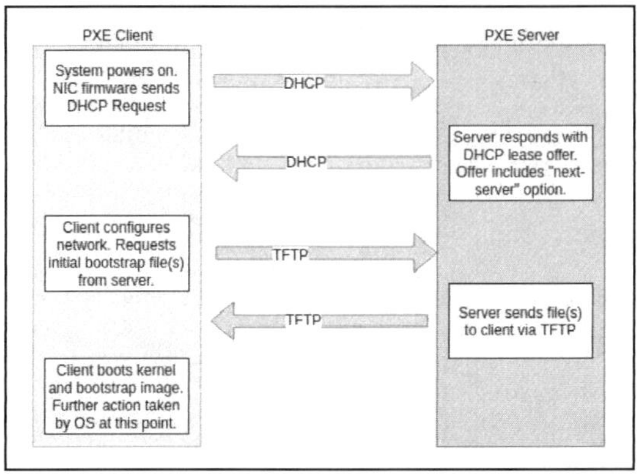

Figure 1. PXE-boot process

2.2 Docker Swarm

Docker is a container-based virtualization solution. In contrast to virtual machines, which use a hypervisor to provide a virtual hardware environment for guest operating systems, containers employ virtualization capabilities of the host kernel to provide a virtual system environment for applications. Scheduling and resource management for all containers is performed by the host kernel, which is also responsible for keeping the containers isolated from each other. This approach makes containers more light-weight compared to virtual machines in several ways. And since no guest operating system needs to be booted or shut down, containers can usually be started and stopped very quickly (Boettiger, C., 2015).

There are multiple tools available for orchestrating these services and Docker contains a built-in swarm mode for service and cluster orchestration. This facilitates functionalities like service scaling and discovery, load balancing, high availability and seamless service versioning (i.e., service updates). Compared to Kurbenetes, swarm comes packaged together with docker and most importantly it is lightweight. For the purposes and needs for this project, this suffices to provide required capabilities within our platform (Pentyala, S.K., 2017; Ahmed, A. and Pierre, G., 2020).

3. METHODS AND MATERIALS

3.1 Materials

Docker supports multiple CPU architectures including our RPi 3 Model B+ and RPi 4 Model B ARM. For device and service orchestration, Docker can be run in swarm mode. According to (Mendki, P., 2018), this facilitates services scaling, discovery, load balancing and updating the swarm and corresponding deployed processes (Pentyala, S.K., 2017; Al-Rakhami, M. et al, 2018).

In our project setup we experimented using the following physical equipment (i) Raspberry Pi 3 Model B+, (ii) Raspberry Pi 4 Model B, (iii) 48-port Gigabit tp-link TL-SG 1048 switch, (iv) WaveShare RPi 2.8 inch LCD screens and (v) SanDisk Extreme Portable SSD 1-TB.

3.2 Methods

The design of the set up in this project consists of a server environment that enables network booting RPis. The server provisions Dynamic Host Configuration Protocol (DHCP), Domain Name System (DNS) services, Trivial File-Transfer Protocol (TFTP) and Network File System (NFS). The server was run on a virtual machine with sufficient computing and memory (the co-called Big-VM environment) for faster operations and mass storage. All the nodes in the rig (except DHCP server) use a common root NFS share, with one of the nodes having privileges to make configuration changes, thus, updates cascade throughout the rest of the nodes in the RPi cluster which we named the 'demo rig'.

With light-weight virtualization, we are able to orchestrate in container environment software-stack to facilitate collection of various metrics from each node in the cluster, and facilitate cluster management. The collected metrics were ingested into data storage and various dashboards created in a visualization application. This software-stack is deployed in a server environment (Big-VM) capable of handling moderate to large datasets and performing efficient data processing. This in turn, is an open setup where other orchestrations can be provisioned. An illustration of physical design and the implementation of the setup are shown in Figures 2 left and Figure 2 right respectively.

4. IMPLEMENTATION

Our physical setup shown in Figure so consists of RPi boards connected with Ethernet cables via the switch. Docker swarm was deployed to manage the cluster and orchestrated Telegraf-InfluxDB-Grafana (TIG-stack) to facilitate collection of hardware metrics. The purpose is to monitor resource utilization of the rig by extracting this information and utilize machine learning and scripts to automate management of the cluster.

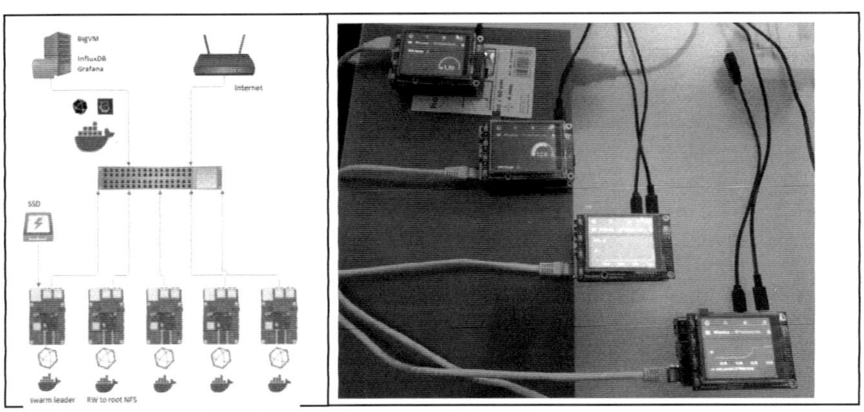

Figure 2. Left: Design of the setup; Right: Implemented setup

Challenges we faced were that 1) Docker initially failed to start on the nodes with read-only rights and managed to make it work by creating temporary mount point that pertain to configurations needed for docker daemon; 2) whenever we restarted the Docker swarm worker nodes, they were not able to authenticate, thus not recognized in the swarm. This was fixed by using extra partitions on the mounted microSD cards to make the configurations more persistent; 3) visualizations and dashboards did not perfectly fit the display on the LCD screens, and we are still exploring options to address this and other challenges that arise. For detailed implementation we refer the reader to our GitHub[1] repository.

5. CONCLUSION AND FUTURE WORK

In this project, we set up a cyber-physical demonstration environment consisting of Raspberry Pi microcomputers (RPi's) with the Preboot eXecution Environment (PXE) also known as the 'demo-rig' or 'CDN rig'. All the nodes in the rig mount to a common root NFS share, thus, all updates cascade throughout the rest of the nodes in the rig. Only one RPi node has privileges to make configuration changes. With such a design, RPi cluster provisioning and maintenance becomes more consolidated and efficient, facilitating automated updates or upgrades. As a result of this project, we created a cloud testbed environment that can be scaled and managed with less administrative effort, facilitates monitoring resources, analytics, deployment of various orchestrations and load balancing. With this demo-rig setup we are able to orchestrate for different use cases and utilize various tools for applications and infrastructure monitoring. Docker in swarm mode was used to unify and manage and orchestrate the cluster of the RPis. The immediate next steps for the demo-rig are to deploy actual use cases, comprehensive comparison with other frameworks and related work, experiment with the Content Delivery Network (CDN) power consumption use cases and provide evaluation results. Additionally, we will prepare a holding box/compartment for the demo-rig which will allow us to transfer it and utilize in different demonstrations conducted in academic or professional settings e.g., student conferences, project demonstrations, cloud app showcase.

REFERENCES

Ahmed, A. and Pierre, G., (2018), July. Docker container deployment in fog computing infrastructures. In *2018 IEEE International Conference on Edge Computing (EDGE)* (pp. 1-8). IEEE.

Ahmed, A. and Pierre, G., (2020). Docker-pi: Docker container deployment in fog computing infrastructures. *International Journal of Cloud Computing*, *9*(1), pp.6-27.

Al-Rakhami, M. et al, (2018), August. Cost efficient edge intelligence framework using docker containers. In *2018 IEEE 16th Intl Conf on Dependable, Autonomic and Secure Computing, 16th Intl Conf on Pervasive Intelligence and Computing, 4th Intl Conf on Big Data Intelligence and Computing and Cyber Science and Technology Congress (DASC/PiCom/DataCom/CyberSciTech)* (pp. 800-807). IEEE.

Bellavista, P. and Zanni, A., (2017), January. Feasibility of fog computing deployment based on docker containerization over raspberrypi. In *Proceedings of the 18th international conference on distributed computing and networking* (pp. 1-10).

Boettiger, C., (2015). An introduction to Docker for reproducible research. *ACM SIGOPS Operating Systems Review*, *49*(1), pp.71-79.

Hajji, W. and Tso, F.P., (2016). Understanding the performance of low power Raspberry Pi Cloud for big data. *Electronics*, *5*(2), p.29.

Johnston, S.J. and Cox, S.J., (2017). The raspberry Pi: A technology disrupter, and the enabler of dreams. *Electronics*, *6*(3), p.51.

Mendki, P., (2018), February. Docker container based analytics at iot edge video analytics usecase. In *2018 3rd International Conference On Internet of Things: Smart Innovation and Usages (IoT-SIU)* (pp. 1-4). IEEE.

Pentyala, S.K., (2017), August. Emergency communication system with Docker containers, OSM and Rsync. In *2017 International Conference On Smart Technologies For Smart Nation (SmartTechCon)* (pp. 1064-1069). IEEE.

Stewart, C., (2021). Network-Based Deployment Using Preboot Execution Environment.

[1] ELTE-DSED/RPi_Demorig: All code, scripts, configurations for our cyber-physical cloud testbed (github.com)

COMPARATIVE ANALYSIS OF CHANGE DETECTION METHODS IN CYBER-PHYSICAL SYSTEMS

Máté Hekfusz, Vrushali Mahajan, Adolf Kamuzora and Imre Lendák
Department of Data Science and Engineering,
Institute of Industry - Academia Innovation, ELTE Eötvös Loránd University,
Pázmány Péter sétány 1/C, H-1117, Budapest, Hungary

ABSTRACT

Networks of sensors and Internet of Things have led to an ever-increasing amount of data that is now more commonly available in streaming settings. Often it is assumed that the process generating these streams is stationary, however, in real world scenarios, systems are evolving and dynamic in nature. Thus, methods and approaches to be able to detect when there is a change or drift in the environment are necessary. In this research project, we analyzed various data distribution-based change detection algorithms and provided a framework for real-time analysis and change detection in Secure Water Treatment, that facilitated gaining insights into the comparative analysis and assessment of performance of the methods. Results show that out of the algorithms we experimented with, the Kolmogorov-Smirnov WINdowing algorithm was more balanced in terms of changes detected and detection delay, albeit with a high false alarm rate.

KEYWORDS

Drift, Change Detection, Stream Processing, Cyber-Physical Systems, Swat, Open-Source Technologies

1. INTRODUCTION

Rapidly changing environment of new products, new markets, customer behaviors and system behaviors result in appearance of concept and data drift problem. Often it is assumed that the process generating data from these environments is a stationary one. It has been observed that systems are dynamic and evolve over time. In the field of machine learning, this creates a problem as accuracy of predictions and decisions suffer and degrade over time. A lot of interest in the field of drift and change detection has increased and approaches to deal with this phenomenon are being researched to be incorporated into machine learning (Halder, B. et al, 2023). In cyber-physical systems, for instance, water treatment systems, quick and accurate identification of changes or drifts and evolving patterns is important for operational security and resilience.

Main contribution of this research work is a stream processing pipeline with the aim of analysis of various data distribution-based change detection methods in Secure Water Treatment cyber-physical system. An array of various open-source technologies was used in this endeavor. Our work aspires to contribute insights into the efficacy of change detection algorithms on the dynamic and evolving water treatment facility while providing a framework that can be used for change detection.

The remainder of this paper is organized as follows: Section 2 provides problem description around concept drift and change detection algorithms, with Section 3 stipulating the approach followed, datasets and performance evaluation metrics used. Section 4 presents results and discussion and Section 5 provides conclusion and future directions.

2. PROBLEM DESCRIPTION

2.1 Concept Drift Definition

According to Gama, J. et al, 2014, drift is a phenomenon in which the underlying statistical properties of the feature/target domain change over time in an arbitrary way. Change/drift is quantified by indicating change-points or small time-intervals of change (Bayram, F. et al, 2022; Halder, B. et al, 2023). In supervised learning aspect, the relationship between the features and the target variable changes over time. Data distribution-based change detection methods can work well also in an unsupervised learning setting (Guo, H. et al, 2022). The following section provides state-of-the-art algorithms applied in this research study.

2.2 Algorithms

CUmulative SUM (CUSUM): This is a sequential analysis technique. The test outputs an alarm when the mean g_t of incoming data significantly deviates from zero. Decision rule is if $g_t > \lambda$ then signal change then followed by setting $g_t = 0$, where λ is a user defined threshold at confidence δ (Callegari, C. et al, 2012; Vaswani, N., 2005; Flynn, T. and Yoo, S., 2019)

Kolmogorov-Smirnov WINdowing: This method is based on the Kolmogorov-Smirnov test. KS-test is a statistical method that quantifies a distance between the empirical distribution function of the instances (sliding window) and the cumulative distribution function of the reference distribution (reference window). When significant deviation is detected, it signals a potential (warning or drift state) change/drift in the data distribution (Hu, H. et al, 2020).

Crámer-von Mises: The idea behind CvM is similar to that described in KSWIN. It is a criterion used for monitoring the goodness-of-fit of a cumulative distribution function of examples compared to a given empirical distribution. With a given threshold, if the distance between the distributions is significant enough, then this signals a change (Zhou, C. et al, 2019).

Principal component analysis-based change detection (PCA-CD): PCA-based change/drift detection signals a change when the *p*-value of a generalized Wilcoxon test statistic W_{BF} is significantly large. This framework detects changes in unlabeled multidimensional data streams by using PCA to project the original stream into a lower dimensional space. Changes in the original stream such as mean, variance and correlation between variables, can be observed on one or several Principal Components, which are orthogonal by their construction (Qahtan, A.A. et al, 2015).

k-dimensional-quad-tree (kdqTree): It is an Information-Theoretic Approach (ITA) density-based change detection method. A kdqTree is used to partition the historical and new data (multi-dimensional) into a set of bins A, and then use Kullback-Leibler divergence to quantify the difference of the density θ in each bin. With a given threshold α, if it is exceeded, then a change/drift is confirmed (Dasu, T. et al, 2006).

3. MATERIALS AND METHODS

3.1 Dataset and Evaluation

We used iTrust Secure Water Treatment (SWaT A4 and A5, 2019) dataset of Singapore University of Technology and Design (SUTD). SWaT testbed is a scaled down version of a real industrial water treatment plant. It is a cyber-physical system (CPS) (Lamshöft, K. et al, 2021; Yoong, C.H. and Heng, J., 2019). We focused on sensor data recorded at a frequency of one observation per second for about 4 hours on July 20, 2019. The data comprises 14,996 records, 78 columns, consisting of values and states of 28 sensors. During operation, the system was subjected to two scenarios, a predominant period 3 hours of normal operations, and

1 hour of series and stages of single point and multipoint attacks, 36 in total (Goh, J. et al, 2017). For a detailed description of the attacks generated, we refer the reader to the dataset website[1].

Performance evaluation metrics used for change detection algorithms were 1) true detection rate characterizing capacity of the system to detect drifts, 2) false detection rate which means detecting a change when there is no change and 3) delay of detection which is an estimate of how many new instances are required to detect a change after the actual occurrence of a change (Gama, J. et al, 2014; Lu, J. et al, 2018).

3.2 Experiments Framework

The architecture consists of open-source technologies utilized in our pipeline which highlights stream processing framework that facilitates analysis of various data distribution-based change/drift detection methods. Docker is the linchpin of this framework as it facilitates quick deployment and virtualization of various technology solutions as lightweight processes in containerized environment. We were able to setup and configure Apache Kafka, Spark, Telegraf-InfluxDB-Grafana and Chronograf. Figure 1 illustrates this framework, and for detailed implementation we refer the reader to our GitHub[2] repository.

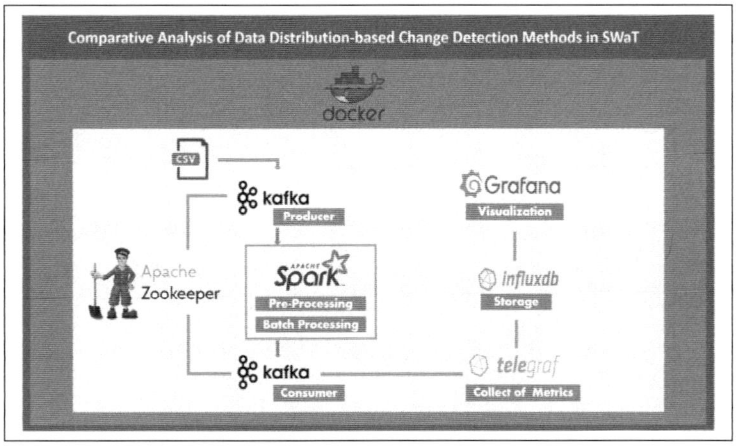

Figure 1. System architecture

Stream processing: We implemented sliding window aggregation on streaming data for real-time analysis. The key steps involved defining a 10-minutes window and a sliding window of 5 minutes. The streaming data frame is grouped by both the specified time window and the attack label, allowing computation of various aggregations (10-minute average for each numeric column, minimum, maximum, standard deviation). The use of watermark on timestamps ensures that late data is appropriately handled. Results of aggregations enable monitoring trends in water treatment at real time.

Phase 1: Data ingestion and preprocessing: A platform for data ingestion was Kafka. At this stage, the dataset is converted into streaming format and relevant attributes (targeted for attack, and the timestamps) are ingested into Kafka, leveraging its resilience and scalability. CMAK (a Kafka manager) was deployed to monitor the flow of data. The dataset undergoes preprocessing in Spark Streaming. Each record, uniquely distinguished by its timestamp is formatted and labelled accordingly (normal or attack) and aggregate values calculated.

Phase 2: Change detection: Five data distribution-based change detection methods were implemented and analyzed: Information Theoretic k-dimensional-quad-Tree (kdqTree), CUmulative SUM (CUSUM), Crámer-von Mises (CvM) test, Principal Component Analysis-Change Detection (PCA-CD), and Kolmogorov-Smirnov WINdowing (KS-WIN) test. The data stream was run through these algorithms, recording true and false detections, and detection lag. The algorithms were implemented using the open-source libraries Menelaus and River.

[1] http://itrust.sutd.edu.sg/research/datasets
[2] https://github.com/mateotis/swat-change-detection

Phase 3: Storage: Telegraf takes the preprocessed Kafka records and pushes into InfluxDB, where this persists as time-series data. This is useful for later stages of visualizing results.

Phase 4: Visualization: Grafana is the technology we used to visualize the data persisted in InfluxDB. We also integrated Chronograf into this part of the pipeline to get another option of more meaningful data visualizations. In this stage we were able to craft insightful dashboards to show sensor values in real-time, highlight correlations between them and visualize monitor performance of change detection methods.

4. RESULTS

We applied the change detection algorithms over the entire SWaT data stream and recorded their performance. We observed that most attacks concentrated on Level Transmitter; UF Feed Tank with LIT-301 sensor and Flow Transmitter; controls the UV dechlorinator with FIT-401 sensor, thus, they were points of interest for monitoring. Figure 2 on the left shows the rate of true change detection (TDR) and rate of false alarms (FAR). And on the right illustrates the detection delay per change detection method. Intuitively we aim to maximize TDR while minimizing FAR and detection delay.

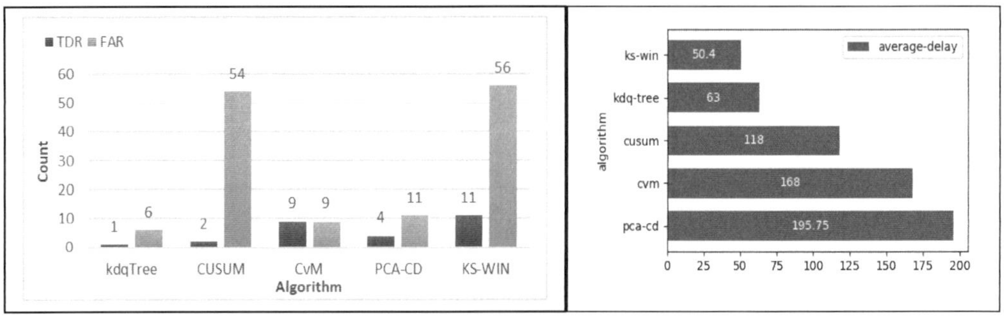

Figure 2. Left: Change detection and false alarms; Right: Change detection delay

Parameters for algorithms; KS-WIN (α=0.0001, window size=200, stat_size=100), PCA-CD (window size=50, divergence="kl", δ=0.1), kdqTree (window size=1000, α=0.05, bootstrap samples=500, count unbound=100), CUSUM (λ=1, δ=0.005), CvM (window size=10, *p*-value=0.05, λ=10). Observing the percentage TDR to FAR ratio as illustrated in Table 1, shows CvM is better-off compared to other algorithms, however with a delay more than 3 times of KS-WIN, with CUSUM being worse-off in false alarms, and KS-WIN being relatively fast in detecting changes, however with relatively low TDR to FAR ratio.

Table 1. Change detection algorithms TDR to FAR ratio with detection delay

Change detection method	TDR to FAR Ratio (%)	Delay
KS-WIN	16%	50
kdqTree	14%	63
CUSUM	4%	118
CvM	50%	168
PCA-CD	27%	196

KS-WIN detected relatively more changes than CvM, but with 6 times more false alarms. With the delay in consideration, we propose that reducing the rate of false alarms for KS-WIN would be a better approach than reducing lag in CvM. It is better to be more sensitive to changes than delay in the order of more than 3 times the magnitude, however annoying it might be; in cyber security, a tiny crack is a flood, no pun intended.

5. CONCLUSION AND FUTURE WORK

In this research project, we analyzed various data distribution-based change detection algorithms and provided a framework for real-time analysis and change detection in Secure Water Treatment. Out of the algorithms we experimented with, the Kolmogorov-Smirnov WINdowing algorithm showed to be more balanced in terms of changes detected and detection delay. It was relatively fast on detecting changes, however limited on the rate of false alarms. We intended to run experiments on more sensors with more algorithms, however we had a limited schedule. This study laid the groundwork for further research in the domain of cyber-physical systems, emphasizing continuous and real-time monitoring in safeguarding critical infrastructure in non-stationary environments. Another prospect is to experiment with other approaches to further improve performance of methods applied, making the cyber-physical system more robust.

REFERENCES

Bayram, F. et al, (2022). From concept drift to model degradation: An overview on performance-aware drift detectors. *Knowledge-Based Systems*, *245*, p.108632.

Callegari, C. et al, (2012). WAVE-CUSUM: Improving CUSUM performance in network anomaly detection by means of wavelet analysis. *computers & security*, *31*(5), pp.727-735.

Dasu, T. et al, (2006). An information-theoretic approach to detecting changes in multi-dimensional data streams. In *Proc. Symposium on the Interface of Statistics, Computing Science, and Applications (Interface)*.

Flynn, T. and Yoo, S., (2019), December. Change detection with the kernel cumulative sum algorithm. In *2019 IEEE 58th Conference on Decision and Control (CDC)* (pp. 6092-6099). IEEE.

Gama, J. et al, (2014). A survey on concept drift adaptation. *ACM computing surveys (CSUR)*, *46*(4), pp.1-37.

Goh, J. et al, (2017). A dataset to support research in the design of secure water treatment systems. In *Critical Information Infrastructures Security: 11th International Conference, CRITIS 2016, Paris, France, October 10–12, 2016, Revised Selected Papers 11* (pp. 88-99). Springer International Publishing.

Guo, H. et al, (2022). Concept drift type identification based on multi-sliding windows. *Information Sciences*, *585*, pp.1-23.

Halder, B. et al, (2023). Autonomic Active Learning Strategy using Cluster-based Ensemble Classifier for Concept Drifts in Imbalanced Data Stream. *Expert Systems with Applications*, p.120578.

Hu, H. et al, (2020). No free lunch theorem for concept drift detection in streaming data classification: A review. *Wiley Interdisciplinary Reviews: Data Mining and Knowledge Discovery*, *10*(2), p.e1327.

Lamshöft, K. et al, (2021), June. Information hiding in cyber physical systems: Challenges for embedding, retrieval and detection using sensor data of the SWAT dataset. In *Proceedings of the 2021 ACM Workshop on Information Hiding and Multimedia Security* (pp. 113-124).

Lu, J. et al, (2018). Learning under concept drift: A review. *IEEE transactions on knowledge and data engineering*, *31*(12), pp.2346-2363.

Qahtan, A.A. et al, (2015), August. A pca-based change detection framework for multidimensional data streams: Change detection in multidimensional data streams. In *Proceedings of the 21th ACM SIGKDD International Conference on Knowledge Discovery and Data Mining* (pp. 935-944).

Vaswani, N., (2005), March. The modified CUSUM algorithm for slow and drastic change detection in general HMMs with unknown change parameters. In *Proceedings.(ICASSP'05). IEEE International Conference on Acoustics, Speech, and Signal Processing, 2005.* (Vol. 4, pp. iv-701). IEEE.

Yoong, C.H. and Heng, J., (2019), September. Framework for continuous system security protection in SWaT. In *Proceedings of the 2019 3rd International Symposium on Computer Science and Intelligent Control* (pp. 1-6).

Zhou, C. et al, (2019). Comparative analysis of nonparametric change-point detectors commonly used in hydrology. *Hydrological sciences journal*, *64*(14), pp.1690-1710.

LEVERAGING AI FOR ENHANCING DISASTER RESILIENCE IN SMART COASTAL CITIES: STRATEGIES FOR TOMORROW'S URBAN CHALLENGES

Rouzbeh Nazari[1, 2, 3], Mujungu Lawrence Museru[1, 2] and Maryam Karimi[3, 1, 2]
[1]Sustainable Smart Cities Research Center, University of Alabama at Birmingham, AL, USA
[2]Department of Civil, Construction, and Environmental Engineering, University of Alabama at Birmingham, AL, USA
[3]Department of Environmental Health Science, School of Public Health, University of Alabama at Birmingham, AL, USA

ABSTRACT

Natural disasters pose significant threats to the sustainability of smart cities, emphasizing the urgent need to re-evaluate and fortify their infrastructural resilience systematically. This necessitates a holistic approach considering the interplay between population dynamics, critical infrastructures, and hazard/disaster risks within communities. To address these challenges, proactive measures, including land use planning, public education, and disaster preparedness, are essential for enhancing resilience. However, there is a critical gap in accessible tools and comprehensive resources for citizens, communities, and decision-makers faced with these complex tasks. This study investigates the economic impacts of hurricanes across small and large counties, aiming to discern disparities in vulnerability and resilience in coastal communities along the Gulf of Mexico and the Atlantic Coast in the United States. Leveraging data from the National Flood Insurance Program (NFIP) and simulated flood hazards, oversampling techniques are employed to enrich flood damage datasets. Ensemble machine learning methods, including random forest, extra tree, extreme gradient boosting, and categorical boosting, are utilized to develop multi-variable flood damage models. Results indicate that extreme gradient boosting outperforms other methods, achieving notable precision, recall, and F1-score metrics, while effectively quantifying relative damage levels. Despite inherent dataset limitations, including SHapley Additive exPlanations offers valuable insights into feature interactions. This research underscores the potential of AI in bolstering disaster resilience strategies and advancing flood damage modeling in coastal regions.

KEYWORDS

Smart Cities, Resilience, Multivariable Flood Damage Model, Ensemble Machine Learning

1. INTRODUCTION

The coastal regions of the United States, accommodating nearly 40% of the population in less than 10% of the land area, face severe flood-related challenges with significant societal, economic, and environmental implications (NOAA, 2013). Instances like Hurricane Katrina, which caused substantial infrastructure damage, and the CoreLogic 2022 report (Corelogic, 2022) projecting approximately 7.8 million homes at risk underscore the pressing need to accurately assess and manage flood risks. Traditional flood damage models, predominantly Depth-Damage Functions (DDFs), have limitations in capturing the complexities of flood damage processes. Recent studies advocate for multivariable models considering diverse factors, recognizing these drawbacks, and demonstrating enhanced accuracy and predictive capability. Yet, limited, inconsistent, and incomplete historical flood data impede the development and reliability of such models. Emerging machine learning (ML) models, particularly ensemble methods, have shown promise in improving predictive flood damage assessments. However, challenges persist, including imbalanced datasets, limited interpretability, and the need for more nuanced modeling techniques. This study applies supervised ML methods to estimate building-level flood damage in vulnerable coastal areas, specifically Baldwin and Mobile counties in Alabama, addressing data limitations in the National Flood Insurance Program (NFIP) claims data. It aims to mitigate data imbalances, explore crucial influencing variables, and enhance model interpretability by employing ensemble ML models alongside Synthetic Minority Oversampling Techniques (SMOTE) variants and SHapley Addictive exPlanations (SHAP), contributing novel insights to multivariable flood damage modeling.

2. STUDY AREA

The coastal region in Mobile and Baldwin counties, Alabama, faces recurrent flooding due to hurricanes and intense rainfall. This area, part of the Mobile Bay watershed system, witnesses a dynamic hydrologic system with the convergence of Mobile and Tensaw rivers. It has been significantly impacted by 28 major hurricanes in the last 50 years, causing extensive damage, particularly from Hurricane Katrina (Figure 1).

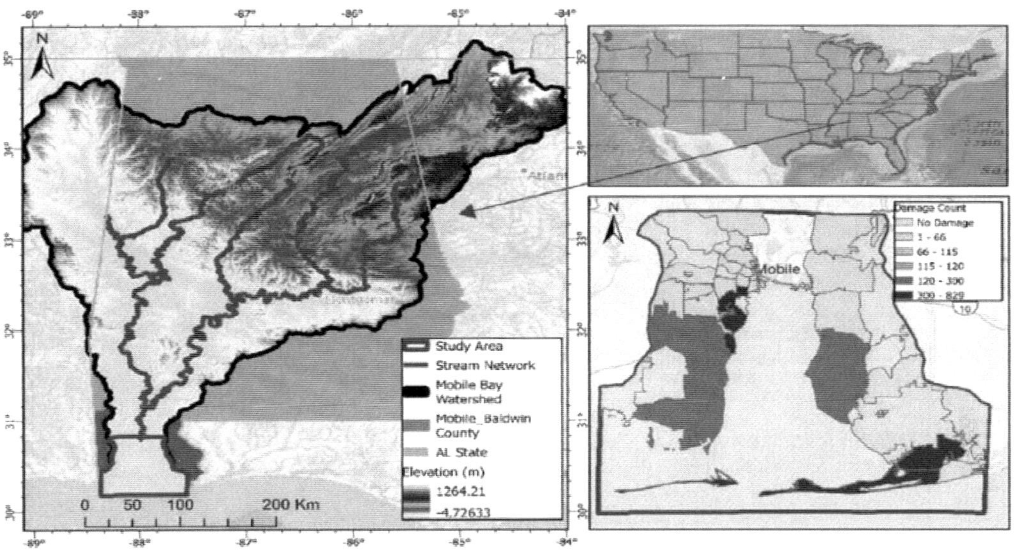

Figure 1. Location of the study area and property damage count during Hurricane Katrina

3. DATA AND METHODOLOGY

The methodology comprises three key steps: (1) integrating NFIP claims and Alabama property records, (2) incorporating simulated flood hazard variables, and (3) constructing multi-variable ML flood damage models with SMOTENC and SMOGN techniques. Classification and regression models were used to predict damaged properties and estimate their relative damage. The NFIP dataset, consisting of over 2 million flood damage claims, lacks specific property details due to confidentiality protocols, presenting challenges for analysis. To address this, properties were filtered based on damage consistency and criteria, reducing the dataset to 3122 properties. Twelve variables were selected and merged with AL property data, including property location and characteristics. Flood zone designations and property values were updated to 2017 prices. The dataset introduced a Relative Damage (RD) variable, calculated by dividing observed damage by property value, following similar approaches in prior studies (Pinelli, et al., 2020 & Paleo-Torres, et al., 2020). This method accommodates data limitations in determining true values.

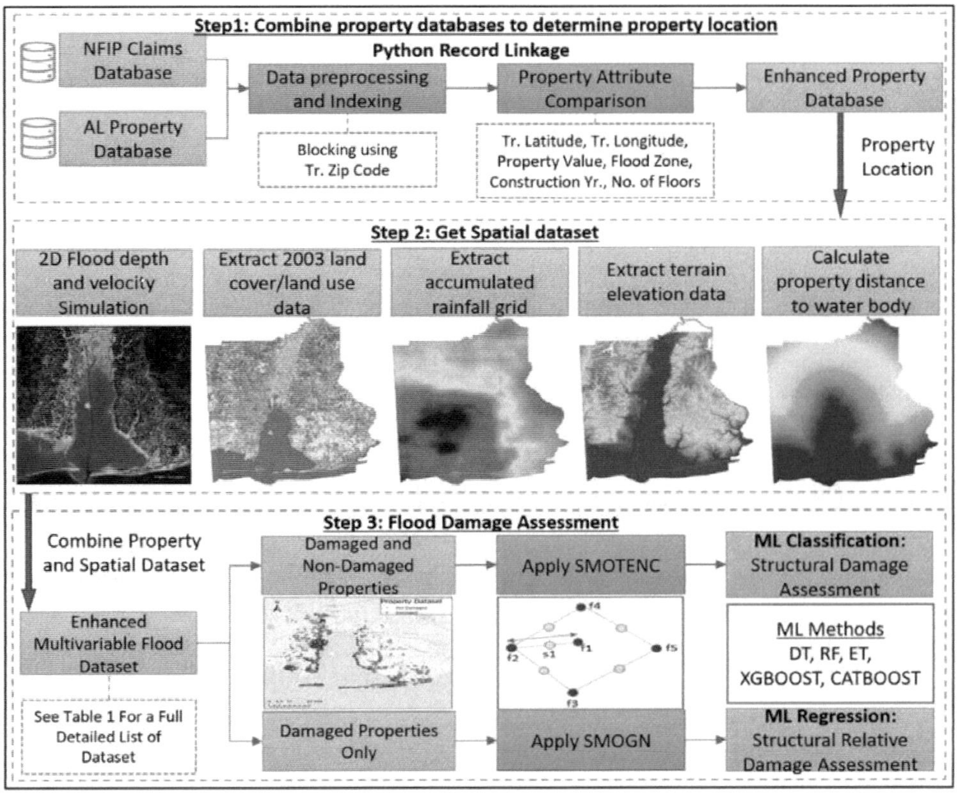

Figure 2. Workflow diagram showing data development process and damage assessment methodology

4. RESULTS

Initially, the dataset encompasses 14,835 properties, with 16.67% representing damaged properties. Model accuracy ranges from 0.8 to 0.85, and ROC AUC values span 0.69 to 0.89, indicating strong performance. However, relying solely on accuracy and ROC AUC might skew results, as these metrics prioritize the majority class, presenting challenges in predicting the minority class accurately. Metrics like precision, recall, and F1 scores were lower for predicting damaged properties (precision: 0.49-0.68, recall: 0.46-0.56, F1: 0.5-0.59) compared to undamaged properties (precision: 0.88-0.89, recall: 0.87-0.95, F1: 0.88-0.91). Without data balancing, models perform better on the majority class but struggle with the minority class. After applying SMOTENC, achieving a balanced 1:1 ratio between classes, all classifiers show significant improvements. Accuracy, ROC AUC, precision, recall, and F1 score improve notably (2.5%-5.88%, 14.45%-28.99%, 30.88%-67.34%, 60.71%-93.49%, and 51.72%-64%, respectively). Balanced datasets enable optimal performance nearing one, which is crucial for effective classification, with XGBoost exhibiting the best performance post-SMOTENC.

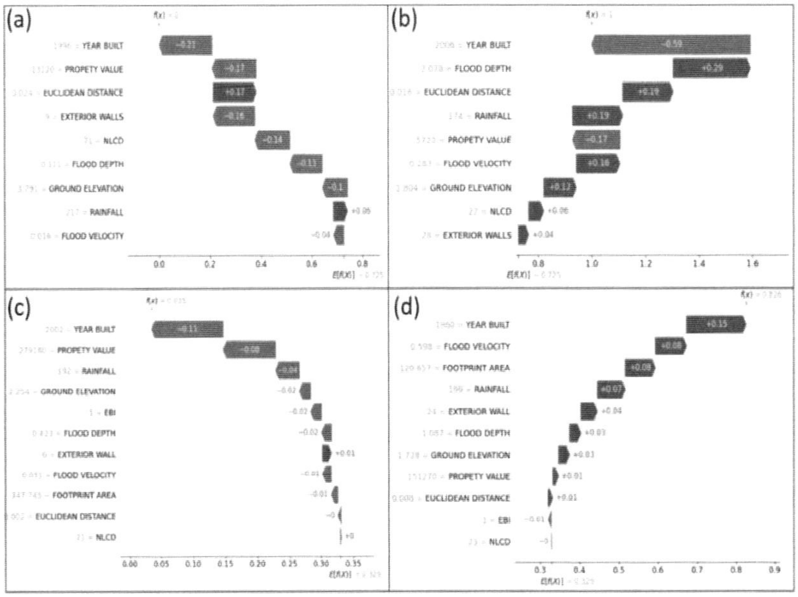

Figure 3. Local analysis for feature importance using waterfall plots for single observation in the XGBoost model for (a) property classified as not damaged, (b) property classified as damaged, (c) property RD predicted as 0.0035, and (d) property RD predicted as 0.826. NLCD = 21,22,23,71 represents developed open space, low and medium intensity Grassland. EBI=1 elevated property

5. DISCUSSION

Over the past decade, numerous studies have emphasized the importance of ensemble models for developing multivariable flood damage models. In this study, we extended this line of research by exploring the performance of various ensemble methods, including Random Forest (RF), Extra Trees (ET), XGBoost, and CATBoost, in predicting flood damage in coastal urban areas. Our analysis, which covered classification and regression models, revealed a consistent superiority of ensemble methods over single Decision Trees (DT), echoing findings from previous research by Merz et al. (2013) and others.

Ensemble models' effectiveness can be attributed to their ability to combine multiple decision trees, creating stronger learners that enhance accuracy, reduce overfitting, and handle complex variable relationships more effectively. For instance, XGBoost, a gradient-boosting ensemble method, demonstrated exceptional performance in our analysis, achieving high precision, recall, and F1 scores for identifying damaged properties in the classification task. Similarly, its performance in the regression task, as measured by metrics like R^2, RMSE, and Spearman correlation, exceeded that of other models.

However, one of the persistent challenges in flood damage modeling is the imbalanced distribution of flood damage observations. This issue can significantly impact the performance of data-driven models, as there may be insufficient data to adequately train the models, particularly for underrepresented classes or scenarios. Our analysis revealed that techniques like Synthetic Minority Over-sampling Technique for Nominal and Continuous (SMOTENC) and Synthetic Minority Over-sampling Technique for Regression (SMOGN) can effectively address imbalanced data distributions, leading to improved model performance.

In addition to addressing data imbalances, our study also explored input variables' global and local influence on prediction outcomes. Flood hazard and property characteristics emerged as the most influential parameters in the prediction models, aligning with previous studies. However, our analysis revealed discrepancies between global and local interpretations of feature importance, highlighting the context-dependent nature of feature contributions. This variability underscores the need for comprehensive model interpretation techniques like SHAP (SHapley Additive exPlanations), which offer insights into global trends and individual prediction explanations.

Overall, our findings underscore the effectiveness of ensemble tree methods in flood damage modeling and highlight the importance of addressing data imbalances and enhancing model interpretability for accurate and reliable predictions in smart coastal cities.

6. CONCLUSION

In conclusion, this study has demonstrated the potential of leveraging AI-based approaches to enhance disaster resilience in smart coastal cities. By developing multivariable flood damage models using ensemble tree methods and incorporating techniques like SMOTENC, SMOGN, and SHAP analysis, we have provided valuable insights into the complex dynamics of flood damage in urban coastal areas. Our research aligns with the core principles of the United Nations Sustainable Development Goal 11 (UNSDG11), which emphasizes the importance of building resilient and sustainable cities. By emphasizing vulnerability analysis as a key element of flood disaster management, our study contributes to global efforts to create disaster-resilient communities. Our findings have practical implications for urban planners, policymakers, and disaster management agencies tasked with mitigating flood risks in coastal cities. By leveraging AI-based models and advanced analytical techniques, stakeholders can better understand and address the multifaceted challenges posed by floods, ultimately fostering more resilient and sustainable urban environments. As smart coastal cities continue to evolve, the integration of AI-based models into disaster risk management strategies will be essential for effectively mitigating the impacts of floods and ensuring the long-term resilience of urban coastal communities. We can build a more resilient future for smart coastal cities and their residents by embracing innovative approaches and leveraging cutting-edge technologies.

REFERENCES

Corelogic. Millions of Home At Risk of Hurricane and Storm-Surge Damages. *Press Release* 2022

Merz, B., Kreibich, H., & Lall, U. (2013). Multi-variate flood damage assessment: a tree-based data-mining approach. Natural Hazards and Earth System Sciences, 13(1), 53-64. https://doi.org/10.5194/nhess-13-53-2013

NOAA. Population Trends from 1970 - 2020. *NOAA Coastal Population Report.* 2013

Pinelli, J.-P., Da Cruz, J., Gurley, K., Paleo-Torres, A. S., Baradaranshoraka, M., Cocke, S., & Shin, D. Uncertainty Reduction Through Data Management in the Development, Validation, Calibration, and Operation of a Hurricane Vulnerability Model. *International Journal of Disaster Risk Science*, 2020

Paleo-Torres, A., Gurley, K., Pinelli, J.-P., Baradaranshoraka, M., Zhao, M., Suppasri, A., & Peng, X. Vulnerability of Florida residential structures to hurricane-induced coastal flood. *Engineering Structures*. 2020

CONDITIONAL DIFFUSION MODEL FOR GENERATING BIOLOGIC DATA

Denis Sidorenko and Anatoly Shalyto
ITMO University
Saint-Petersburg, Kronverksky Pr., 49, Russia

ABSTRACT

Diffusion models have shown remarkable success in generating high-quality data across various domains. However, applications in the biomedical field often require conditioning the generative process on additional information to obtain relevant and controlled outputs. This paper presents a conditional diffusion model tailored for multimodal data fusion, with a focus on integrating categorical features and continuous variables to generate biologically plausible patterns of gene expression and methylation. The model architecture builds upon the U-Net with self-attention mechanisms and employs techniques to effectively incorporate categorical conditions via learnable embeddings and continuous conditions through transformation networks. To represent the gene data as images for the diffusion model, self-organizing maps are used to construct a unified coordinate system based on expression or methylation profiles. Experimental results on the GTEx and CNCB datasets demonstrate the model's promising performance in tasks such as tissue classification and age prediction from generated methylation patterns. However, there is room for improvement in handling continuous conditions for generating more accurate expression patterns. The conditional diffusion approach shows strong potential for generating biologically relevant data conditioned on multiple factors, with key areas for future work including enhanced continuous condition modeling and capturing intricate details in the generated patterns.

KEYWORDS

Conditional Diffusion Models, Biomedical Data Generation, Self-Organizing Maps

1. INTRODUCTION

Diffusion models have (Chang, Koulieris and Shum, 2023) demonstrated impressive performance in generating high-quality samples across various domains, such as images, audio, and text. However, many applications in the biomedical field require conditioning the generative process on additional information, such as categorical or continuous features, to obtain more controlled and specific outputs. In this work, we present a conditional diffusion model tailored for multimodal data fusion, with a particular focus on the integration of categorical features (tissue, gender, type of omics) and continuous variables (age) for generating gene expression and methylation data. Understanding gene regulation through the analysis of gene expression and methylation data is of paramount importance in biomedical research. Our approach aims to leverage the power of conditional diffusion models to generate realistic and biologically plausible patterns of gene expression and methylation, conditioned on relevant factors.

2. MODEL

The core of our conditional diffusion model is built upon the U-Net (Ronneberger, Fischer and Brox, 2015) architecture, which consists of a series of downsampling and upsampling layers. Between these layers, we introduce self-attention blocks, allowing the model to capture long-range dependencies and better understand the global context of the input data. Central to the diffusion process is the time step tensor, which controls the noising and denoising operations at each step. This tensor is initialized using a sinusoidal positional embedding (Vaswani *et al.*, 2017), enabling the model to encode temporal information effectively.

To condition the diffusion model on categorical features, such as gender, tissue, omics, we leverage PyTorch's nn.Embedding layer. This learnable embedding layer maps each class of a categorical condition (represented by unique integers) to a tensor with the same shape as the time step tensor. These embeddings are then added to the time step, effectively conditioning the diffusion process on the categorical information.

For continuous features as an age, a direct embedding approach is not applicable. Instead, we transform the continuous value using a small network consisting of three linear layers with ReLU activation functions. This transformation maps the scalar age value to a tensor with the same dimensions as the time step tensor, allowing it to be fused into the diffusion process by addition.

When incorporating multiple conditions (both categorical and continuous), we observed that the optimizer tends to prioritize conditions with higher embedded values, leading to an imbalance in the conditioning process. To mitigate this issue, we normalize all condition embeddings before adding them to the time step tensor, ensuring a more balanced contribution from each condition.

To represent gene expression and methylation data as images suitable for the diffusion model, we employ Self-Organizing Maps (SOMs) (Sagir and Sathasivam, 2016). SOMs are unsupervised neural networks that project high-dimensional data onto a lower-dimensional grid, preserving the topological structure of the input data. In our approach, we use SOMs to construct a unified coordinate system for genes based on their expression or methylation profiles. Each gene is assigned a coordinate on the SOM grid, and its expression or methylation value is mapped to an intensity value, resulting in a set of image-like representations. These image-like representations serve as the input to our conditional diffusion model, allowing us to leverage the model's generative capabilities to synthesize novel patterns of gene expression and methylation, conditioned on the relevant categorical and continuous features. Figure 1 shows the architecture of the model from the model transfer stage, to the diffusion process with the transfer of conditions to generation.

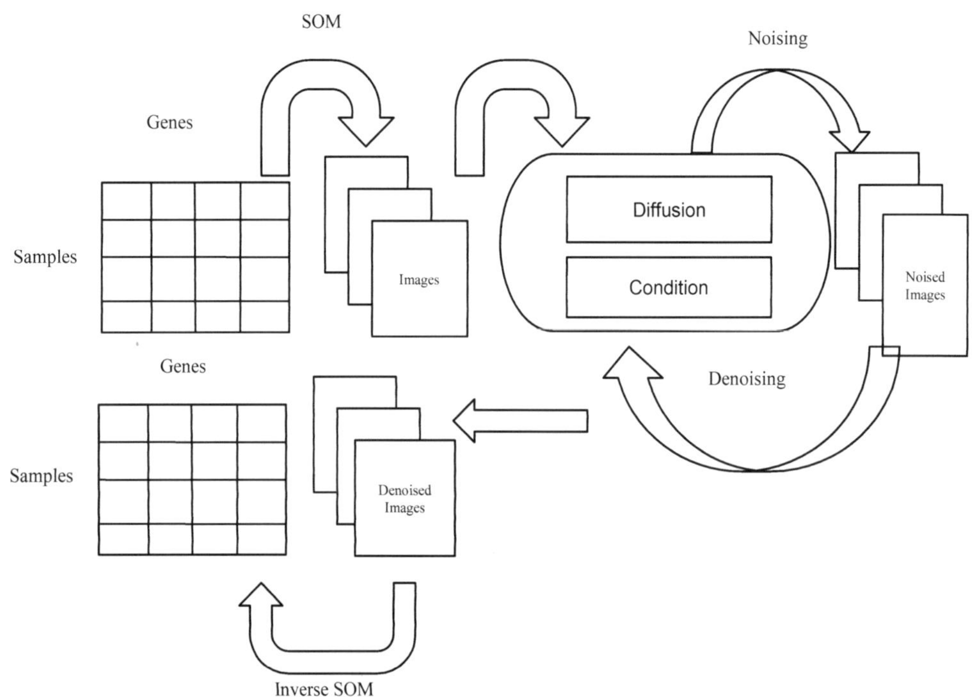

Figure 1. Model Architecture

2.1 Experiments

To evaluate the performance of our conditional diffusion model, we employed two datasets: GTEX (Lonsdale *et al.*, 2013) and CNCB (Xiong *et al.*, 2020). . We assessed the model's performance for tissue classification and age regression for the generated data. Our experiments were conducted using PyTorch 2.0 and trained on T4. We used the following hyperparameters: 4 attention head, 130x130 SOM size. The model was trained for 1000 epochs using the RAdam (Liu *et al.*, 2019) optimizer with a learning rate of 0.05.

Two experiments were conducted, one with a tissue condition and another with an age condition. Each experiment was carried out in two variants: with a uniformly generated label and with a real label. For classification, two standard classification metrics, Accuracy (ACC) and F-measure (F1) were applied, while for regression on the generated data, Mean Average Error (MAE) and R-squared (R^2) were used.

The conditional diffusion model demonstrated promising performance in generating biologically relevant patterns of gene expression and methylation. In the task of age prediction from generated methylation data (expectation that the predicted age on the generated data will be as close as possible to the given age condition), our model achieved strong results, accurately estimating age from the generated methylation patterns. Table 1 shows F1 and accuracy between the predicted and actual tissues (31 tissues were represented in the expression data, 37 tissues in the methylation data). In the case of the categorical tissue condition, both omics data show that the data generated by our model are well classified, suggesting potential applications in the arbitrary tissue generator for methylation or expression. Table 2 shows MAE and R2 metrics between the predicted and actual ages. However, the model's performance in generating gene expression patterns (for the age condition) was moderate, suggesting room for improvement.

Table 1. Tissue classification experiment

Model	Expression		Methylation	
Metrics	F1	ACC	F1	ACC
CDiffusion+real label	0.95	0.99	0.93	0.95
CDiffusion+generated label	0.9	0.97	0.87	0.93

Table 2. Age prediction experiment

Model	Expression		Methylation	
Metrics	MAE	R^2	MAE	R^2
CDiffusion+real label	11.2	0.26	6.3	0.87
CDiffusion+generated label	11.9	-0.07	7.1	0.83

3. CONCLUSION

In this paper, we presented a conditional diffusion model capable of effectively fusing categorical and continuous features into the generative process. By leveraging learnable embeddings for categorical conditions and transforming continuous values into embeddings, our model can seamlessly incorporate multimodal information. Furthermore, we addressed the issue of optimizer bias towards conditions with higher embedded values by normalizing all condition embeddings before adding them to the time step tensor.

Our experimental results on GTEX and CNCB demonstrate the efficacy of our approach. However, there are still limitations and room for improvement. One key aspect that requires further attention is the handling of continuous conditions, such as age. While our model performed well in age prediction from methylation data, the quality of generated expression patterns conditioned on age could be enhanced. Developing more sophisticated techniques for incorporating continuous features into the diffusion process may lead to improved performance in such scenarios.

In summary, our conditional diffusion model demonstrated strong capabilities in generating biologically plausible patterns of gene expression and methylation, conditioned on relevant categorical and continuous features. However, there is scope for improvement, particularly in handling continuous conditions more effectively and refining the model's ability to capture intricate details in the generated patterns.

ACKNOWLEDGEMENT

This work was supported by Russian Science Foundation, Grant 23-11-00346.

REFERENCES

Chang, Z., Koulieris, G. and Shum, H. (2023) On the Design Fundamentals of Diffusion Models: A Survey.

Liu, L. et al. (2019) 'On the Variance of the Adaptive Learning Rate and Beyond', ArXiv, abs/1908.03265. Available at: https://api.semanticscholar.org/CorpusID:199528271.

Lonsdale, J. et al. (2013) 'The Genotype-Tissue Expression (GTEx) project', Nature Genetics, 45(6), pp. 580–585. Available at: https://doi.org/10.1038/ng.2653.

Ronneberger, O., Fischer, P. and Brox, T. (2015) 'U-Net: Convolutional Networks for Biomedical Image Segmentation'.

Sagir, A. and Sathasivam, S. (2016) 'The Self Organizing Map as a Tool for Cluster Analysis', 38, pp. 44–53.

Vaswani, A. et al. (2017) 'Attention Is All You Need'. https://arxiv.org/abs/1706.03762

Xiong, Z. et al. (2020) 'EWAS Data Hub: a resource of DNA methylation array data and metadata', Nucleic Acids Research, 48(D1), pp. D890–D895. Available at: https://doi.org/10.1093/nar/gkz840.

MANAGEMENT BY DIGITAL RESPONSIBILITY IN HEALTH ORGANIZATIONS

Jutta Juliane Meier[1], Christian Weigand[2], Jochen Bauer[3] and Björn Eskofier[4]
[1]*Identity Valley® Research gGmbH, Unkel, Germany*
[2]*Mobile Health Lab, Fraunhofer IIS, Erlangen, Germany*
[3]*Lehrstuhl für Fertigungsautomatisierung und Produktionssystematik, Friedrich-Alexander Universität Erlangen-Nürnberg, Erlangen, Germany*
[4]*Prof., Machine Learning and Data Analytics (MaD) Lab Friedrich-Alexander Universität Erlangen-Nürnberg, Erlangen, Germany*

ABSTRACT

This paper describes the Digital Responsibility Goals, their purpose, and their exemplary relevance for digital innovations in the health sector. In addition, the document makes a first proposal for implementing digital responsibility through a management system based on the Digital Responsibility Goals.

KEYWORDS

Digital Health, Digital Responsibility, Framework, ISMS, Patient Participation, Management Approaches, Trust, Responsible AI, Digital Sovereignty, European Digital Rights and Principles

1. INTRODUCTION

Digital technologies have the potential to improve peoples' lives, but technological innovations and the use of innovative technologies must be geared more to taking responsibility for the well-being of people and society - especially in the sector of healthcare provision. Guidelines and laws are indispensable in this regard, but the dynamics of technological development, especially in the age of Artificial Intelligence (AI), also challenge social developments and the ethical dimension in dealing with digital technologies.

Likewise, the internet and digital technologies can bring negative side effects: for example, in many places in the world, especially in totalitarian states, the internet is restricted, regulated, monitored, and used for their propaganda; also, fake news and hate speech poison the atmosphere and make social discourse more difficult.

Oftentimes, critical decisions about the future of digital developments are made without a clear framework. Trustworthy, ethically sensitive, and sustainable guidelines that focus on the benefits for people are missing. The Digital Responsibility Goals (DRGs) aim to define this framework and work towards a truly human-centered digital transformation.

This paper describes the DRGs, their purpose, the possibilities, and their relevance for digital innovations in the health sector, aiming to enable patient participation and digital sovereignity. In addition, the paper investigates an innovative implementation and innovative management system, that is currently being developed within the research project TEAM-X - Trusted Ecosystem of Applied Medical Data eXchange, aimed at developing a trustworthy data ecosystem for health and care data.

2. FRAMEWORK: THE DIGITAL RESPONSIBILITY GOALS

To an increasing extent leading organizations and companies are committed to the United Nations' 17 Sustainable Development Goals (SDGs) (IEA, IRENA, UNSD, WB, WHO, 2019). Following the same logic, addressing the digital dimension, going beyond a purely "corporate perspective" as proposed by the Corporate Digital Responsibility approach (Lobschat, L. et al, 2021), the 7 DRGs aim to guide decision makers, companies, and other stakeholders, such as researchers and users, to develop and use trustworthy digital products and services by getting orientation and information.

The DRGs provide an opportunity for various stakeholders and decision-makers from businesses, regulators, academia, and civil society to form a common agenda and plan a common course of action to deal with a human-centered digital transformation (Meier, J. et al, 2022).

DRG #1 "Digital literacy" shapes the forefront of the DRGs, as only knowledge, education, and comprehensive information can be the basis of all self-determination and informed decision-making. This includes access to the technological infrastructure itself. Goals 2 to 6 of the DRGs are oriented along the data value chain. They cover security of the system (DRG #2 "Cybersecurity") as technological prerequisite against manipulation, protection of personal data (DRG #3 "Privacy") as part of digital self-determination, fair handling of data and a new understanding of how to deal with non-personal information (DRG #4 "Data Fairness"), responsible collection, processing and trustworthy evaluation of data (DRG #5 "Trustworthy Algorithms") and transparent communication of principles of digital processes (DRG #6 "Transparency").

Finally, the DRGs form a social frame and contribute to the protection of human identity and ensure human responsibility in the digital space (DRG #7 "Human Agency & Identity").

In the sense of the DRGs, people including their identity are elevated from a previous position as marginal figures to sovereign, self-determined individuals in the center of digital transformation.

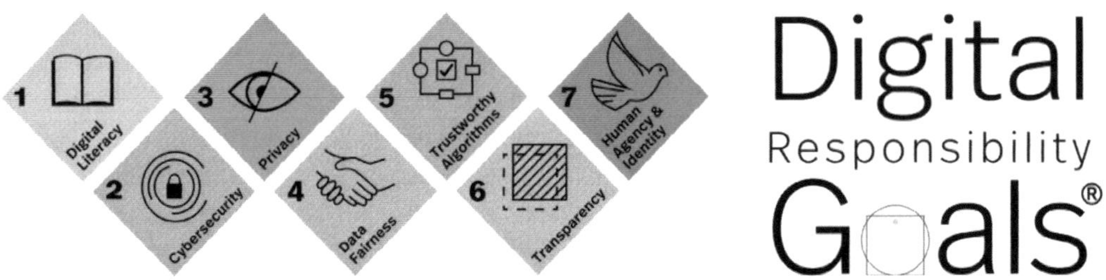

Figure 1. The 7 Digital Responsibility Goals by Identity Valley®

3. CHALLENGE: IMPLEMENTATION SYSTEMS AND MEASUREMENT PROSSESSES

Many challenges are associated with the current state of digital transformation. Critical decisions about the future of digital developments are oftentimes made without a clear framework and trustworthy, ethically sensitive, and sustainable guidelines that focus on the benefits for people. Furthermore, power to amass and use data is concentrated within few companies. Nowadays, organisations often develop their own AI & Ethics policies, being aware of the challenges but missing a global framework, similar to the SDG approach, in the digital space.

OECD AI Policy Observatory provides a live repository (OECD.AI, n.d.) of over 1000 AI policy initiatives from 69 countries, territories and the European Union. Thus, Orientation is difficult, and innovations are not comparable due to non-existent uniform measurement criteria of AI ethics, regardless of whether it is local or global.

The DRGs aim to define this missing measurable framework for emerging technologies – including Artificial Intelligence - and therefore work towards a trustworthy, ethically sensitive, sustainable, and truly human-centered digital transformation – aligned with the European Digital Rights and Principles, that have been signed by all European Member States in 2022 (European Commission, n.d.).

For the health vertical, patients' possibilities regarding digital access, digital education, and integration need to be improved to enhance trust. This is easily understandable since medical data are among the most critical data of all, as they are always personal and usually highly sensitive (Liu, V. et al, 2009). Significant ethical questions also typically arise in the medical field.

The trustworthiness of digital products, services, and processes is not just a question of technology, but one in which human behaviour and procedures within and between organizations are equally important. In the definition of the DRGs, a system perspective is therefore combined with a process perspective: The system perspective refers to the requirements for an artefact, the digital solutions themselves. The process perspective specifies the requirements for the design, implementation, and operational processes of the organizations, manufacturers, or service providers behind them. Both perspectives are independent, have their strengths and weaknesses, but are somehow ground-breaking when combined. The way a digital solution is set up should be given equal weight to how it was conceived, created, implemented, and how it is operated. Overall, the method for dealing with responsible digitization in practice requires not just a "best of both worlds" approach that incorporates both the system and process perspectives. Our holistic approach adds yet another perspective – the human-team perspective – in order to guarantee all assessments and developments are verifiably human-centered. Therefore, the DRGs not only combine the systems- and process-perspective, but also add the human dimension through a human centered perspective.

Giving orientation and providing information to enable informed decisions will become crucial. Measuring this responsible behaviour will be possible through an overall holistic approach along all DRG Criteria that then enables the creation of a DRG Index - empowering a truly human-centered digital transformation. The DRG Index shall then be incorporated in a Dashboard tool for companies, resp. organisations in which the evaluation of digitally responsible activities is continuously monitored, and the current level of digital responsibility is permanently updated.

4. SOLUTION: INVESTIGATING A MANAGEMENT SYSTEM

Each of the seven DRGs represents an urgent need for action in the digital space. Comprehensible, innovative, and applicable measures are needed and most of all, need to be implemented and actively pursued within and be activated by an organization.

To achieve this change towards a responsible behaviour, a responsible leadership, a comprehensive solution to get the DRGs into the organization is needed. As a reference, the authors of this paper regard Quality Management (QM) systems (Bacoup, P. et al, 2017), like the ISO 9001 or ISO 13485, Information Security Management System (ISMS) systems (Haufe, K. et al, 2016) according to the ISO 27001, or to a lesser extend corporate social responsibility approaches.

The common denominators of such systems are their systemic, continuous improvement approaches (i.e., Plan-Do-Check-Act (PDCA) cycle (Moen, R. et al, 2009), as well as their focus on establishing a custom-fit regime of processes and controls to guide entire organizations. They also include the idea of regular reviews and audits to seek for areas of improvement.

These management systems also address the tension between strategically implementing improvements, while ensuring short-term achievements of important project milestones. Moreover, they have demonstrated to be effective especially in environments that frequently do not possess the resources or focus to implement improvement systems like small and medium enterprises (SME) or municipalities.

An adaptation of best practice approaches, such as ISIS12 (Information security management system) (Bauer, J. et al, 2021), might be suitable as a method to apply the DRG framework towards implementation. This could be achieved in two ways:

Being embedded in existing management systems, for example as an additional module in an ISMS. Alternatively, the DRG framework can also establish itself as a stand-alone management by objectives system in the organization and refer to existing systems, such as a specific ISMS or project management method, as part of its 7 objectives, the 7 DRGs.

4.1 An ISMS Based Approach

An ISMS is a promising framework, because of its focus on systemic, continuous improvement taking a holistic view at a given challenge as argued above (Raza, M., 2023). In addition, with ISIS-12 being suited for smaller organizations, it fits well in the rather fragmented and small-scale environment of health and care.

The investigated approach works in three broad phases: (a) the initialization phase (LEARN) with steps 1 and 2, (b) the structural and process phase (ENGAGE) with steps 3 through 6 and (c) the implementation phase (ADAPT) with steps 7 through 12.

Initialization phase (LEARN)

Step 1: Establish guideline

Step 2: Train Employees

Organizational structure and processes (ENGAGE)

Step 3: Establish governance

Step 4: Analyse the System

Step 5: Analyse the Processes

Step 6: Analyse the Team

Implementation (ADAPT)

Step 7: Derive measures & link them to existing systems

Step 8: Compare target picture/actual

Step 9: Plan measure implementation

Step 10: Implement measures

Step 11: Perform self-audit and revision

Step 12: Report progress, and possibly audit and certify

4.2 Testing the Concept in Gaia-X

Gaia-X is a European initiative to create a secure and open data infrastructure that enables data sovereignty and interoperability (Braud, A. et al, 2021). Within the research project of TEAM-X, funded by the German Federal Ministry for Economic Affairs and Climate Action, research is ongoing within two Gaia-X use cases in the areas of breast cancer treatment and in the geriatric care sector. As part of this research project, the DRGs are put into practice through a dedicated management system, responsible leadership trainings, assessments, monitoring, reports, and other methods for the first time in the health vertical: DRG4Health.

The DRGs are anchored in the project consortium, they are committed to by the team, are included actively in use case definition and are key part of the project management for example by being integrated in tools (e.g., GitLab-based issues tracking) and templates (e.g., project-one-pagers) (see Figure 2).

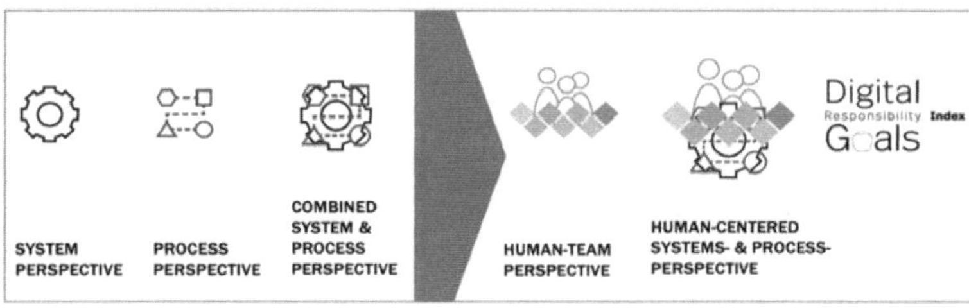

Figure 2. Depiction of integration of Digital Responsibility Goal Approach to TEAM-X.

The DRGs not only combine the systems- and process-perspective within TEAM-X, but also add the human dimension through a human centered perspective. This is how the DRG Index will be generated in an overall holistic approach along all DRG Criteria.

5. DISCUSSION

As a target picture to shape a sustainable human-centered digital transformation, the DRGs offer an opportunity to promote greater responsibility in the digital space across sectors. Responsible behaviour and responsible leadership all along the data life cycle is at the core of establishing trust. By adhering to the framework of the DRGs, and implementing it with a dedicated management system, building trust will no longer be a random by-product, but a pro-active and targeted achievement.

The DRGs provide a framework to mobilize companies and organizations to invest in digital trust in a continuous and scalable way while pursuing their business interests, sustainably and responsibly. By doing so, they go beyond "traditional" headquarter-centric Corporate Social Responsibility (CSR) and Corporate Digital Responsibility (CDR) approaches, that mitigate otherwise unjust business models and are not necessarily addressed within digital strategies, as they are more focused on "corporate social and corporate digital" activities. The aim of the DRG approach is to be ingrained in both business model and organization from end-to-end. The associated management system proposes clear governance mechanisms to do so (e.g., clear roles, integration in processes, education measures) and corresponding tools such as a DRG Index and a corresponding DRG Dashboard for continuous automated monitoring.

Especially the need for healthcare companies to implement a QM-System according to the ISO 13485 makes it easy for them to integrate the DRGs in their processes. Applying for a CE mark with their product should make it also necessary to not only go through the Conformity Assessment Procedure but also to declare DRG conformity.

In addition, to assess trustworthiness, it is crucial to communicate and convey the responsibility enhancing characteristics of a given digital solution in a way that citizens, users, and consumers, as well as policymakers, business leaders, academia, regulators, and/or standardization organisations (like IEEE/ISO) can easily understand and comprehend (Blöbaum, B., 2016). This importance of clear communication is especially true in the doctor-patient relationship (Gopichandran, V. et al, 2021).

We propose to measure and represent the complexity of these demonstrated behaviours in the overarching Digital Responsibility Index. Sufficient differentiation between different levels of maturity in value fulfilment becomes visible according to the granularity of the observation data (see Figure 3).

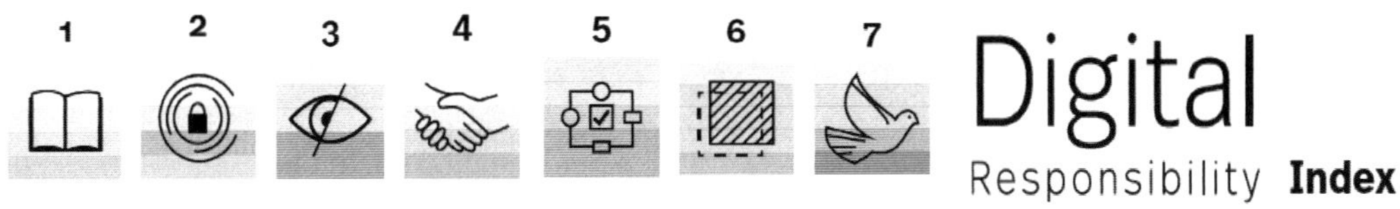

Figure 3. The granularity of the observation data leads to a DRG Index.

Finally, any development towards more digital responsibility does not happen in isolation inside an organization only, but also takes place in an ecosystem if not the wider environment and on a societal level.

The European Union has taken a first positive stride with its publication of the European Digital Rights and Principles, yet further development for implementation in organisations is needed. Based on the DRGs in line with responsible and human-centered digital transformation, more and more actors come together. They jointly shape the living organism of the DRGs, bring them into action, and further develop them towards measurability.

6. CONCLUSION

Integration of Digital Responsibility in a sector like healthcare is challenging albeit its enormous potential in enhancing patient-doctor-researcher interaction. To further incite change, there are more considerations to enhance the impact of the management system based on the DRGs. A next step is to depict a brief overview of the status quo of existing regulations, standards, and initiatives for each DRG and to derive the resulting need for action. This will be one key input towards extending and enhancing the catalogue of measures underlying the management system. In addition, responsible leadership trainings, a meaningful sustainable structure for consultants, best practices and software tools to help with implementation needs to be established.

Furthermore, the incorporation of the DRG Index and DRG Dashboard into the framework will enhance the evaluation. The Gaia-X-related implementation and evaluation is highly influenced by the participating companies, so more and other companies should be considered to verify the insights.

All of these activities could ultimately help to further depict this view of the DRGs and their concrete application to build trust for example for a European Health Data Space (European Commision, n.d.) or an extension for existing data spaces like smart living.

ACKNOWLEDGEMENT

The project TEAM-X consists of a consortium of dedicated research institutions, non-profit organisations, and SMEs. Funding is supported by the German Federal Ministry for Economic Affairs and Climate Action.

REFERENCES

Bacoup, P. et al (2017). From a Quality Management System (QMS) to a Lean Quality Management System (LQMS). The TQM Journal. 30. 00-00. 10.1108/TQM-06-2016-0053.

Bauer, J. et al (2021). ForeSight Approach to improve Privacy and Security in the Smart Living Domain. Current Directions in Biomedical Engineering, 7(2), 903-906.

Blöbaum, B. (2016). Trust and communication in a digitized world. Models and Concepts of Trust Research. Heidelberg et al.: Springer.

Braud, A. et al (2021). The road to European digital sovereignty with Gaia-X and IDSA. IEEE Network, 35(2), 4-5.

European Commision (n.d.). European Health Data Space. https://ec.europa.eu/health/ehealth-digital-health-and-care/european-health-data-space_en

European Commission (n.d.). European Digital Rights and Principles. https://digital-strategy.ec.europa.eu/en/policies/digital-principles

Gopichandran, V. et al (2021). Doctor-patient communication and trust in doctors during COVID 19 times - A cross sectional study in Chennai, India. Plos one, 16(6), e0253497.

Haufe, K. et al (2016). ISMS Core Processes: A Study. Procedia Computer Science. 100. 339-346. 10.1016/j.procs.2016.09.167.

IEA, IRENA, UNSD, WB, WHO (2019), Tracking SDG 7: The Energy Progress Report 2019, Washington DC

Liu, V. et al (2009). Privacy and security in open and trusted health information systems. In Proceeding of the Third Australasian Workshop on Health Informatics and Knowledge Management (pp. 25-30). Australian Computer Society

Lobschat, L. et al (2021). Corporate digital responsibility. Journal of Business Research, 122, 875-888.

Meier, J. et al (2022). Digital Responsibility Goals – a Framework for a Human-Centred Sustainable Digital Economy with a Focus on Trusted Digital Solutions, dHealth 2022 publication, Vienna

Moen, R. et al (2009). Evolution of the PDCA cycle. https://rauterberg.employee.id.tue.nl/lecturenotes/DG000%20DRP-R/references/Moen-Norman-2009.pdf

OECD.AI (n.d.). National AI policies & strategies. https://oecd.ai/en/dashboards/overview

Raza, M. (2023). ISMS: Information Security Management Systems Explained. https://www.splunk.com/en_us/blog/learn/isms-information-security-management-systems.html

AI-ENHANCED ANALYSIS OF BRAIN MORPHOMETRY LINKED TO FACTORS INFLUENCING MENTAL AND NEUROLOGICAL HEALTH OUTCOMES

Paula Arandia[2,3], Christian Núñez[2,3], Christian Stephan-Otto[2,3], Raul Benitez[1,2,3] and Christian Mata[1,2,3]
[1]*Automatic Control Department. Universitat Politècnica de Catalunya, Barcelona, Spain*
[2]*Institut de Recerca Hospital Sant Joan de Déu Barcelona, Esplugues de Llobregat, Spain*
[3]*Pediatric Computational Imaging Center, Institut de Recerca Sant Joan de Déu, Esplugues de Llobregat, Spain*

ABSTRACT

This work introduces a platform aimed at connecting clinicians and data scientists, providing clinicians the opportunity to engage with machine learning models without the need for a background in data science. Moreover, the tool facilitates the exploration of predictive models, with a particular emphasis on analyzing risk and protective factors influencing brain diseases and overall well-being health. Initially conceived for the UKBiobank Project, this initiative has expanded its scope through collaboration with Sant Joan de Déu Hospital in Barcelona, augmenting its capacity to generate psychopathologically relevant models. This innovative approach represents a significant stride in harnessing the full potential of the UKBiobank database for advancements in healthcare research. The primary goal of the project is to identify the top features, underscoring the platform's potential for upcoming studies. The emphasis lies in developing an interpretable model to assist clinicians in pinpointing the most influential lifestyle and brain morphometry features in the development of neurodegenerative disorders.

KEYWORDS

UKBiobank, Supervised Machine Learning, Principal Component Analysis, Decision Trees, Neurological Health

1. INTRODUCTION

For decades, much research has gone into assessing the associations between risk factors, protective factors, and brain structure metrics. There is ample proof that the development of psychopathological disorders is heavily related to combinations of these factors. As an example, some intraindividual traits, such as neurotic personality, have been established as a causal factor in Major Depressive Disorder (MDD) (Navrady et al. 2017).

The use of supervised machine learning (ML) to handle large datasets offers a significant advantage in capturing intricate relationships between features, especially as datasets increase in size and variables become more complex (Tate et al. 2020). Notably, ensemble learning techniques (ELTs) have proven highly efficient in identifying neurobiological markers for neurodevelopmental disorders, such as attention-deficit/hyperactivity disorder (ADHD) (Luo et al. 2020). Unsupervised machine learning approaches, including principal component analysis (PCA) and Lasso regression, successfully distinguish between different bipolar disorder phenotypes, providing a more accurate diagnosis (Wu et al. 2017). Various machine learning techniques have been applied to analyze frontotemporal lobe structure, insula structure, gyrification patterns, cortical thickness, brain geometry, white and grey matter volume anomalies, hippocampus structure, and whole brain models (Kim et al. 2019, Koutsouleris et al. 2022). These applications aim to contribute to the diagnosis, prognosis, classification, or prediction of neurological and psychopathological disorders, showcasing the exceptional tools that machine learning offers in the analysis of clinical and psychopathologically relevant data.

In this context, the UK Biobank has been utilized in this proposed project (Sudlow et al. 2015). Presently, it stands as a valuable resource for researchers in areas like genetics, epidemiology, medicine, and public health, playing a significant role in advancing scientific knowledge and enhancing our understanding of factors that impact human health. This proposed work is an exploration of the possibilities of decision trees as interpretable classification models, as they will be applied to a select group of features from the UKBiobank in order to verify their ability to distinguish between healthy participants and those diagnosed with a neurological disorder.

Research on this specific neurological disorder using this database is lacking, underscoring that this study serves as an initial illustration of its potential usefulness in this field. As such, the primary objective of this project is to pinpoint the most promising features, accentuating the platform's potential for subsequent investigations.data, Methodology and results.

2. METHODOLOGY

2.1 UKBiobank Database

The UK Biobank is a large-scale ongoing research initiative involving more than 500,000 individuals from various regions of the United Kingdom. This prospective cohort study was conducted between 2006 and 2010, during which participants between the ages of 40 and 69 were invited to one of 22 centers located across the country. The study involved the collection of diverse biological samples, including blood, urine, and saliva, as well as the gathering of physical measurements. Furthermore, the initiative has expanded to include follow-up assessments, inviting a subset of the original participants for brain, heart, and body imaging. These imaging examinations offer valuable insights into the structural and functional aspects of these vital organs.

2.2 Preparation of the Testing and Training Datasets

Given that there is almost no pathological sample in the UK Biobank, preprocessing had to be done reducing the 40,000 participants to 368. The dataset, with 18,234 columns and 39,566 rows corresponding to individual participants, poses challenges for developing an all-encompassing machine learning model due to computational constraints. To address this, experts manually narrowed down the features by assessing their relevance, documenting the chosen variables for further processing.

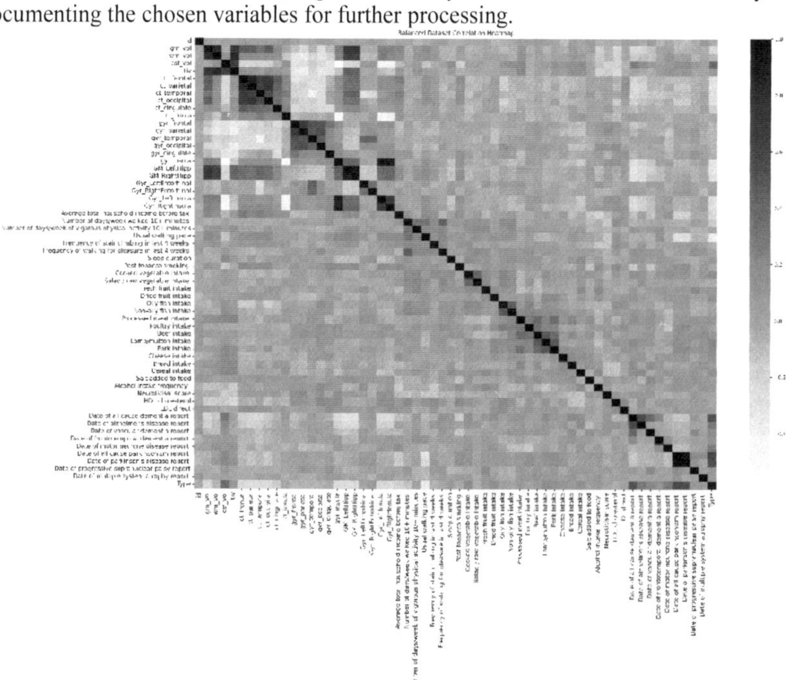

Figure 1. Heatmap of the correlation matrix displaying all model variables, with correlation coefficient mapping to hue on the right scale

Optimal classification problem resolution involves initiating with an extensive Exploratory Data Analysis (EDA) to extract insights and identify potential issues. The correlation matrix heatmap proved valuable for gaining quick insights into dataset characteristics during EDA (Figure 1). After applying an 80/20 split, the training dataset consists of 31,652 samples and the testing dataset consists of 7,914 samples, and both have 18,234 columns which will be reduced later.

2.3 Selecting Features and Building a Decision Tree

Addressing the imbalanced class distribution, a common challenge in classification datasets, was the first step. After balancing the dataset by means of undersampling the majority class, the final study population comprises a total of 184 pathological participants –labeled as 'pathological' (1) – and 184 healthy participants –labeled as 'healthy' (0) –, resulting in a total of 368 participants. Several tests were carried out by varying the imbalance ratio to some extent and checking its effect on classification accuracy, and a 1:1 proportion appeared to perform the best.

Afterwards, a feature selection process is performed by which the input variables of the current predictive model are reduced to only those that are believed to be the most relevant for predicting the target variable, while all the other non-informative variables are removed. This reduction is beneficial in lowering computational costs associated with modeling and, in certain scenarios, can enhance the model's performance. In this sense, a series of feature selection techniques were used. Some methods – wrapper methods, such as Recursive Feature Elimination (RFE) – eliminate irrelevant variables by assessing their relevance to the target, while others – embedded methods, such as Random Forests (RFs) – perform feature selection automatically as part of the learning pipeline (Solorio-Fernandez et al. 2019). After the application of the feature selection techniques, a reduced set of only eight features remains.

The next step in the methodology involves constructing and training the final decision tree based on the selected features. This phase of the study is centered on acquiring an inference tool with an emphasis on high interpretability. In our context, the objective is to infer the correlation between risk and protective factors in the development of neurodegenerative diseases, specifically dementia and its subtypes. To accomplish this, a highly interpretable model is essential for evaluating the underlying relationships between the target variable and the predictors. This inherently requires a simple model with an acceptable level of accuracy, and decision trees are the perfect tool to check both boxes. The objective is to create a random forest algorithm capable of categorizing each patient as 'healthy' (0) or 'pathological' (1). This entails using data from UKBiobank variables along with variables related to brain morphological metrics obtained from diverse studies conducted by the Hospital Sant Joan de Déu Barcelona. In this sense, three random forest models were generated: the baseline model, the optimized model with hyperparameters tuned via random search, and the optimized model with hyperparameters tuned through grid search (Nembrini et al. 2018). Moreover, this method aims to pinpoint the essential variables within these datasets to forecast the likelihood of a patient developing a neurodegenerative disease in subsequent stages, potentially including the evaluation of user perspectives to ensure ease of use and the attainment of valid and reliable outcomes by non-experts.

3. RESULTS

In summary, the methodology evaluated the accuracy of the decision tree classifier by first fitting the model with default parameters (Zahedi et al. 2022). This established a baseline performance, representing the initial expectation of how well the model was anticipated to perform on a given task before any optimization or fine-tuning took place. This benchmark served as a starting point for evaluating and comparing the effectiveness of the model. The primary objective of this project is to identify the top features, demonstrating the potential of this platform for future studies. The next steps will be summarized as follows:

1. The partition between training and testing datasets is done by applying a 75/25 split to total the 368 samples, which results in a training set of 276 participants and a testing set of 92 participants. These datasets only include the eight features that were selected as the most influential by the previously trained random forest.

2. A basic decision tree model with default parameters is evaluated to compute the baseline accuracy in order to measure the improvement after the optimization.

3. Once again, a grid search is carried out to find the best combination of hyperparameters. Random search is not required because decision trees have a small and manageable pool of hyperparameters, so its optimization can be done using grid search directly.

4. The 10-fold cross-validated accuracy of the classifier is computed and compared to the baseline accuracy to visualize the improvement after the hyperparameter tuning.

The eight features that were deemed as the most relevant by the random forest algorithm were, ordered by importance: *GM_LeftHipp, GM_RightHipp, ct_temporal, ct_occipital, Gyr_RightInsula, Gyr_LeftInsula, gyr_temporal and Usual walking pace*. A confusion matrix provides a summary of a classifier's classification performance in relation to a set of test data using a set of metrics. In this case, the positive class will be the healthy participants (labeled 0) and the negative class will be the pathological participants (labeled 1). The confusion matrix obtained from the decision tree classifier is depicted in Figure 2.

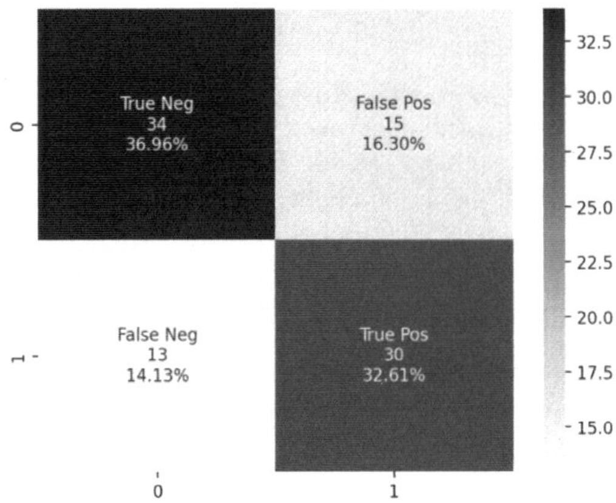

Figure 1. Confusion matrix of the decision tree classifier

Out of the 92 participants that conform the test set, 64 were correctly classified (34 were true negatives, 30 were true positives) and 28 were not (there were 15 false positives and 13 false negatives), which results in a $(64/92)*100\cong70\%$ classification accuracy. This seems in line with the 72.83% obtained through cross-validation: due to the randomness of the split between training/testing datasets, and due to the small size of the testing dataset, some iterations of the classification might obtain worse results than the cross-validated average. Another useful tool to assess the performance of the model is the classification report (Figure 3). Three performance metrics are employed: precision, recall and the F1-score. 'Support' refers to how many samples of each label were in the testing dataset originally. Where "Precision" also known as positive predictive value, is the ratio of correctly predicted positive instances to the total predicted positive instances (Sharma et al. 2022) [199]. In simple terms, precision represents the proportion of accurately identified positive values, which in this case signifies participants correctly diagnosed with a neurodegenerative disease. "Recall", often called sensitivity or TP rate (TPR), represents the proportion of correctly identified positive values out of the total actual positive instances. The F1 score is the harmonic mean of precision and sensitivity, which provides equal weight to both factors. These results have been validated by two neurology experts from HSJD.

	Precision	Recall	F1 score	Support
Negative **(Healthy, 0)**	0.72	0.69	0.71	49
Positive **(Diseased, 1)**	0.67	0.70	0.68	43
Accuracy			0.70	92

Figure 2. Classification report of the optimized decision tree

4. CONCLUSION

Considering this study as an initial version of a highly functioning platform, the objectives set for this part of the project can be deemed successfully accomplished, as the top promising features related to neurological health outcomes were obtained from the UKBiobank. The machine learning models built on the UKBiobank database not only functioned as intended but also provided clinically relevant results. Consequently, the platform received professional validation from clinical experts. The delicate balance in the tradeoff between accuracy, complexity, and interpretability has been successfully achieved, as evidenced by the clinically logical diagnosis workflow generated by the decision tree.

Ultimately, the primary objective of this project has been accomplished by identifying the most significant features. This showcases the platform's potential for future research endeavors, emphasizing its effectiveness and the potential impact it could have on studies in the field.

In future work, the emphasis will be on utilizing the acquired features for an exploratory analysis of general brain status indicators, such as the cortical thickness of the insula or the temporal lobe, and examining their associations with cognition. Furthermore, a second study will be undertaken, focusing on the development of an interpretable model to assist clinicians in identifying the most influential lifestyle and brain morphometry features in the development of neurodegenerative disorders.

ACKNOWLEDGEMENTS

The authors wish to extend their gratitude to the UKBiobank for granting permission to use the dataset in this proposed work.

REFERENCES

Kim, J. P., Kim, J., Park, Y. H. (2019). Machine learning based hierarchical classification of frontotemporal dementia and Alzheimer's disease. *NeuroImage: Clinical*. Vol. 23, p. 101811.

Koutsouleris, N., Pantelis, C. (2022). Exploring Links Between Psychosis and Frontotemporal Dementia Using Multimodal Machine Learning. *JAMA Psychiatry*. Vol. 79, Issue 9, p. 907.

Luo, Y., Alvarez, T. L., Halperin, J. M., Li, X. (2020). Multimodal neuroimaging-based prediction of adult outcomes in childhood-onset ADHD using ensemble learning techniques. *NeuroImage: Clinical*. Vol. 26, p. 102238.

Navrady, L. B., Ritchie, S. J. (2017). Intelligence and neuroticism in relation to depression and psychological distress: Evidence from two large population cohorts. *European Psychiatry* Vol. 43, pp. 58–65.

Nembrini, S., König, I. R., Wright, M. N. (2018). The revival of the Gini importance? In A. Valencia (Ed.), *Bioinformatics* Vol. 34, Issue 21, pp. 3711–3718.

Sharma, D. K., Chatterjee, M., Kaur, G., Vavilala, S. (2022). Deep learning applications for disease diagnosis. *Deep Learning for Medical Applications with Unique Data*, pp. 31–51.

Solorio-Fernández, S., Carrasco-Ochoa, J. A., Martínez-Trinidad, J. Fco. (2019). A review of unsupervised feature selection methods. *Artificial Intelligence Review*. Vol. 53, Issue 2, pp. 907–948.

Sudlow C, Gallacher J, Allen N, Beral V, Burton P, Danesh J, et al. (2015). UK Biobank: An Open Access Resource for Identifying the Causes of a Wide Range of Complex Diseases of Middle and Old Age. *PLoS Med* Vol.12, num. 3, e1001779.

Tate, A. E., McCabe, R. C., Larsson, H., Lundström, S., Lichtenstein, P., Kuja-Halkola, R. (2020). Predicting mental health problems in adolescence using machine learning techniques. *PLOS ONE. Public Library of Science (PLoS)*. Vol. 15, Issue 4, p. e0230389.

Wu, M.-J., Mwangi, B. (2017). Identification and individualized prediction of clinical phenotypes in bipolar disorders using neurocognitive data, neuroimaging scans and machine learning. *NeuroImage*. Vol. 145, pp. 254–264.

Zahedi, M., & Golchin, S. (2022). Prediction of blast loading on protruded structures using machine learning methods. *International Journal of Protective Structures*. p. 204141962211440.

KNOWLEDGE GRAPH CONSTRUCTION VIA PRETRAINED END-TO-END TRANSFORMER MODEL

Denis Sidorenko and Anatoly Shalyto
ITMO University
Saint-Petersburg, Kronverksky Pr., 49, Russia

ABSTRACT

This work presents a novel end-to-end deep learning model for extracting and classifying interactions between biological entities from scientific biomedical literature texts. Additionally, a visualization tool has been developed for visualizing the extracted interactions in the form of intuitive knowledge graph structures. The proposed model is based on the pre-trained BioBERT language model and utilizes an entity masking approach. In the first stage, biological entities (genes,, diseases, drugs, biological processes) are identified in the texts and then masked with special tokens. Next, a multi-classification neural network, trained on a large set of labeled data, classifies the relationships between the masked entities into one of four interaction categories: activation, inhibition, increase amount, and decrease amount. The visualization tool allows visualizing the extracted interactions in the form of intuitive knowledge graph structures, facilitating the analysis of complex biological processes. The developed model and tool open up new opportunities for accelerating scientific research and drug development by effectively extracting and visualizing valuable knowledge from unstructured biomedical text data using modern deep learning and natural language processing methods.

KEYWORDS

Name Entity Recognition, Relation Extraction, Knowledge Graph, Bioinformatics, Deep Learning, Natural Language Processing

1. INTRODUCTION

Scientific biomedical literature texts contain extensive knowledge about complex interactions between various biological entities, such as genes, diseases, drugs, and biological processes. Extracting and systematizing this information is crucial for accelerating scientific research, developing new treatment methods, and understanding fundamental biological processes. However, a significant portion of this knowledge is encoded in unstructured text data, making it challenging to analyze and apply effectively.

In recent years, deep learning and natural language processing (NLP) methods have demonstrated impressive results in extracting structured information from texts (J. Li et al., 2022). In particular, pre-trained language models, such as BERT (Bidirectional Encoder Representations from Transformers) (Devlin et al., 2018), allow for the creation of highly efficient solutions for tasks such as entity extraction, relation classification, and other NLP tasks. In the field of bioinformatics and biomedicine, specialized models such as BioBERT (Lee et al., 2019) have been developed, pre-trained on a large corpora of biomedical texts and demonstrating high accuracy on relevant tasks.

However, existing approaches for extracting interactions from biomedical texts still have several limitations. Firstly, many of them are based on rules (Soomro et al., 2017) or the use of contextual features (Cho and Lee, 2019), which constrains their ability to extract complex relationships between biological entities. Secondly, the extracted interactions are often presented in unstructured lists or tables, making their analysis and interpretation difficult. Thirdly, most existing tools are focused on specific areas of biomedicine (Trieu et al., 2020) and do not provide a comprehensive solution for extracting and visualizing interactions from various data sources.

In this work, we present a novel scalable deep learning model for extracting entities from texts and classifying interactions between biological entities from scientific biomedical literature texts. The model is based on the pre-trained BioBERT language model, utilizing an entity masking approach, which involves modifying original sentences by replacing tokens with masked entities identified by the NER model. Additionally, was developed the visualization tool for visualizing the extracted interactions in the form of intuitive filtered graph structures, facilitating the analysis of complex biological processes.

2. DATA

In the field of biomedical NER, high-quality annotation is crucial (J. Li et al., 2022). For this reason, there is a significant lack of labeled text data in biomedicine. For the present model, all current open-labeled datasets were pre-collected for both the entity extraction task and the task of classifying relationships between them. For the task NER were used NCBI Disease (Dogan et al., 2014), BC5CDR (Li et al., 2016), GNormPlus (Wei et al., 2015), GO (Carbon and Mungall, 2024), for the task RE used BioRED dataset (Luo et al., 2022).

3. METHOD

3.1 NER

NER algorithms typically use a sequence learning approach, with methods based on Conditional Random Fields (CRF) or Recurrent Neural Networks (RNN). For example, we have a sentence S consisting of words $\{w_1, w_2, \dots, w_n\}$ and named entities $\{e_1, e_2, \dots, e_k\}$ that we are interested in. CRF defines the probability of a single sequence of entities E given the sentence S as: $P\,[E|S] = (1/z) \times exp(\sigma(\lambda f \times f(S,E) + \mu_g * g(S,E)))$, where f, g are feature functions of words and entities, σ is the sum function, λ, μ are the weights of these functions, and z is a normalization constant that prevents probabilities from exceeding 1. The CRF model is trained on labeled data, adjusting the weights λ and μ for each input sentence and its corresponding entities.

3.2 Proposed Model

The proposed approach consists of two main components: 1) the deep learning model for extracting and classifying interactions between biological entities from texts; 2) the visualization tool for visualizing the extracted interactions in the form of graph structures. The model is based on the pre-trained BioBERT language model (Lee et al., 2019) and uses an entity masking approach. In the first stage, biological entities such as genes, diseases, and biological processes are identified in the texts using a specialized tool for extracting entities from biomedical texts. The identified entities are then masked with special tokens (e.g., [GENE], [DISEASE], [DRUG], [Biological Process]).

Next, the masked texts are fed into a multi-classification neural network trained on a large set of pre-labeled interaction data from BioRED (Luo et al., 2022). The neural network classifies the relationships between the masked entities into one of four categories:

1. Activation: one biological entity activates or induces another;
2. Inhibition: one biological entity suppresses or blocks another;
3. Increase amount: one biological entity increases the expression, concentration, or quantity of another;
4. Decrease amount: one biological entity decreases the expression, concentration, or quantity of another;
5. No relation: when there is no relationship.

The use of the pre-trained BioBERT model and entity masking allows the model to more effectively extract contextual information and recognize complex relationships between biological entities in texts. Additionally, the multi-class architecture provides a more fine-grained classification of interactions compared to binary approaches (interaction/no relation).

3.3 Filtration

A unique filtering algorithm was also developed, as presented in Figure 1. First, a key entity is selected, then nodes are filtered based on their vertex degree. Next, the key entity is enriched from the initial graph proportionally to each entity type. Clinical trials are added, a single connected component is filtered, and meta-information about the publications from which the relationships were extracted is incorporated. Figures 2-3 show examples of graphs before and after filtering for visualization purposes. This filtering process aims to reduce the complexity of the initial graphs, highlight the most relevant information for the selected key entity, and provide additional context from clinical trials and source publications. The resulting filtered graphs enable more effective visual analysis and interpretation of the extracted biological interactions and their context.

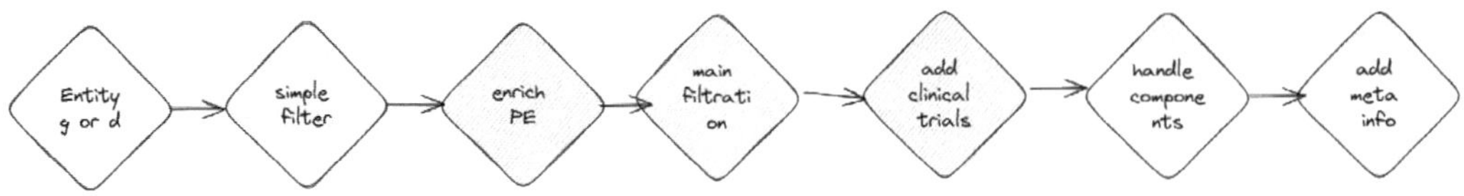

Figure 1. Filtration process

3.4 Visualization

The tool is designed to visualize the interactions extracted by the model in the form of intuitive graph structures. The nodes of the graph represent biological entities (genes, diseases, drugs, biological processes), and the edges represent classified interactions between them.

Each type of interaction (activation, inhibition, increase/decrease in quantity) is visualized using different colors and edge styles, facilitating information perception. Additional interactive features are provided, such as filtering interactions by type, searching for and highlighting specific entities or paths.

The tool is also integrated with external databases and resources, allowing users to obtain additional information about biological entities, including their functions, structures, relevant publications, and clinical aspects.

Figures 2 and 3 show the model results before and after filtering for the EGFR gene. It can be seen that through visual analysis directly without filtering, it is difficult to draw any conclusions about the biological entities associated with it. However, after filtering, the graph becomes convenient for analysis. Colors represent different types of relationships and various node types. Using this model visualization tool, medical professionals, biologists, or chemists can search for genes, biological processes, or diseases of interest, with connections confirmed by scientific articles. Each article is represented by a sentence from a scientific paper, containing entities between which a relationship has been formed and predicted by the developed model.

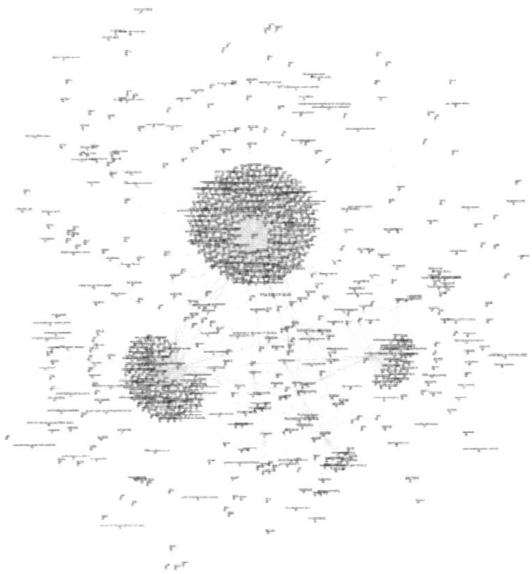

Figure 2. Knowledge graph visualization before filtration

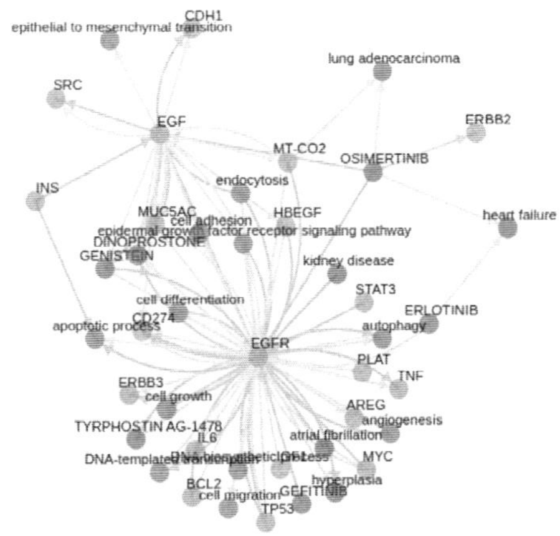

Figure 3. Knowledge graph visualization after filtration

4. CONCLUSION

In this work, we presented a new scalable deep learning model for extracting and classifying interactions between biological entities from scientific biomedical literature texts. The model is based on the pre-trained BioBERT language model and utilizes an entity masking approach for more effective recognition of complex relationships in texts. Additionally, we developed the visualization tool for visualizing the extracted interactions in the form of intuitive graph structures, facilitating the analysis of biological processes.

The developed model and tool demonstrate the effectiveness of applying modern deep learning and natural language processing methods for extracting and visualizing valuable knowledge from unstructured biomedical text data. They open up new opportunities for accelerating scientific research, developing new treatment methods, and gaining a deeper understanding of fundamental biological processes.

In the future, we plan to continue enhancing the model, particularly by expanding the set of interaction types and integrating additional data sources.

ACKNOWLEDGEMENT

This work was supported by Russian Science Foundation, Grant 23-11-00346.

REFERENCES

Carbon, S., Mungall, C., 2024. Gene Ontology Data Archive. https://doi.org/10.5281/ZENODO.10536401

Cho, H., Lee, H., 2019. Biomedical named entity recognition using deep neural networks with contextual information. BMC Bioinformatics 20. https://doi.org/10.1186/s12859-019-3321-4

Devlin, J., Chang, M.-W., Lee, K., Toutanova, K., 2018. BERT: Pre-training of Deep Bidirectional Transformers for Language Understanding.

Dogan, R., Leaman, R., lu, Z., 2014. NCBI Disease Corpus: A Resource for Disease Name Recognition and Concept Normalization. J. Biomed. Inform. 47. https://doi.org/10.1016/j.jbi.2013.12.006

J. Li, A. Sun, J. Han, C. Li, 2022. A Survey on Deep Learning for Named Entity Recognition. IEEE Trans. Knowl. Data Eng. 34, 50–70. https://doi.org/10.1109/TKDE.2020.2981314

Lee, J., Yoon, W., Kim, Sungdong, Kim, D., Kim, Sunkyu, So, C., Kang, J., 2019. BioBERT: a pre-trained biomedical language representation model for biomedical text mining. Bioinforma. Oxf. Engl. 36. https://doi.org/10.1093/bioinformatics/btz682

Li, J., Sun, Y., Johnson, R., Sciaky, D., Wei, C.-H., Leaman, R., Davis, A.P., Mattingly, C., Wiegers, T., lu, Z., 2016. BioCreative V CDR task corpus: a resource for chemical disease relation extraction. Database 2016, baw068. https://doi.org/10.1093/database/baw068

Luo, L., Lai, P.-T., Wei, C.-H., Arighi, C.N., Lu, Z., 2022. BioRED: a rich biomedical relation extraction dataset. Brief. Bioinform. 23, bbac282. https://doi.org/10.1093/bib/bbac282

Soomro, D., Banbhrani, S., Shaikh, A., Raj, H., 2017. Bio-NER: Biomedical Named Entity Recognition using Rule-Based and Statistical Learners.

Trieu, H.-L., Tran, T., Duong, K., Nguyen, A., Miwa, M., Ananiadou, S., 2020. DeepEventMine: End-to-end neural nested event extraction from biomedical texts. Bioinforma. Oxf. Engl. 36. https://doi.org/10.1093/bioinformatics/btaa540

Wei, C.-H., Kao, H.-Y., Lu, Z., 2015. GNormPlus: An Integrative Approach for Tagging Genes, Gene Families, and Protein Domains. BioMed Res. Int. 2015.

Reflection Paper

ADVANCING USER-CENTRIC DESIGN AND TECHNOLOGY ADOPTION FOR AGING POPULATIONS: A MULTIFACETED APPROACH

Andreea Stamate[1*], Magda Velciu[1†], Cosmina Paul [1‡], Luiza Spiru[2§] and Mircea Dan Marzan[3]**

[1] *Dr., Researcher, Research Department, Ana Aslan International Foundation, Bucharest, Romania*
[2]*President, Ana Aslan International Foundation; Head, Saint Luca's Chronic Diseases Clinical Hospital; Prof., Department of Geriatrics, Gerontology, Old Age Psychiatry and Longevity Medicine, Carol Davila University of Medicine and Pharmacy, Bucharest, Romania*
[3]*MD, PhD Student, Department of Geriatrics, Gerontology, Old Age Psychiatry and Longevity Medicine, Carol Davila University of Medicine and Pharmacy, Bucharest, Romania*

ABSTRACT

The global aging trend requires a nuanced approach to assistive technology for older adults. This paper summarizes challenges and strategies for technology adoption among aging populations. User-centric design and co-creation are crucial for effective technologies. Challenges like participant selection, feedback attitudes, and trust-building are vital for meaningful user involvement. Accurate assessment of technological literacy in older adults is crucial. Objective measures are needed due to unreliable self-reported proficiency. The fragmented digital ecosystem and digital divide hinder technology adoption. Integrating caregivers into adoption models is crucial. Collaboration and research standards are vital to address societal inequalities. User-centric design, caregiver integration, and interdisciplinary collaboration can improve technology acceptance and older adults' quality of life in the digital era.

KEYWORDS

Older Adults, Digital Products, User Experience, Best Practices, Co-Creation

1. BACKGROUND AND KNOWLEDGE TRANSFER

The relevance of Healthy and Independent Living at Home has increased due to aging populations, rising healthcare costs, technological advancements, and events like the COVID-19 pandemic. Despite efforts in health literacy, digital health, age-friendly environments, and integrated care, few products for older adults are on the market, and even fewer are widely accepted. Many older adults struggle to use healthcare technology without assistance. Factors contributing to digital product failures for this segment include:1. Lack of User-Centric Design - products often don't consider older adults' needs, preferences, and capabilities, leading to complex or frustrating interfaces.; 2. Underestimation or overestimation of Technological Literacy – older users have varied digital skills and may not accurately assess their abilities.; 3. Fragmented Ecosystem - numerous products targeting aging create confusion for users.; 4. Lack of Integrated Technology Adoption Models - caregiver roles are overlooked, impacting technology acceptance.; 5. Ignoring Social Innovation - not considering social aspects widens the digital gap.; 6. Lack of Standardized Literature - there's a need for collaborative research to establish norms.

* https://orcid.org/0000-0002-3385-9714
† https://orcid.org/0000-0003-1457-004X
‡ https://orcid.org/0000-0003-3827-2290
§ https://orcid.org/0000-0002-5308-205X
** https://orcid.org/0009-0005-8456-1300

Sharing knowledge and best practices can lead to more user-friendly technologies for older adults. Collaborative efforts enhance innovation and adoption rates. We humbly share our experiences from about 10 years of working on ICT solutions for Healthy Aging projects in Europe, though examples of best practices that we have aided in developing innovative products to enhance older adults' quality of life through tested digital solutions.

1.1 Lack of User-Centric Design

The growing aging population emphasizes the need for research involving older adults in co-design processes due to the challenges of representing diverse opinions and target audiences across countries, as highlighted by von Hippel (2005). Participant diversity is crucial for generating ideas that appeal to a wide consumer base, benefiting companies through insights gained from co-creation and influenced by participants' personality traits and interests (Mahr et al., 2014).

1.1.1 The Individual Characteristics Bias

Older adults exhibit varied characteristics, including socio-cultural backgrounds, income levels, ICT skills, and experiences with health technologies. These factors shape their perceptions and feedback during co-creation activities (Moschis, 2019; Sthienrapapayut et al, 2018; Nimrod, 2017).

1.1.2 Differences in Sharing their Opinions and Giving Feedback

Life experiences influence how older adults express opinions, with cultural backgrounds impacting openness. Former authoritarian regimes' legacies, seen in Eastern Europe, may lead to reluctance in voicing critical views. Conversely, Western cultures promote open dialogue, enhancing idea exchange during co-creation. Our organization (based in Romania – a former communist country) encourages participants to provide honest feedback by emphasizing the prototype's imperfections and the benefits of constructive criticism. Building trust fosters open communication and diverse perspectives, crucial for successful co-creation efforts.

Beside building trust, researchers and evaluators need to be aware of the types of biases which affect data collection with respect to the satisfaction and interest in the new technologies. Hence, the opinions and feedback of the older adults and their formal and informal caregivers should be collected in ways which are able to deconstruct biases, such as social desirability bias, the need for social/ emotional connection bias (older adults involve into projects for combating loneliness), the need to help bias (older adults successfully aging try to help others, and therefore involve into testing though they have not an interest/ need into the project as such). Practical recommendations concern discourse analyses and a more qualitative undertaken of the methodologies employed.

1.1.3 Differences in Willingness to Explore New Things, Healthcare Technology and Continuous Learning

The willingness of older adults to explore healthcare technology varies significantly, affecting adoption rates. Tailored outreach, flexible engagement options, personalized support, and continuous feedback are essential for inclusive participation and meaningful contributions in co-creation endeavours (Hunsaker, 2019).

To address older adults' varying engagement with healthcare technology in co-creation, tailor outreach, offer flexible participation, provide personalized support and training, empower participants, and seek continuous feedback. Segmentation based on technology comfort, varied engagement channels, individualized assistance, empowerment, and open communication ensure inclusive and meaningful involvement in shaping healthcare technologies. Nevertheless, success is not assured, as it involves a complex, multi-task, and multi-stakeholder process that surpasses traditional citizen participation efforts (Jarke, 2021). Establishing strict guidelines and recommendations in the inherently open and complex co-creation process can be challenging; however, knowledge transfer insights can provide valuable evidence for effectively co-creating more user-centric public services with and for older adults.

1.2 Underestimation or Overestimation of Technological Literacy

It is crucial to evaluate the computer proficiency of older adults when designing new applications and promoting co-creation engagement. Understanding their unique needs and limitations in digital technology design could improve user experience and drive advancements in assessment and diagnosis technologies, aiming to optimize patient benefits while minimizing associated risks (Robillard et al., 2018). A recent study that we conducted (currently under review), suggested that self-reported digital proficiency may not reliably reflect actual competence in older adults, highlighting the importance of incorporating objective measures in assessing digital competence. Additionally, considering older adults' digital skills, especially in digital cognitive testing, is critical to avoid overdiagnosis or under diagnosis based on varying levels of computer experience (Lee et al., 2021). These insights can guide the development of more effective assessment and diagnosis technologies for older adults.

1.3 Fragmented Ecosystem

The fragmented digital ecosystem complicates accurate measurements in devices for older adults, hindering adoption and contributing to product failures due to the difficulty in finding suitable solutions. Ensuring marketed digital devices provide reliable measurements is crucial, as inconsistencies can worsen navigating this ecosystem. For instance, our recent reliability assessment of the Squegg Smart Grip Trainer, which showed good agreement with the gold standard Jamar dynamometer (ICC = .912), highlights the importance of thorough testing before endorsing devices (Stamate et al., 2023). Wearable devices in AAL consortia projects offer benefits like simultaneous data collection but require rigorous testing for reliability, especially concerning the intended population and purpose of use.

1.4 Lack of Integrated Technology Adoption Models

The use of assistive technology by older adults often relies on support from formal or informal caregivers, encompassing technical assistance, device setup, and troubleshooting (Wei, 2023; Zhou, 2024). Caregivers also play a role in introducing older adults to new technology and interpreting health data. Despite their significant influence on technology acceptance, there is a gap in understanding how caregivers justify and structure their acceptance of new products and services. Studies highlight the importance of technical support and social influence from caregivers in older adults' acceptance of assistive technologies (Li et al., 2018). However, the integration of caregivers into Technology Adoption Models for older adults remains insufficient in the field of assistive technologies. Neglecting caregiver roles hinders the development of comprehensive models, leading to increased burdens on caregivers and perpetuating digital divides among the older adult population. Therefore, a holistic approach is necessary to advance Technology Adoption Models, considering all stakeholders to promote successful technology adoption and reduce caregiver burdens and digital disparities.

However, concerted efforts towards a unified approach in theory acceptance of technologies have been made (Venkatesh, 2003; Venkatesh and Davis, 2000; Venkatesh et al, 2012). And all these efforts heavily impact the health care sector, as the theories are adapted, identifying variables which affect the acceptability, acceptance, and adoption of new technologies in health care sector (Wutz et al, 2023; Renaud and Biljon, 2008).

1.5 Ignoring the Social Innovation Dimension

The digital gap among older adults, particularly in countries like Romania, stems mainly from economic disparities (known as the 'first digital divide') and seniors' limited digital skills. Despite initial engagement with technology before retirement, many older adults find new technologies irrelevant post-retirement, leading to decreased usage, especially in those aged 75 and older. This trend perpetuates societal marginalization and deepens broader inequalities.

The development of assistive technologies has attracted attention from both business and government sectors, with businesses seeing the growing elderly population as a lucrative market and governments viewing it as a financial challenge, particularly in healthcare. This convergence has led to a focus on commercial gains, neglecting the social dimension and exacerbating the gap between affluent and marginalized groups. Addressing these challenges requires political action beyond business interests.

1.6 Lack of Systematized Literature

Since the digital revolution, there has been increased exploration of older adults' knowledge, attitudes, and behaviours as users of assistive technologies. Nonetheless, there is still a lack of organized literature, research protocols, and standards in this field. Despite the development of gerontographics segmentation and technology adoption models tailored for older adults, they are often perceived as a homogeneous group based on chronological age. Interdisciplinary collaboration and coordinated efforts are needed to establish comprehensive research and practice standards in this area.

2. CONCLUSIONS

Addressing challenges in assistive technology adoption among aging populations necessitates a comprehensive approach. User-centric design through co-creation is vital, addressing participant selection, feedback attitudes, and fostering trust for meaningful involvement. Assessing technological literacy is key for effective solutions, while integrating caregivers into Technology Adoption Models enhances acceptance and support. The fragmented digital ecosystem underscores the urgency of tackling economic disparities and improving digital literacy. Collaborative interdisciplinary efforts are crucial for developing robust research standards and advancing technology adoption, ultimately reducing societal inequalities and enhancing older adults' quality of life.

ACKNOWLEDGEMENT

This work was supported by a grant from the Ministry of Research, Innovation, and Digitization under project code PN-III-P3-3.6-H2020-2020-0174 within PNCDI III, titled "Increasing research capacity and visibility of the Ana Aslan International Foundation (AAIF) for better promotion of results".

REFERENCES

Hunsaker, A., Nguyen, M.H., Fuchs, J., Djukaric, T., Hugentobler, L., Hargittai, E.: He explained it to me and I also did it myself: how older adults get support with their technology uses. Socius, *5*, 2378023119887866 (2019)

Jarke, J. (2021). Co-creating digital public services for an ageing society: Evidence for user-centric design (p. 228). Springer Nature.

Mahr, F. D., Lievens, A., & Blazevic, V. (2014). The value of customer cocreated knowledge during the innovation process. *Journal of Product Innovation Management*, *31*(3), 599-615. https://doi.org/10.1111/jpim.12116Mitzner, 2019Mitzner, 2019

Moschis, G.P.: Consumer Behavior over the Life Course: Research Frontiers and New Directions. Springer, Cham (2019). https://doi.org/10.1007/978-3-030-05008-5

Nimrod, G.: Older audiences in the digital media environment. Inf. Commun. Soc., *20*(2), 233–249 (2017)

Renaud, K., Van Biljon, J.: Predicting technology acceptance and adoption by the elderly: a qualitative study. In: Proceedings of the 2008 Annual Research Conference of the South African Institute of Computer Scientists and Information Technologists on IT Research in Developing Countries: Riding the Wave of Technology (2008)

Robillard, J. M., Lai, J. A., Wu, J. M., Feng, T. L., & Hayden, S. (2018). Patient perspectives of the experience of a computerized cognitive assessment in a clinical setting. Alzheimer's & Dementia: Translational Research & Clinical Interventions, 4, 297-303. https://doi.org/10.1016/j.trci.2018.06.003

Stamate, A., Bertolaccini, J., Deriaz, M., Gunjan, S., Marzan, M. D., & Spiru, L. (2023). Interinstrument Reliability Between the Squegg® Smart Dynamometer and Hand Grip Trainer and the Jamar® Hydraulic Hand Dynamometer: A Pilot Study. *The American Journal of Occupational Therapy*, 77(5), 7705205150. https://doi.org/10.5014/ajot.2023.050099

Sthienrapapayut, T., Moschis, G.P., Mathur, A.: Using gerontographics to explain consumer behaviour in later life: evidence from a Thai study. J. Consum. Mark., *35*(3), 317–327 (2018). https://doi.org/10.1108/JCM-02-2017-2083

Venkatesh, M., et al.: User acceptance of information technology: toward a unified view. MIS Q. 27(3), 425–478 (2003). https://doi.org/10.2307/30036540

Venkatesh, V., Davis, F.D.: A theoretical extension of the technology acceptance model: four longitudinal field studies. Manag. Sci., *46*(2), 186–204 (2000)

Venkatesh, V., Thong, J.Y.L., Xu, X.: Consumer acceptance and use of information technology: extending the unified theory of acceptance and use of technology. MIS Q., *36*, 157–178 (2012).

Von Hippel, W., Von Hippel, C., Conway, L., Preacher, K. J., Schooler, J. W., & Radvansky, G. A. (2005). Coping with stereotype threat: denial as an impression management strategy. *Journal of personality and social psychology*, *89*(1), 22. https://psycnet.apa.org/doi/10.1037/0022-3514.89.1.22

Wei, W., Gong, X., & Tian, K. (2023). A study on community older people's willingness to use smart home—an extended technology acceptance model with intergenerational relationships. *Frontiers in Public Health*, 11, 1139667. doi:10.3389/fpubh.2023.1139667

Wutz M, Hermes M, Winter V, Köberlein-Neu J. Factors Influencing the Acceptability, Acceptance, and Adoption of Conversational Agents in Health Care: Integrative Review. J Med Internet Res. 2023 Sep 26;25:e46548. doi: 10.2196/46548. PMID: 37751279; PMCID: PMC10565637

Zhou, C., Qian, Y., & Kaner, J. (2024). A study on smart home use intention of elderly consumers based on technology acceptance models. *Plos one*, 19(3), e0300574.doi: 10.1371/journal.pone.0300574

Posters

USING EYE-GAZE TO EVALUATE E-HEALTH DECISION SUPPORT SYSTEMS

Elizabeth A. Krupinski
PhD, Department of Radiology & Imaging Sciences Emory University
1364 Clifton Rd NE Atlanta, GA 30032 USA

ABSTRACT

Eye-tracking technologies have been used for decades to assess the impact of technologies on clinical decision making in medical imaging. eHealth and telemedicine present new opportunities to integrate images and other visual data into routine clinical care. Artificial intelligence (AI) tools can work synergistically with these data to improve the efficacy and efficiency of data analysis and interpretation by the human decision maker. However, we know little about the optimal ways to present AI output to providers. This study used radiology image interpretation with AI-based decision support to explore the impact of different types of AI output on reader performance. Four different outputs (plus no AI feedback) were evaluated with experienced radiologists and radiology residents. Results reveal that the rates of decision changes prior to and after receiving the AI output differ as a function of the output format and reader experience. More complex output formats tend to increase reading time.

KEYWORDS

Eye-Tracking, Decision Support Systems, Radiology, Artificial Intelligence

1. INTRODUCTION

Our eyes and brains are amazingly good at what they do. They are, however, often easily fooled (e.g., visual illusions). Medical imaging (e.g., radiology, pathology, dermatology, ophthalmology) relies on human visual perception and cognitive reasoning to render accurate and efficient diagnoses. Unfortunately, errors can occur (false negatives and false positives) that impact patient care and outcomes. Errors occur for many reasons (e.g., poor image quality, satisfaction of search, experience level, cognitive biases), leading to variability in observer performance and decisions (Degnan, 2019).

For over 30 years, researchers have been exploring the potential of computers and image analysis techniques to aid providers in the decision-making process (Krupinski, 2022), with artificial intelligence (AI) emerging as the most recent platform. None of the AI schemes developed to date (and likely never will be) 100% accurate and error free. Thus, we must improve our understanding of the human-computer interface and how we can optimize the synergy between human decision makers and computer-based decision aids.

Since medical imaging is such a visual modality, it makes sense to capitalize on the capabilities of the human visual system. Can we use eye gaze or search data in AI evaluation? How can eye gaze be used in assessing the impact of AI on clinical decision making? Can eye gaze reveal anything about the best way to present AI information to users? These are some emerging areas of research that have significant potential to further promote and enhance the use of AI tools in medical imaging.

Evaluating new technologies in medical imaging has incorporated eye-gaze data for over 30 years to better understand exactly how the technology impacts the way clinicians access and process the information contained in the images – and that information can change as a function of several factors throughout the imaging chain, including image acquisition, pre- and post-processing, display, storage, and retrieval (Krupinski, 2022; Krupinski, 2019; Auffermann et al., 2018; Zhan et al., 2021; Hanna et al., 2018).

With the advent of AI these investigations continue, but also explore ways that gaze can inform AI and vice-versa. With respect to assessing the impact of AI on clinical decision-making, one key question is what form should the AI take to best convey information. For example, should the AI output be displayed as a heatmap, a probability graph, an arrow pointing to the potential abnormalities, a bounding box or outline? The options are many but what is optimal – what will aid in rendering more accurate and efficient decisions? Efficiency is critical – if it takes more time to use an AI tool without much gain in accuracy it will not be used in the clinic. However, if an AI tool can improve efficiency and at least maintain current accuracy levels, it will be used. This study explores the impact of the form of the AI output on observer performance.

2. METHODS

Eight radiologists and radiology residents participated as observers in this IRB approved study. Each participated in 5 sessions separated by at least 3 weeks to promote forgetting. In each session two series of 40 chest x-ray images with no COVID or mild, moderate, or severe COVID were viewed in random order. Each series provided a different form of AI feedback from an in-house developed AI tool: no AI provided, a heat map, a probability graph, heat map plus probability graph, or a single word (normal, mild, moderate, severe). In every trial the original image was displayed on a computer screen and the reader was required to render a decision (normal, mild moderate, severe) plus their confidence in that decision (possible, probable, definite). After providing a decision the AI output was displayed and they could either keep their original decision and confidence or change either/both decisions. Throughout the trial, a Tobii eye-tracking system was used to record their visual search patterns.

3. RESULTS

Overall, the residents had more decision ($p < 0.0001$) and confidence ($p < 0.0001$) changes with the provision of the AI output than the experienced radiologists, although the format did not appear to have a significant impact. The form of the AI output did however affect the visual search patterns. The single word did not add significantly to viewing time and overall readers simply glanced over to the word and then back to the image. The heat maps, probability graphs, heatmap plus graphs resulted in significantly more time spent looking at the output as well as back and forth between the chest image and the AI output. Figures 1 and 2 show examples of a radiologist searching a chest x-ray image for COVID with an AI heatmap (Figure 1) and heatmap plus probabilty graph (Figure 2) provided (these are actually color heatmaps but black and white for publication) while eye gaze is recorded. The superimposed circles and lines represent fixations (circles) or where the eye lands with foveal or high-resolution vision, and saccades (lines) or how the eyes jump from one location to the next.

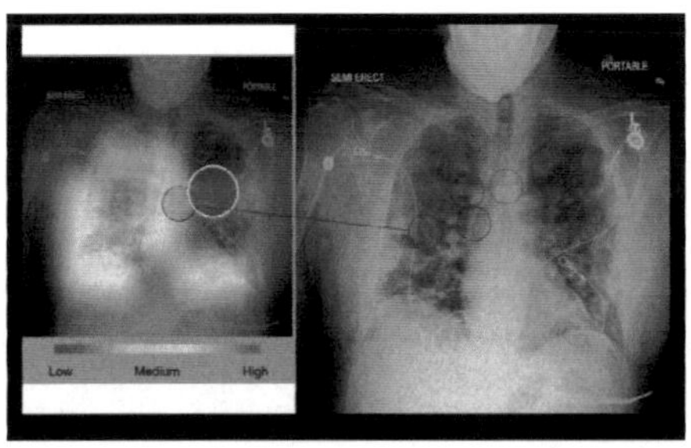

Figure 1. Example of a radiologist searching a chest x-ray image for COVID with an AI heatmap (these are actually color heatmaps but black and white for publication) while eye gaze is recorded. The superimposed circles and lines represent fixations (circles) and saccades (lines)

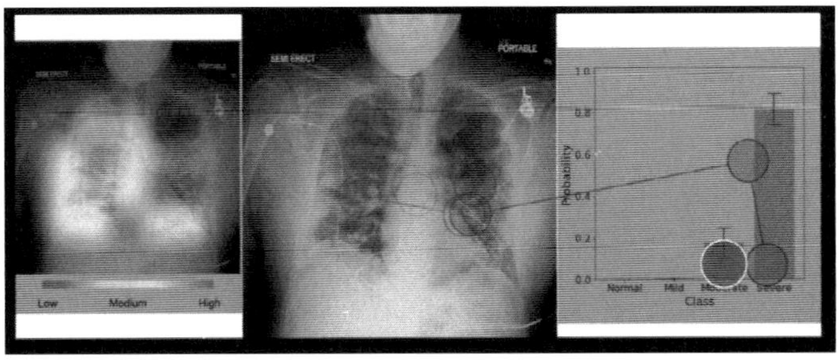

Figure 2. Example of a radiologist searching a chest x-ray image for COVID with an AI heatmap plus probabilty graph while eye gaze is

4. DISCUSSION

The potential for eye gaze to be incorporated into AI and other technologies is immense. The key to the success of AI tools in medical imaging will be to incorporate the human into the overall process to optimize and synergize the human-computer dyad, as at least for the foreseeable future, the human is and will be the ultimate decision maker. Our results demonstrate that the form of the AI output is important as it can impact clinical decision making and efficiency. Perhaps of equal importance is that it may impact different users in different ways, especially trainees versus those with more experience. To date, there have been no other eye-tracking studies like this that have been published. There were some eye-tracking studies done in prior decades examining the types of prompts used for computer-aided diagnoses (CAD), but they differ in that CAD prompts typically appeared directly on the images (e.g., arrows, bounding boxes) as opposed to additional images or information presented outside the context of the images being searched. Some important limitations of this study include use of only one type of image and disease, one AI tool, and not all forms that AI output could take (although they were the most common). Future work needs to expand into different application areas and possibly different clinical decision tasks (e.g., tumor measurement).

REFERENCES

Auffermann W. F., et al. (2018). Search pattern training for evaluation of central venous catheter positioning on chest radiographs. *J Medical Imaging*, 5, 031407-1 – 031407-5.

Degnan A. J., et al. (2019). Perceptual and interpretative error in diagnostic radiology – causes and potential solutions. *Acad Radiol*, 26, 833-845.

Hanna T.N., et al. (2018). The effects of fatigue from overnight shifts on radiology search patterns and diagnostic performance. *J Am Coll Radiol*, 15, 1709-1716.

Krupinski E. A. (2019). Impact of patient photos on detection accuracy, decision confidence and eye-tracking parameters in chest and abdomen images with tubes and lines. *J Digital Imaging*, 32, 827-831.

Krupinski E. A. (2022). In between are the doors of perception. *Proc SPIE Medical Imaging*, 12035, 120350K.

Zhan H., et al. (2021). The impact of fatigue on complex CT case interpretation by radiology residents. *Acad Radiol*, 28, 424-432.

ENHANCING HEALTH CARE THROUGH REMOTE SERVICES AND INTERACTIVE TECHNOLOGIES

Ssu-Hsuan Lu
Department of Computer Information and Network Engineering
Lunghwa University of Science and Technology
Taoyuan, Taiwan

ABSTRACT

With the aging population, health issues have become a significant challenge for society. The increasing demand for remote medical services stems from changing interests in reducing healthcare costs and the sudden outbreak of COVID-19 globally. As the Internet of Things (IoT) and cloud technology advance rapidly worldwide, this trend has driven the digital transformation of biotechnology and pharmaceutical-related industries. Health care is shifting from a focus on medical institutions to a decentralized approach, emphasizing care facilities and home-based services. However, remote medical care brings about certain challenges, such as limited appointment availability due to network constraints, concerns about internet security, data encryption, privacy issues, and lack of awareness among people in remote areas. With the continuous development and widespread adoption of technology, smart home systems have become an integral part of modern life. In this project, the system emphasizes interaction with the elderly and utilizes the multifunctionality of touch screens. It provides various features through a simple and user-friendly interface, including health monitoring, outpatient services, life assistance, and social interaction functions. In the future, the system can incorporate more interactive features for the elderly, such as mini-games to enhance cognitive abilities, hand-eye coordination, or rehabilitation training.

KEYWORDS

Smart Home, Telemedicine, Remote Monitoring

1. INTRODUCTION

With the rapid advancement of global Internet of Things (IoT) and cloud technology, it has propelled the biotechnology and pharmaceutical-related industries towards digitization. As the healthcare market continues to grow steadily, the scope of healthcare coverage has expanded from disease prevention, diagnosis, and monitoring to include long-term care management. Technology giants are also actively investing in key areas of digital healthcare development. In response to the shift in healthcare focus from traditional medical institutions to care facilities and home-based services, biopharmaceutical companies are expanding into the healthcare market by deploying in vitro diagnostic reagents and smart medical technologies. As the healthcare market continues to grow steadily, the scope of healthcare has expanded from disease prevention, diagnosis, and monitoring to long-term care management. Technology giants are actively investing in key areas of digital healthcare development. In response to the shift in healthcare focus from traditional medical institutions to care facilities and home-based services, biopharmaceutical companies are expanding into the healthcare market with in vitro diagnostic reagents and smart medical technologies. The most prominent investment targets for technology companies include healthcare, telemedicine, and medical diagnostics. The three key components of telemedicine are remote monitoring, information transmission, and interactive communication. Both current investment trends and future developments are closely tied to the challenges of an aging society and healthcare workforce shortages. However, telemedicine presents several issues, such as prolonged consultation times due to network limitations, internet security, data encryption, identity authentication, and the lack of awareness among people in remote areas. These limitations are expected to constrain the telemedicine services market.

Against this backdrop, this paper aims to design and develop an intelligent home care system specifically for the elderly. The system is intended to provide comprehensive health monitoring, outpatient services, life assistance, and social interaction functions, thereby promoting the health, happiness, and independence of elderly individuals. By thoroughly understanding the needs and usage habits of the elderly and integrating the latest smart home technologies and design concepts, the system will enable them to enjoy a better quality of life in their later years.

2. SMART HOME SYSTEMS FOR THE ELDERLY

This system's features focus on interaction with the elderly and the multifunctional use of touch screens. Interaction with the elderly—in an era where digital devices are widely used, this system aims to help seniors easily learn and adapt, providing them with friendly interactive feedback, so they are no longer overlooked by the technology market. The system will facilitate interaction with the elderly in various ways, including voice medication reminders, medication image displays, and outpatient appointment inquiries, enabling seniors to use the system effortlessly while enjoying the benefits of smart home functionalities. Moreover, the multifunctional use of touch screens is another key feature of this system. In addition to basic functions similar to those of everyday apps, such as browsing the internet, video calling with family, and watching videos, the system's app offers more applications. It can integrate with wearable devices to display health information, such as measuring blood pressure, heart rate, and weight through wearable devices and inputting this data into the system, which then reminds seniors to monitor their health. The screen can also display practical functions like weather information and calendar reminders, helping seniors better organize their daily lives.

The system is expected to include two subsystems: healthcare and active aging. The system will collaborate with medical units, utilizing interactive health check software to record users' health conditions and transmit the data in real-time back to the medical units. The system will offer customized daily health plans, designing rehabilitation exercises based on individual needs, and recording activity and responses. Additionally, the system will include a series of interactive games specifically designed for the elderly to promote physical activity and cognitive function. In addition to games, the system can also offer online courses from internet communities, allowing users to participate in various learning activities from home.

3. CONCLUSION

In recent years, advancements in information and telecommunication technologies have made telemedicine possible. Using the latest sensor technology, it is possible to collect an individual's vital data within a smart home environment. This biometric information can be transmitted wirelessly or via the internet to medical databases and healthcare professionals. By using appropriate sensing devices in a smart home environment, users can monitor their health signals and information in real-time and can regularly chart their progress. The system is currently under construction. In the future, the system will incorporate more interactive features to achieve the goal of smart home health services.

REFERENCES

Ahmed Shihab, Albahri, et al, 2021. "IoT-based telemedicine for disease prevention and health promotion: State-of-the-Art."*In Journal of Network and Computer Applications,* Volume 173.

Haleem, Abid, et al, 2022. "Medical 4.0 technologies for healthcare: Features, capabilities, and applications." *In Internet of Things and Cyber-Physical Systems*, No. 2, pp 12-30.

Khoshrounejad, Farnaz, et al, 2021. “Telehealth-based services during the COVID-19 pandemic: a systematic review of features and challenges.” *In Frontiers in public health*, Volume 9.

Javaid, Mohd, et al, 2023. "5G technology for healthcare: Features, serviceable pillars, and applications." *In Intelligent Pharmacy*, Volume 1, Issue 1, pp 2-10.

Doctoral Consortium

IMAGE SEGMENTATION APPROACHES: A COMPARATIVE STUDY

Rajaa Miftah, Mostafa Hanoune and Mohssine Bentaib
LTIM laboratory. Faculty of Science Ben M'sik , Hassan II University of Casablanca
Casablanca 20000, Morocco

ABSTRACT

Object detection is an intriguing and pivotal topic in the field of computer vision. Meanwhile, segmentation plays a crucial role and stands as one of the oldest problem statements that researchers have extensively investigated in the realm of Artificial intelligence (AI). Within computer vision systems, image segmentation poses significant challenges. Consequently, numerous methods have been developed to address this complex task. Given the vast array of options available, this research focuses on exploring and comparing the frequently employed classical techniques for image segmentation. The study aims to shed light on the strengths, limitations, and performance characteristics of each approach. By comprehensively examining these traditional segmentation methods, we strive to provide valuable insights into their applicability, potential use cases, and overall effectiveness in addressing the challenges posed by image segmentation tasks. Ultimately, this research seeks to contribute to the advancement of computer vision systems and inspire further exploration in the field of image segmentation.

KEYWORDS

Image Segmentation, Computer Vision, Segmentation Techniques, Artificial Intelligence, Object Detection, Scene Understanding

1. INTRODUCTION

This paper explores the operational differences among three types of image segmentation: semantic segmentation, instance segmentation, and panoptic segmentation. Semantic segmentation labels every pixel in an image, providing detailed scene understanding. Instance segmentation goes further by distinguishing individual object instances within the same class. Panoptic segmentation combines both, offering a unified representation of all scene objects.

A detailed examination of fundamental segmentation techniques such as thresholding, region growing, edge-based methods, clustering algorithms, and watershed transformation is provided. By comparing their strengths, limitations, and suitability for different tasks, the paper aims to aid researchers and practitioners in choosing the most appropriate method for their needs.

Despite the prominence of deep learning, classical segmentation techniques remain relevant, especially where data availability or computational resources are limited. The paper contributes insights into segmentation methods and applications, providing a valuable resource for researchers and practitioners in the field.

2. TYPES OF SEGMENTATION

Image segmentation takes an input image and outputs regions or segments, as seen in Figure 1 (Minaee et al., 2022). Segmentation includes semantic, instance, and panoptic types, each serving unique purposes.

2.1 Semantic Segmentation

Semantic segmentation labels each pixel in an image (Figure 1) with a class, providing detailed scene understanding. While demanding, it offers valuable insights. However, it may struggle with instance differentiation, limiting its effectiveness (Minaee et al., 2022).

Figure 1. Semantic segmentation

2.2 Instance Segmentation

Identifies individual object instances with masks or contours (Figure 2) (Minaee et al., 2022), aiding precise identification but not labeling every pixel.

Figure 2. Instance segmentation

2.3 Panoptic Segmentation

Masks all objects, including the background, for a comprehensive scene representation (Figure 3) (Elharrouss et al., 2021), merging semantic and instance methods.

Figure 3. Panoptic segmentation

3. CLASSICAL IMAGE SEGMENTATION TECHNIQUES

Image segmentation techniques, shown in figure 4 have evolved from classical methods to deep learning (Sharma et al., 2010).

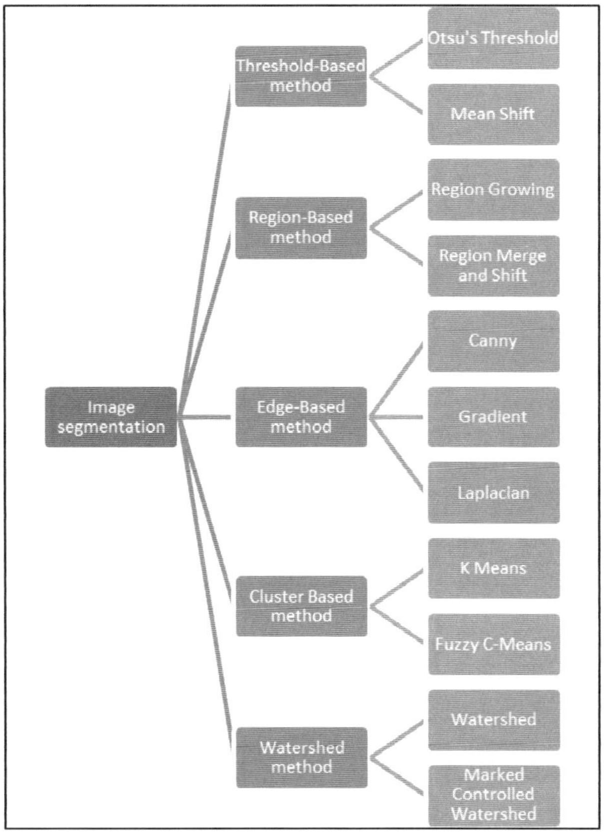

Figure 4. Image segmentation classical techniques

3.1 Thresholding

Thresholding partitions an image based on intensity values (Planckt & Mueller, 1977) (Kaur & Kaur, 2014), distinguishing foreground from background using threshold T and spawning three common segmentation algorithms.

Global threshold: Uses a single threshold T for light objects on a dark background.

Variable threshold: Includes local and adaptive thresholding for enhanced accuracy (Bhargavi & Jyothi, 2014).

Multiple threshold: Uses thresholds like T1 and T2 for objects with different intensity levels (Hu et al., 2016).

3.2 Region Based Segmentation

These methods group pixels into regions based on similarity. Region growing starts with seed points and expands based on similarity (Kaur & Kaur, 2014). Region split and merge divides and combines regions to enhance segmentation (Kaur & Kaur, 2014) (Chaudhuri & Agrawal, 2010). These techniques offer effective ways to segment images into coherent regions.

3.3 Edge Based Segmentation

Detects image edges, highlighting differences in gray-level, color, and texture (Farag, 1992). Utilizes operators like Sobel, Canny, and Roberts to reveal image structure (Kaur & Kaur, 2014) (Al-Amri et al., 2010).

3.4 Clustering Based Segmentation

Clustering organizes data into classes with high intra-class similarity and low inter-class similarity (Yambal & Gupta, 2013) (Wiharto & Suryani, 2020). Hierarchical methods create a tree-like structure, while partition-based methods refine cluster assignments. In image processing, clustering includes hard methods like K-means and soft methods like fuzzy c-means, accommodating noise (Singh, 2015) (Dehariya et al., 2010) (Chuang et al., 2006).

3.5 Watershed Based Methods

Watershed-based methods treat a grayscale image as a 3D surface, segmenting it into regions resembling basins. Each intensity level forms a basin, merging when water reaches boundaries. Dilation creates dams to prevent merging (Kaur & Kaur, 2014) (Kang et al., 2009).

4. COMPARISON

Table 1 compares segmentation techniques from (Kaur & Kaur, 2014), aiding application decisions by outlining their characteristics and limitations, enhancing understanding of segmentation utility.

Table 1. Comparison of traditional segmentation techniques

Segmentation technique	Description	Advantages	Disadvantages
Threshold method	Find the specific threshold values using the histogram peaks of the image.	Simple technique, no requirement for prior information.	Largely dependent on peaks, with scant consideration for spatial details.
Region based method	Dividing the input image into homogeneous regions.	Highly resistant to noise, helpful when defining similarity criteria is simple.	Cost more time and memory.
Edge based method	Detection of discontinuity.	Decent technique for images having better contrast between objects.	Unsuitable for incorrect or excessive edges.
Clustering method	Partitioning the input image into homogeneous regions.	More beneficial for real-world issues.	Difficult to determine membership function.
Watershed method	Topological concept.	Results are more consistent, and continuous boundaries are found.	Complicated gradient calculations.

5. CONCLUSION

In conclusion, this paper explored classical image segmentation approaches, highlighting their significance across various domains. The comparative table offered insights into their strengths and limitations, aiding in technique selection. While traditional methods remain effective, future research should leverage emerging technologies like deep learning to enhance segmentation accuracy and efficiency. Integration of deep learning promises to address segmentation complexities, pushing boundaries in computer vision. Researchers must stay updated on advancements to drive breakthroughs in image segmentation applications.

REFERENCES

Al-amri et al. (2010) S. S. Al-amri, N. V Kalyankar, and K. S.D, "Image segmentation by using edge detection," Int. J. Comput. Sci. Eng., vol. 2, no. 3, pp. 804–807, 2010.

Bhargavi and Jyothi (2014) K. Bhargavi and S. Jyothi, "A Survey on Threshold Based Segmentation Technique in Image Processing," Int. J. Innov. Res. Dev., vol. 3, no. 12, pp. 234–239, 2014, [Online]. Available: https://www.researchgate.net/publication/309209325.

Chaudhuri and Agrawal (2010) D. Chaudhuri and A. Agrawal, "Split-and-merge procedure for image segmentation using bimodality detection approach," Def. Sci. J., vol. 60, no. 3, pp. 290–301, 2010, doi: 10.14429/dsj.60.356.

Chuang et al. (2006) K. S. Chuang, H. L. Tzeng, S. Chen, J. Wu, and T. J. Chen, "Fuzzy c-means clustering with spatial information for image segmentation," Comput. Med. Imaging Graph., vol. 30, no. 1, pp. 9–15, 2006, doi: 10.1016/j.compmedimag.2005.10.001.

Dehariya et al. (2010) V. K. Dehariya, S. K. Shrivastava, and R. C. Jain, "Clustering of image data set using k-means and fuzzy k-means algorithms," Proc. - 2010 Int. Conf. Comput. Intell. Commun. Networks, CICN 2010, pp. 386–391, 2010, doi: 10.1109/CICN.2010.80.

Elharrouss et al. (2021) O. Elharrouss, S. Al-Maadeed, N. Subramanian, N. Ottakath, N. Almaadeed, and Y. Himeur, "Panoptic Segmentation: A Review," 2021, [Online]. Available: http://arxiv.org/abs/2111.10250.

Farag (1992) A. A. Farag, "Edge-based image segmentation," Remote Sens. Rev., vol. 6, no. 1, pp. 95–121, 1992, doi: 10.1080/02757259209532148.

Hu et al. (2016) M. H. Hu, Q. L. Dong, B. L. Liu, L. Q. Pan, and J. Walshaw, "Image Segmentation of Bananas in a Crate Using a Multiple Threshold Method," J. Food Process Eng., vol. 39, no. 5, pp. 427–432, 2016, doi: 10.1111/jfpe.12233.

Kang et al. (2009) W. X. Kang, Q. Q. Yang, and R. P. Liang, "The comparative research on image segmentation algorithms," Proc. 1st Int. Work. Educ. Technol. Comput. Sci. ETCS 2009, vol. 2, pp. 703–707, 2009, doi: 10.1109/ETCS.2009.417.

Kaur and Kaur (2014) Kaur D and Kaur Y, "Various image segmentation techniques: A review," Int. J. Comput. Sci. Mob. Comput., vol. 3, no. 5, pp. 809–814, 2014.

Minaee et al. (2022) S. Minaee, Y. Boykov, F. Porikli, A. Plaza, N. Kehtarnavaz, and D. Terzopoulos, "Image Segmentation Using Deep Learning: A Survey," IEEE Trans. Pattern Anal. Mach. Intell., vol. 44, no. 7, pp. 3523–3542, 2022, doi: 10.1109/TPAMI.2021.3059968.

Planckt and Mueller (1977) S. R. Planckt and G. C. Mueller, "DNA Chain Growth in Isolated HeLa Nuclei," Biochemistry, vol. 16, no. 12, pp. 2778–2782, 1977, doi: 10.1021/bi00631a030.

Sharma et al. (2010) N. Sharma et al., "Automated medical image segmentation techniques," J. Med. Phys., vol. 35, no. 1, pp. 3–14, 2010, doi: 10.4103/0971-6203.58777.

Singh (2015) A. Singh, "Current Image Segmentation Techniques-A Review," vol. 6, no. 2, pp. 1940–1942, 2015.

Wiharto and Suryani (2020) W. Wiharto and E. Suryani, "The comparison of clustering algorithms K-means and fuzzy C-means for segmentation retinal blood vessels," Acta Inform. Medica, vol. 28, no. 1, pp. 42–47, 2020, doi: 10.5455/AIM.2020.28.42-47.

Yambal and Gupta (2013) M. Yambal and H. Gupta, "Image Segmentation using Fuzzy C Means Clustering: A survey," Int. J. Adv. Res. Comput. Commun. Eng., vol. 2, no. 7, p. 2927, 2013, [Online]. Available: www.ijarcce.com.